A Course in Large-Sample and High-Dimensional Theory

This book provides a systematic treatment of two central regimes in statistical theory: classical large-sample theory for M- and Z-estimation with a fixed number of parameters, and high-dimensional theory where the number of parameters can be comparable to or larger than the sample size. While the former was developed earlier and remains fundamental, high-dimensional statistical theory has become an indispensable part of modern statistics.

Classical large-sample theory and high-dimensional theory are typically compartmentalized into separate books and courses, which can make it difficult for readers to see how they relate. To foster learning, this book brings them together in a compact and integrated manner, highlighting both their differences and their shared underlying structures.

Assuming basic knowledge of mathematics and statistics, the book is intended primarily as a graduate textbook for students and researchers in Statistics, Data Science, and related fields. It serves as a useful resource for those wishing to study classical asymptotics and modern high-dimensional theory as cohesive parts of a broader statistical framework.

Key Features:

- Focuses on core, representative topics in classical and modern statistical theory, emphasizing essential ideas that help readers extend their understanding to related areas.
- Treats important results that are otherwise scattered across research papers and monographs in a coherent and carefully organized manner.
- Provides direct, self-contained proofs of main results while assuming only basic concepts and results from probability and real analysis.
- Reinforces learning with end-of-chapter exercises as well as questions and exercises integrated into the main text.

Zhiqiang Tan is a Distinguished Professor in the Department of Statistics at Rutgers University. His research and teaching interests include Monte Carlo methods, causal inference, statistical learning, and related areas. He is a Fellow of the American Statistical Association, a Fellow of the Institute of Mathematical Statistics, and an Elected Member of the International Statistical Institute.

CHAPMAN & HALL/CRC
Texts in Statistical Science Series

Recently Published Titles

Statistical Inference, Second Edition
George Casella and Roger Berger

Nonparametric Statistical Methods Using R, Second Edition
John Kloke and Joseph McKean

Generalized Linear Mixed Models
Modern Concepts, Methods and Applications, Second Edition
Walter W. Stroup, Marina Ptukhina, and Julie Garai

Analysis of Categorical Data with R, Second Edition
Christopher R. Bilder, Thomas M. Loughin

Applied Nonparametric Statistical Methods, Fifth Edition
Nigel C. Smeeton, Neil H. Spencer, and Peter Sprent

Linear Models with R, Third Edition
Julian J. Faraway

Statistics in Survey Sampling
Jae Kwang Kim

Time Series: A Data Analysis Approach Using R
Robert H. Shumway and David S. Stoffer

Bayesian Statistical Methods: With Applications to Machine Learning
By Brian J. Reich, Sujit K. Ghosh

Experimental Design for Data Science and Engineering
By V. Roshan Joseph

Survival Analysis: Principles and Applications in Clinical Trials and Beyond
Song Yang

Foundations of Bayesian Statistics for Data Scientists: With R and Python
Alan Agresti, Maria Kateri, Ranjini Grove, and Antonietta Mira

A Course in Regression and Smoothing Methods
Zhiqiang Tan

A Course in Large-Sample and High-Dimensional Theory
Zhiqiang Tan

For more information about this series, please visit: https://www.routledge.com/Chapman--HallCRC-Texts-in-Statistical-Science/book-series/CHTEXSTASCI

A Course in Large-Sample and High-Dimensional Theory

Zhiqiang Tan

CRC Press
Taylor & Francis Group
Boca Raton London New York

CRC Press is an imprint of the
Taylor & Francis Group, an **informa** business
A CHAPMAN & HALL BOOK

Designed cover image: Zhiqiang Tan

First edition published 2026
by CRC Press
2385 NW Executive Center Drive, Suite 320, Boca Raton FL 33431

and by CRC Press
4 Park Square, Milton Park, Abingdon, Oxon, OX14 4RN

CRC Press is an imprint of Taylor & Francis Group, LLC

ISBN: 978-1-041-15357-3 (hbk)
ISBN: 978-1-041-15687-1 (pbk)
ISBN: 978-1-003-68073-4 (ebk)

DOI: 10.1201/9781003680734

Typeset in CMR10 font
by KnowledgeWorks Global Ltd.

Publisher's note: This book has been prepared from camera-ready copy provided by the authors.

Contents

Preface

The subject of this book consists of two parts, corresponding to the following grouping of the six chapters of the book:

(i) Classical large-sample theory for M- and Z-estimation (Chapters 1–2), where the number of parameters is fixed and the sample size tends to infinity, and

(ii) High-dimensional theory in linear regression, generalized linear regression, and more broadly M-estimation (Chapters 3–6), where the number of parameters can be close to or exceed the sample size.

Although the first topic was developed earlier both in time and concept, high-dimensional statistical theory such as Lasso-penalized estimation has become an essential and integral part of modern statistics. In fact, each topic is covered by numerous books and is taught across undergraduate and graduate courses. However, the levels of rigor or completeness of such courses may often be constrained by various factors including the background and interest of students and instructors and the availability of suitable textbooks. For example, the two topics, classical large-sample theory and high-dimensional theory, are usually treated in separate books and courses, and their connection and distinction may not be sufficiently explained in a way that fosters learning. Moreover, the literature of these topics continues to evolve, and new material or a better understanding of older results has emerged relatively recently and seems to be scattered in specialized research papers and books. Examples include the classical asymptotic theory of convex M-estimation among the first topic and high-dimensional inference based on debiased Lasso estimation among the second topic. For these reasons, it is valuable to present both classical large-sample theory and modern high-dimensional theory in an accessible, compact, and integrated manner.

To provide an accessible and concise account of large-sample and high-dimensional theory, the book is designed with the following features, while assuming basic knowledge of mathematics and statistics.

(i) The book focuses on main representative topics in classical and modern statistical theory, and attempts to explain essential ideas which, once understood, may enable readers to understand other related topics. In other words, the book aims to illustrate the most representative ideas rather than the most general results.

(ii) The book covers, in a coherent and careful manner, various topics which are important but scattered among research papers and books such as random- vs fixed-design regression, model misspecification, convex nonsmooth M-estimation, and debiased Lasso inference.

(iii) The book presents direct and self-contained proofs of main results, while assuming basic concepts and results from probability and real analysis. To preserve the text flow, technical discussions are sometimes placed in the appendix sections of their respective chapters.

(iv) The book not only provides exercises at the end of each chapter, but also includes questions and exercises alongside regular discussions.

As facilitated by the four features, the book is mainly intended as a graduate textbook (used for teaching and studying representative topics), instead of as a research book (used for documenting and disseminating research). The book includes pedagogical discussions that would normally be omitted in research books. The primary audience for the book is graduate students and researchers in Statistics, Data Science, and related fields. A secondary audience may include anyone who is interested in studying the book material, i.e., both classical large-sample and modern high-dimensional theory.

This book can be studied together with a sister book, “A Course in Regression and Smoothing Methods” (Tan, 2025). The method book aims to provide a coherent and concise account of both parametric regression and nonparametric smoothing methods at an advanced level. A considerable part of the method book is devoted to showing how to put numerical algorithms directly in computer code such as R. I hope that the two books may provide an up-to-date account of statistical theory and methods, moving from elementary, conventional material to advanced, modern material. Errata and updates of the books will be posted on my book webpage:

`https://statweb.rutgers.edu/ztan/books.html`.

This book is based on lecture notes used for several years in a first-year PhD course, titled “Advanced Modern Statistical Inference II”, at the Department of Statistics, Rutgers University. From my teaching experience, the book provides sufficient material to be covered in a semester of 14 weeks with 2 classes each of 80 minutes per week. The method book mentioned above is based on my teaching of another first-year PhD course, titled “Advanced Applied Statistics I”. I am grateful to the Department for the opportunity to design and teach these courses. I am also grateful to students in my classes over the years for their questions and comments. Our discussions have helped to shape the two books.

I thank Cunhui Zhang for helpful discussions on high-dimensional statistics. I also thank my editor, David Grubbs, for encouragement and support.

Zhiqiang Tan, December 2025

Author Biography

Zhiqiang Tan is a Distinguished Professor in the Department of Statistics at Rutgers University. His research and teaching interests include Monte Carlo methods, causal inference, statistical learning, and related areas. He is a Fellow of the American Statistical Association, a Fellow of the Institute of Mathematical Statistics, and an Elected Member of the International Statistical Institute.

1

Basic convergence theory

In this chapter, we review basic convergence theory in probability through the following topics: convergence of random variables, o_p and O_p notation, and convergence of sample means with independent and identically distributed observations. The concepts and results are instrumental to classical large-sample theory in statistics, where the number of parameters is fixed and the sample size tends to infinity. The discussion assumes knowledge of probability at the undergraduate or Master's levels.

We review basic material of convergence theory in probability, oriented toward its applications to classical large-sample theory in statistics, where the sample size tends to infinity but the number of parameters is fixed. The discussion is not comprehensive but selective by design, and proofs of the results are mostly omitted and may be taken as exercises. Among many relevant books, Ferguson (1996) provides a systematic and direct coverage of classical topics, and van der Vaart (1998) provides a broad and succinct discussion of important topics including those in our discussion and other additional topics.

1.1 Convergence of random variables

We review basic concepts and results about convergence of random variables. Proofs of the results are omitted and may be taken as exercises, except for demonstration purposes. By convention, a random variable takes only finite values (hence is finite). A random variable with possible values $\pm\infty$ is called an *extended* random variable. In general, a random variable can be one- or multi-dimensional. For clarity in certain contexts, we adopt a narrower terminology: a random variable takes values in the real line $\mathbb{R}$ and a random vector takes values in a multi-dimensional space $\mathbb{R}^d$ with $d > 1$.

DOI: 10.1201/9781003680734-1

Definition 1.1.

(i) A sequence of random variables X_n, $n = 1, 2, \ldots$, is said to converge almost surely (or with probability one) to X, denoted as $X_n \xrightarrow[\text{a.s.}]{} X$, if

$$P(\lim_{n\to\infty} X_n = X) = 1$$
$$\Longleftrightarrow P(\lim_{n\to\infty} |X_n - X| = 0) = 1.$$

(ii) A sequence of random variables X_n, $n = 1, 2, \ldots$, is said to converge in probability to X, denoted as $X_n \xrightarrow[\mathcal{P}]{} X$, if for every $\epsilon > 0$,

$$\lim_{n\to\infty} P(|X_n - X| \leq \epsilon) = 1$$
$$\Longleftrightarrow \lim_{n\to\infty} P(|X_n - X| > \epsilon) = 0.$$

(iii) A sequence of random variables X_n, $n = 1, 2, \ldots$, is said to converge in distribution to X, denoted as $X_n \xrightarrow[\mathcal{D}]{} X$, if

$$\lim_{n\to\infty} F_{X_n}(x) = F_X(x) \tag{1.1}$$

for all $x \in \mathbb{R}$ at which $F_X(\cdot)$ is continuous, where $F_{X_n}(\cdot)$ or $F_X(\cdot)$ denotes the cumulative distribution function (CDF) of X_n or X, respectively.

We make two remarks. First, for the convergence almost sure or in probability, the random variables X_n, $n \geq 1$, and X are defined in the same probability space. [For the convergence in probability, an exception is that when X is degenerate and treated as a real constant (for example, 0), X_n may be defined in a different probability space for different n.] This may not be true for the convergence in distribution, which is fundamentally a concept about the convergence of distributions (or CDFs). For example, if $X_1, X_2, \ldots$ are independent and identically distributed (IID) as X, then $X_n \xrightarrow[\mathcal{D}]{} X$, but $X_n \not\xrightarrow[\mathcal{P}]{} X$ and $X_n \not\xrightarrow[\text{a.s.}]{} X$ unless X is degenerate (i.e., taking some value with probability one).

Second, to understand the choice of only requiring (1.2) to hold at x such that F_X is continuous, consider the following example. Let X_n be a degenerate random variable at $\frac{1}{n}$, i.e., $P(X_n = \frac{1}{n}) = 1$, and let X be a degenerate random variable at 0, i.e., $P(X = 0) = 1$. Since $\frac{1}{n} \to 0$ as $n \to \infty$, we expect that $X_n \xrightarrow[\mathcal{D}]{} X$. It can be calculated that

$$F_{X_n}(x) = \begin{cases} 0, & x < \frac{1}{n}, \\ 1, & x \geq \frac{1}{n}, \end{cases} \qquad F_X(x) = \begin{cases} 0, & x < 0, \\ 1, & x \geq 0. \end{cases}$$

and

$$\lim_{n\to\infty} F_{X_n}(x) = \begin{cases} 0, & x \leq 0, \\ 1, & x > 0. \end{cases}$$

Hence $\lim_{n\to\infty} F_{X_n}(x) = F_X(x)$ for $x \neq 0$ at which F_X is continuous, but $\lim_{n\to\infty} F_{X_n}(x) \neq F_X(x)$ for $x = 0$ at which F_X is discontinuous.

The following result describes the basic relationship between the three types of convergence.

Theorem 1.1 (Convergence relations).

(i) If $X_n \xrightarrow[\text{a.s.}]{} X$*, then* $X_n \xrightarrow[\mathcal{P}]{} X$.

(ii) If $X_n \xrightarrow[\mathcal{P}]{} X$*, then* $X_n \xrightarrow[\mathcal{D}]{} X$.

(iii) If $X_n \xrightarrow[\mathcal{D}]{} c$ *for a constant* $c \in \mathbb{R}$*, then* $X_n \xrightarrow[\mathcal{P}]{} c$.

The following result allows us to use almost sure (pointwise) convergence to prove results about convergence in probability.

Theorem 1.2 (Subsequence theorem).

(i) If $X_n \xrightarrow[\mathcal{P}]{} X$*, then there exists a subsequence* $(X_{n_k})_{k\geq 1}$ *such that* $X_{n_k} \xrightarrow[\text{a.s.}]{} X$.

(ii) $X_n \xrightarrow[\mathcal{P}]{} X$ *if and only if every subsequence* $(X_{n_k})_{k\geq 1}$ *contains a further subsequence* $(X_{n_{k_l}})_{l\geq 1}$ *such that* $X_{n_{k_l}} \xrightarrow[\text{a.s.}]{} X$.

The following result, a special case of Skorokhod representation theorem on the real line, allows us to use almost sure (pointwise) convergence to prove results about convergence in distribution. See, for example Pollard (1984), Section III.6, for a direct proof using quantile transformations. For two random variables X and Y, we write $X \sim Y$ to indicate that X and Y have the same distribution.

Theorem 1.3 (Skorokhod representation theorem). *If* $X_n \xrightarrow[\mathcal{D}]{} X$*, then there exists on some probability space a sequence of random variables* $\tilde{X}_1, \tilde{X}_2, \ldots$ *and* $\tilde{X}$ *such that* $\tilde{X}_n \sim X_n$ *for* $n \geq 1$, $\tilde{X} \sim X$*, and* $\tilde{X}_n \xrightarrow[\text{a.s.}]{} \tilde{X}$.

The following result is known as the continuous mapping theorem. The proof of the almost-sure version is straightforward, using $\{f(X_n) \to f(X)\} \supset \{X_n \to X\} \cap \{X \in C(f)\}$. From the almost-sure result, the in-probability version can be shown using Theorem 1.2, and the in-distribution version can be shown using Theorem 1.3 [Exercise].

Theorem 1.4 (Continuous mapping theorem). *For a function* $f : \mathbb{R} \to \mathbb{R}$*, the continuity set of* f *is* $C(f) = \{x \in \mathbb{R} : f \text{ is continuous at } x\}$.

(i) If $X_n \xrightarrow[\text{a.s.}]{} X$ *and* $P(X \in C(f)) = 1$*, then* $f(X_n) \xrightarrow[\text{a.s.}]{} f(X)$.

(ii) The same result as (i) holds with $\xrightarrow[\text{a.s.}]{}$ *replaced by* $\xrightarrow[\mathcal{P}]{}$ *or* $\xrightarrow[\mathcal{D}]{}$.

For the convergence of random vectors in $\mathbb{R}^d$, let $\|\cdot\|$ be a norm in $\mathbb{R}^d$. For example, the L_p norm with $p \geq 1$ is defined as $\|x\|_p = (x^{(1)^p} + \cdots + x^{(d)^p})^{1/p}$ for $x = (x^{(1)}, \ldots, x^{(d)})^{\mathrm{T}}$. For $p = \infty$, the L_∞ norm is defined as $\|x\|_\infty = \max(x^{(1)}, \ldots, x^{(d)})$. See Figure 1.1 for an illustration.

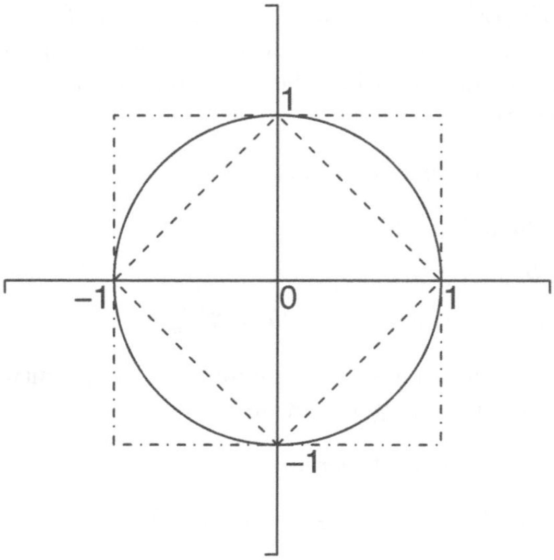

FIGURE 1.1
Illustration of the contours defined as $\{x \in \mathbb{R}^2 : \|x\|_1 \leq 1\}$ (dashed), $\{x \in \mathbb{R}^2 : \|x\|_2 \leq 1\}$ (solid), and $\{x \in \mathbb{R}^2 : \|x\|_\infty \leq 1\}$ (dot-dash).

Definition 1.2.

(i) A sequence of random vectors X_n, $n = 1, 2, \ldots$, is said to converge almost surely (or with probability one) to X, denoted as $X_n \xrightarrow[\text{a.s.}]{} X$, if $P(\lim_{n\to\infty} \|X_n - X\| = 0) = 1$.

(ii) A sequence of random vectors X_n, $n = 1, 2, \ldots$, is said to converge in probability to X, denoted as $X_n \xrightarrow[\mathcal{P}]{} X$, if for every $\epsilon > 0$, $\lim_{n\to\infty} P(\|X_n - X\| \leq \epsilon) = 1$.

(iii) A sequence of random vectors X_n, $n = 1, 2, \ldots$, is said to converge in distribution to X, denoted as $X_n \xrightarrow[\mathcal{D}]{} X$, if

$$\lim_{n\to\infty} F_{X_n}(x) = F_X(x) \tag{1.2}$$

for all $x \in \mathbb{R}$ at which $F_X(\cdot)$ is continuous, where $F_{X_n}(\cdot)$ or $F_X(\cdot)$ denotes the multivariate CDF of X_n or X respectively, for example, $F_X(x) = P(X^{(1)} \leq x^{(1)}, \ldots, X^{(d)} \leq x^{(d)})$, with $X = (X^{(1)}, \ldots, X^{(d)})^\intercal$ and $x = (x^{(1)}, \ldots, X^{(d)})^\intercal$.

Definition 1.2 appears to extend Definition 1.1 in a natural manner. We make two remarks. First, with fixed d, all the norms on $\mathbb{R}^d$ are equivalent to each other. See Section 1.4.1 for details. In particular, the L_p norms are equivalent to L_2 norm in the following sense. For any $p \geq 1$, there exist two constants $c_1, c_2 > 0$ (depending on p and d) such that for any $x \in \mathbb{R}^d$,

$$c_1\|x\|_2 \leq \|x\|_p \leq c_2\|x\|_2. \tag{1.3}$$

By the equivalence of norms, the convergence properties $\underset{\text{a.s.}}{\longrightarrow}$ and $\underset{\mathcal{P}}{\to}$ are not affected by which norm $\|\cdot\|$ is used in Definition 1.2. For example, X_n converges in probability to X under one choice of $\|\cdot\|$ if and only if the same holds under another choice of $\|\cdot\|$.

Second, from Definitions 1.1 and 1.2, convergence of random vectors almost sure or in probability is equivalent to the convergence of component sequences. Convergence of random vectors in distribution implies the convergences of component sequences, but the converse statement may not hold.

Corollary 1.1 (Convergence of components in probability or almost sure). *A sequence of random vectors $X_n = (X_n^{(1)}, \dots, X_n^{(d)})^{\mathrm{T}}$, $n \geq 1$, converges in probability to $X = (X^{(1)}, \dots, X^{(d)})^{\mathrm{T}}$ if and only if for each $j = 1, \dots, d$, the sequence $X_n^{(j)}$, $n \geq 1$, convergence in probability to $X^{(j)}$. The result also holds for almost sure convergence.*

Proof. Take the norm $\|\cdot\|$ in Definition 1.2 to be $\|\cdot\|_\infty$. [Question: How would the proof proceed if another choice of $\|\cdot\|$, for example, $\|\cdot\|_2$ is used?]

[Proof of "$\Longrightarrow$"] Suppose that $X_n \underset{\mathcal{P}}{\to} X$. We show that for $j = 1, \dots, d$, $X_n^{(j)} \underset{\mathcal{P}}{\to} X^{(j)}$. In fact, for any $j = 1, \dots, d$ and $\epsilon > 0$,

$$\{|X_n^{(j)} - X^{(j)}| > \epsilon\} \subset \{\|X_n - X\|_\infty > \epsilon\}, \tag{1.4}$$

i.e., the event on the left-hand side implies that on the right-hand side. Then $P(|X_n^{(j)} - X^{(j)}| > \epsilon) \leq P(\|X_n - X\|_\infty > \epsilon)$. Letting $n \to \infty$, we obtain $P(|X_n^{(j)} - X^{(j)}| > \epsilon) \to 0$ because $P(\|X_n - X\|_\infty > \epsilon) \to 0$ by assumption. Hence $X_n^{(j)} \underset{\mathcal{P}}{\to} X^{(j)}$.

[Proof of "$\Longleftarrow$"] Suppose that for $j = 1, \dots, d$, $X_n^{(j)} \underset{\mathcal{P}}{\to} X^{(j)}$. We show that $X_n \underset{\mathcal{P}}{\to} X$. In fact, for any $\epsilon > 0$,

$$\{\|X_n - X\|_\infty > \epsilon\} \subset \cup_{j=1}^d \{|X_n^{(j)} - X^{(j)}| > \epsilon\}, \tag{1.5}$$

because if $\|X_n - X\|_\infty > \epsilon$, then $|X_n^{(j)} - X^{(j)}| > \epsilon$ for some $j = 1, \dots, d$. [The two sets in (1.5) are equal to each other; the reverse direction $\cup_{j=1}^d \{|X_n^{(j)} - X^{(j)}| > \epsilon\} \subset \{\|X_n - X\|_\infty > \epsilon\}$ is used through (1.4) for $j = 1, \dots, d$ in the proof of "$\Longrightarrow$".] By the union bound, $P(\|X_n - X\|_\infty > \epsilon) \leq \sum_{j=1}^d P(|X_n^{(j)} - X^{(j)}| > \epsilon)$. Letting $n \to \infty$, we obtain $\sum_{j=1}^d P(|X_n^{(j)} - X^{(j)}| > \epsilon) \to 0$ and then $P(\|X_n - X\|_\infty > \epsilon) \to 0$ because d is fixed and $P(|X_n^{(j)} - X^{(j)}| > \epsilon) \to 0$ for each j. Hence $X_n \underset{\mathcal{P}}{\to} X$. □

The following result gives a special but useful situation where the marginal convergences in distribution imply the joint convergence, together with a restatement from Corollary 1.1 about the convergence in probability.

Theorem 1.5 (Stacked convergence in probability or in distribution). *Let X_n and Y_n, $n \geq 1$ be two sequences of random vectors.*

(i) If $X_n \xrightarrow[\mathcal{P}]{} X$ *and* $Y_n \xrightarrow[\mathcal{P}]{} Y$*, then*

$$\begin{pmatrix} X_n \\ Y_n \end{pmatrix} \xrightarrow[\mathcal{P}]{} \begin{pmatrix} X \\ Y \end{pmatrix}.$$

(ii) If $X_n \xrightarrow[\mathcal{D}]{} X$ *and* $Y_n \xrightarrow[\mathcal{D}]{} c$ *for a constant vector* c*, then*

$$\begin{pmatrix} X_n \\ Y_n \end{pmatrix} \xrightarrow[\mathcal{D}]{} \begin{pmatrix} X \\ c \end{pmatrix}.$$

Theorems 1.1–1.4 remain valid in the case of random vectors, where f in Theorem 1.4 can be a function from one vector space to another vector space possibly of different dimensions. Combining Theorem 1.5 with the vector version of Theorem 1.4 (Continuous mapping theorem) leads to various specific results. For example, if $X_n \xrightarrow[\mathcal{P}]{} X \in \mathbb{R}^d$ and $Y_n \xrightarrow[\mathcal{P}]{} Y \in \mathbb{R}^d$, then $X_n + Y_n \xrightarrow[\mathcal{P}]{} X + Y$ and $X_n^{\mathrm{T}} Y_n \xrightarrow[\mathcal{P}]{} X^{\mathrm{T}} Y$. For another example, the following result is known as Slutsky's theorem.

Corollary 1.2 (Slutsky's theorem). *Suppose that* $X_n \xrightarrow[\mathcal{D}]{} X \in \mathbb{R}^d$ *and* $Y_n \xrightarrow[\mathcal{P}]{} c$ *(or equivalently* $Y_n \xrightarrow[\mathcal{D}]{} c$*) for a constant* $c \in \mathbb{R}$*. Then*

(i) $X_n + Y_n \xrightarrow[\mathcal{D}]{} X + c$*;*

(ii) $X_n Y_n \xrightarrow[\mathcal{D}]{} cX$*;*

(iii) $X_n / Y_n \xrightarrow[\mathcal{D}]{} X/c$ *provided* $c \neq 0$*.*

Convergence of random matrices (i.e., random variables taking matrix values) can also be defined, similarly to that of random vectors, provided that a matrix norm is used. All the previous results, including Theorems 1.1–1.5, can be extended in a corresponding manner. In particular, the following multivariate version of Slutsky's theorem holds.

Corollary 1.3 (Slutsky's theorem). *Suppose that* $X_n \xrightarrow[\mathcal{D}]{} X \in \mathbb{R}^d$*,* $Y_n \xrightarrow[\mathcal{P}]{} c$ *for a constant vector* $c \in \mathbb{R}^d$*, and* $Z_n \xrightarrow[\mathcal{P}]{} A$ *for a constant matrix* $A \in \mathbb{R}^{d \times d}$*. Then*

(i) $X_n + Y_n \xrightarrow[\mathcal{D}]{} X + c$*;*

(ii) $Z_n X_n \xrightarrow[\mathcal{D}]{} AX$*;*

(iii) $Z_n^{-1} X_n \xrightarrow[\mathcal{D}]{} A^{-1} X$ *provided that* A *is invertible.*

The following result is known as the delta method.

Theorem 1.6 (Delta method). *Suppose that* $a_n(X_n - \mu) \xrightarrow[\mathcal{D}]{} X$ *for a constant vector* $\mu \in \mathbb{R}^d$ *and a real sequence* a_n *such that* $a_n \to \infty$ *as* $n \to \infty$*. Let* $g : \mathbb{R}^d \to \mathbb{R}^k$ *be a mapping which is differentiable at* μ*, and denote* $D = \frac{\partial g}{\partial x^{\mathrm{T}}}(\mu)$*, which is a* $k \times d$ *matrix. Then* $a_n(g(X_n) - g(\mu)) \xrightarrow[\mathcal{D}]{} DX$*. In particular, if* $\sqrt{n}(X_n - \mu) \xrightarrow[\mathcal{D}]{} \mathrm{N}(0, \Sigma)$*, then* $\sqrt{n}(g(X_n) - g(\mu)) \xrightarrow[\mathcal{D}]{} \mathrm{N}(0, D\Sigma D^{\mathrm{T}})$*.*

Proof. For simplicity, we prove the case of $k = 1$. By the definition of g being differentiable at μ, there exists a scalar function ϕ defined in a neighborhood of $0 \in \mathbb{R}^d$ such that $\phi(0) = 0$ and $\lim_{h \to 0} \phi(h) = 0$ (i.e., ϕ is continuous at 0), and

$$g(x) - g(\mu) = D(x - \mu) + \phi(x - \mu)\|x - \mu\|.$$

[Exercise: Explicitly define the function ϕ.] Then

$$a_n(g(X_n) - g(\mu)) = D \cdot a_n(X_n - \mu) + \phi(X_n - \mu) \cdot \|a_n(X_n - \mu)\|.$$

From $a_n(X_n - \mu) \xrightarrow[\mathcal{D}]{} X$ and $a_n \to \infty$ as $n \to \infty$, we have $a_n(X_n - \mu) = O_p(1)$ by Example 1.1(iii) and $X_n - \mu \xrightarrow[\mathcal{P}]{} 0$ by Corollary 1.5 later. From the continuity of ϕ at 0, we have $\phi(X_n - \mu) \xrightarrow[\mathcal{P}]{} \phi(0) = 0$ by the continuous mapping theorem, and hence $\phi(X_n - \mu) \cdot \|a_n(X_n - \mu)\| \xrightarrow[\mathcal{D}]{} 0 \cdot X = 0$ and $a_n(g(X_n) - g(\mu)) \xrightarrow[\mathcal{D}]{} DX + 0 = DX$ by Slutsky's theorem.

Note: Ferguson (1996), Section 7, presents the delta method and a proof using the mean value theorem, under the slightly stronger condition that the partial derivatives in $\frac{\partial g}{\partial x^{\mathrm{T}}}(x)$ are continuous in a neighborhood of μ (which implies that g is differentiable at μ). □

In Section 1.4.2, we discuss another type of convergence of random variables, called convergence in mean (or in the L_1 norm), and related convergence results.

1.2 o_p and O_p notation

The small-o_p and big-O notation is helpful in large-sample theory, in a similar manner to the small-o and big-O notation for sequences of real numbers or vectors.

Definition 1.3. Let a_n, $n \geq 1$, be a sequence of real vectors in $\mathbb{R}^d$ with norm $\|\cdot\|$.

(i) We write $a_n = o(1)$ if $\lim_{n\to\infty} a_n = 0$, i.e., for any $\epsilon > 0$, there exists $N \geq 1$ such that $\|a_n\| \leq \epsilon$ for $n \geq N$.

(ii) We write $a_n = O(1)$ and say $(a_n)_{n\geq 1}$ is asymptotically bounded if there exist $M > 0$ and $N \geq 1$ such that $\|a_n\| \leq M$ for $n \geq N$.

By definition, $a_n = o(1)$ is just another way to indicate $\lim_{n\to\infty} a_n = 0$. Moreover, the above definition for $O(1)$ is equivalent to a seemingly stronger condition of being bounded: $a_n = O(1)$ if and only if there exists $M > 0$ such that $\|a_n\| \leq M$ for $n \geq 1$, not just for $n \geq N$ with some selected N. [Exercise: Prove this statement.] In other words, $(a_n)_{n\geq 1}$ being *asymptotically bounded* as indicated by Definition (ii) is equivalent to $(a_n)_{n\geq 1}$ being *bounded*. However, the two conditions are no longer equivalent in a more general situation where a_n is allowed to be an extended real vector, i.e., $a_n \in \overline{\mathbb{R}}^d$, with $\overline{\mathbb{R}} = \mathbb{R} \cup \{-\infty\} \cup \{\infty\}$. [Exercise: Explain why.]

Definition 1.4. Let X_n, $n \geq 1$, be a sequence of random vectors in $\mathbb{R}^d$ with norm $\|\cdot\|$.

(i) We write $X_n = o_p(1)$ if $X_n \xrightarrow[\mathcal{P}]{} 0$, i.e., for any $\epsilon, \eta > 0$, there exists $N \geq 1$ such that $P(\|X_n\| \leq \epsilon) \geq 1 - \eta$ for $n \geq N$.

(ii) We write $X_n = O_p(1)$ and say $(X_n)_{n\geq 1}$ is asymptotically bounded in probability if for any $\eta > 0$, there exist $M > 0$ and $N \geq 1$ such that $P(\|X_n\| \leq M) \geq 1 - \eta$ for $n \geq N$.

Similar observations can be made about Definition 1.4 as earlier about Definition 1.3. First, $X_n = o_p(1)$ is just another way to indicate $X_n \xrightarrow[\mathcal{P}]{} 0$. Moreover, the above definition for $O_p(1)$ is equivalent to a seemingly stronger condition of being bounded in probability: $X_n = O_p(1)$ if and only if for any $\eta > 0$, there exists $M > 0$ such that $P(\|X_n\| \leq M) \geq 1 - \eta$ for $n \geq 1$, not just for $n \geq N$ with some selected N (Exercise 1.4). However, the two conditions are no longer equivalent in a more general situation where X_n is allowed to be an extended random vector, taking values in $\overline{\mathbb{R}}^d$. [Exercise: Explain why.]

To clarify an important difference between $O(1)$ and $O_p(1)$, consider the concepts of boundedness by a specific constant $M > 0$.

Definition 1.5.

(i) A sequence of real vectors $(a_n)_{n\geq 1}$ is said to be asymptotically bounded by a constant $M > 0$ if there exists $N \geq 1$ such that $\|a_n\| \leq M$ for $n \geq N$.

(ii) A sequence of random vectors $(X_n)_{n\geq 1}$ is said to be asymptotically bounded in probability by a constant $M > 0$ if for any $\eta > 0$, there exists $N \geq 1$ such that $P(\|X_n\| \leq M) \geq 1 - \eta$ for $n \geq N$.

(iii) A sequence of random vectors $(X_n)_{n\geq 1}$ is said to be *strongly* asymptotically bounded in probability if there exists a constant $M > 0$ such that $(X_n)_{n\geq 1}$ is asymptotically bounded in probability by M.

A sequence of real vectors $(a_n)_{n\geq 1}$ is $O(1)$, i.e., asymptotically bounded if and only if there exists $M > 0$ such that $(a_n)_{n\geq 1}$ is asymptotically bounded by M. In contrast, it may occur that $(X_n)_{n\geq 1}$ is $O_p(1)$, i.e., asymptotically bounded in probability, but there exists no $M > 0$ such that $(X_n)_{n\geq 1}$ is asymptotically bounded in probability by M. In other words, a $O_p(1)$ sequence may not be strongly asymptotically bounded in probability. For example, a sequence of $\mathrm{N}(0, 1)$ variables are $O_p(1)$, but not asymptotically bounded in probability by any constant $M > 0$. Conversely, a sequence strongly asymptotically bounded in probability is always $O_p(1)$.

The difference between $O(1)$ and $O_p(1)$ in terms of boundedness by a constant can be explained from the following observation: fixing M in Definition 1.3(ii) leads to asymptotic boundedness by M, whereas fixing M in Definition 1.4(ii) is meaningless, because M in general depends on η.

For completeness, we note that the asymptotic boundedness by $M > 0$ and that in probability in Definition 1.5(i)–(ii) can be related to $o(1)$ and $o_p(1)$ definitions in a similar manner: $a_n = o(1)$ if and only if $(a_n)_{n\geq 1}$ is asymptotically bounded by any $\epsilon > 0$, and $X_n = o_p(1)$ if and only if $(X_n)_{n\geq 1}$ is asymptotically bounded in probability by any $\epsilon > 0$ (Exercise 1.5).

Similarly as in Corollary 1.1, a sequence of random vectors being $O_p(1)$ is equivalent to the component sequences being $O_p(1)$.

Corollary 1.4 ($O_p(1)$ components). *A sequence of random vectors $X_n = (X_n^{(1)}, \dots, X_n^{(d)})^{\mathrm{T}}$, $n \geq 1$, satisfies $X_n = O_p(1)$ if and only if $X_n^{(j)} = O_p(1)$ for each $j = 1, \dots, d$.*

The following example gives simple cases of $O_p(1)$.

Example 1.1. Let X_n, $n \geq 1$, be a sequence of (finite) random variables.

(i) If $X_n \equiv X$ for $n \geq 1$, with X a (finite) random variable, then $X_n = O_p(1)$.

(ii) If $X_n \underset{\mathcal{P}}{\to} X$, with X a (finite) random variable, then $X_n = O_p(1)$. If further $|X| \leq M$ almost surely for a constant $M > 0$, then $(X_n)_{n\geq 1}$ is asymptotically bounded in probability by $M + c$ for any constant $c > 0$.

(iii) If $X_n \underset{\mathcal{D}}{\to} X$, with X a (finite) random variable, then $X_n = O_p(1)$. If further $|X| \leq M$ almost surely for a constant $M > 0$, then $(X_n)_{n\geq 1}$ is asymptotically bounded in probability by $M + c$ for any constant $c > 0$.

(iv) If $E|X_n| \leq K$ for $n \geq 1$ and a constant K, then $X_n = O_p(1)$.

Similar results hold in the case where X_n and X are random vectors.

Proof.

(i) By definition, $P(|X| < \infty) = 1$ for a finite random variable X. Note that $P(|X| < \infty) = \lim_{M\to\infty} P(|X| \leq M)$ by the continuity of probability. Hence $\lim_{M\to\infty} P(|X| \leq M) = 1$, i.e., for any $\eta > 0$, there exists $M > 0$ such that $P(|X| \leq M) \geq 1 - \eta$.

(ii) Suppose that $X_n \underset{\mathcal{P}}{\to} X$. From the proof of (i), for any $\eta > 0$, there exists $M > 0$ such that $P(|X| > M/2) \leq \eta/2$. By $X_n \underset{\mathcal{P}}{\to} X$, there exists $N \geq 1$ such that $P(|X_n - X| > M/2) \leq \eta/2$ for $n \geq N$. By the triangle inequality, $\{|X_n| > M\} \subset \{|X| > M/2\} \cup \{|X_n - X| > M/2\}$. Hence by the union bound, for $n \geq N$,

$$P(|X_n| > M) \leq P(|X| > M/2) + P(|X_n - X| > M/2) \leq \eta.$$

This indicates $X_n = O_p(1)$.

Suppose further that $|X| \leq M$ almost surely, i.e., $P(|X| > M) = 0$. By $X_n \xrightarrow[\mathcal{P}]{} X$, for any $c > 0$ and $\eta > 0$, there exists $N \geq 1$ such that for $n \geq N$, we have $P(|X_n - X| > c) \leq \eta$ and hence

$$P(|X_n| > M + c) \leq P(|X| > M) + P(|X_n - X| > c) \leq \eta.$$

This indicates that $(X_n)_{n\geq 1}$ is asymptotically bounded in probability by $M+c$.

Note: The asymptotic boundedness in probability by $M+c$ for $c > 0$ cannot be improved to that by M. For example, let $X_n = M + \frac{1}{n}$. Then $X_n \xrightarrow[\mathcal{P}]{} M$ but $P(|X_n| \leq M) = 0$ for $n \geq 1$.

(iii) The proof is left as Exercise 1.6.

(iv) For any $\eta > 0$, take $M = K/\eta$. Then for $n \geq 1$, by the Markov inequality (Theorem 3.1),

$$P(|X_n| > M) \leq \frac{E|X_n|}{M} \leq \frac{K}{M} = \eta.$$

□

The following proposition gives basic properties for $o_p(1)$ and $O_p(1)$ (Exercise 1.8).

Proposition 1.1 (Properties of o_p and O_p). *For sequences of random vectors of the same dimensions,*

(i) $o_p(1) + o_p(1) = o_p(1)$;

(ii) $o_p(1) + O_p(1) = O_p(1)$;

(iii) $o_p(1)O_p(1) = o_p(1)$.

The interpretation of these properties, for example, (ii) and (iii) is that if $X_n = o_p(1)$ *and* $Y_n = O_p(1)$, *then* $X_n + Y_n = O_p(1)$ *and* $X_n^{\mathrm{T}} Y_n = o_p(1)$.

The small-o and big-O notation and small-o_p and big-O_p notation can also be defined with respect to an arbitrary sequence of positive numbers.

Definition 1.6. For a sequence of real vectors $(b_n)_{n\geq 1}$ and a sequence of positive numbers $(a_n)_{n\geq 1}$, we write

$$b_n = o(a_n) \text{ if } \frac{b_n}{a_n} = o(1),$$

$$b_n = O(a_n) \text{ if } \frac{b_n}{a_n} = O(1).$$

For a sequence of random vectors $(X_n)_{n\geq 1}$, we write

$$X_n = o_p(a_n) \text{ if } \frac{X_n}{a_n} = o_p(1),$$

$$X_n = O_p(a_n) \text{ if } \frac{X_n}{a_n} = O_p(1).$$

As a special case of Proposition 1.1(iii), the following relationship can be deduced.

Corollary 1.5 (From O_p to o_p). *If $a_n = o(1)$, i.e., $\lim_{n\to\infty} a_n = 0$, then $O_p(a_n) = o_p(1)$.*

A simple but useful example of $O_p(a_n)$ is as follows.

Example 1.2. For a sequence of random variables $(X_n)_{n\geq 1}$, if $E(X_n^2) = O(\frac{1}{n})$, then $E(nX_n^2) = O(1)$ and hence, by Example 1.1(iv), $nX_n^2 = O_p(1)$ or equivalently $X_n = O_p(\frac{1}{\sqrt{n}})$. Similarly, for a sequence of random vector $(X_n)_{n\geq 1}$, if $E(\|X_n\|_2^2) = O(\frac{1}{n})$, then $X_n = O_p(\frac{1}{\sqrt{n}})$.

1.3 Convergence of sample means

For (statistical) large-sample theory, the convergence of sample means is of fundamental importance. Let X_n, $n \geq 1$, be a sequence of IID random vectors as X. If exist, the mean vector and variance matrix, or simply the mean and variance, of X are denoted as

$$\mu = E(X), \quad \Sigma = \text{var}(X) = E\{(X-\mu)(X-\mu)^{\mathrm{T}}\}.$$

The sample mean is defined as $\bar{X}_n = \frac{1}{n}\sum_{i=1}^n X_i$. If $E(\|X\|_2^2) < \infty$, then μ and Σ are finite and by standard calculation,

$$E(\bar{X}_n) = E(X) = \mu,$$
$$\text{var}(\bar{X}_n) = \frac{1}{n}\text{var}(X) = \frac{1}{n}\Sigma.$$

As a consequence, if $E(\|X\|_2^2) < \infty$, then

$$E(\|\bar{X}_n - \mu\|_2^2) = \frac{1}{n}\text{tr}(\Sigma) = O\left(\frac{1}{n}\right),$$

and hence by Example 1.2, $\bar{X}_n = \mu + O_p(\frac{1}{\sqrt{n}})$. This indicates that not only $\bar{X}_n = \mu + o_p(1)$ or equivalently $\bar{X}_n \xrightarrow[\mathcal{P}]{} \mu$ (by Corollary 1.5) but also $\sqrt{n}(\bar{X}_n - \mu)$ is $O_p(1)$.

The convergence of the sample mean can also be established under weaker conditions or in a more refined manner.

Theorem 1.7 (Law of large numbers (LLN)). *Let X_n, $n \geq 1$, be a sequence of IID random vectors as X, with finite $\mu = E(X)$. Then $\bar{X}_n \xrightarrow[\mathcal{P}]{} \mu$.*

Theorem 1.8 (Central limit theorem (CLT)). *Let X_n, $n \geq 1$, be a sequence of IID random vectors as X, with finite $\mu = E(X)$ and $\Sigma = \text{var}(X)$. Then*

$$\sqrt{n}(\bar{X}_n - \mu) \xrightarrow[\mathcal{D}]{} \text{N}(0, \Sigma),$$

where $\text{N}(0, \Sigma)$ *is the multivariate normal distribution with mean* 0 *and variance* Σ.

As a simple but important application, we show the asymptotic normality of the t statistic: if $\Sigma = \text{var}(X)$ is invertible, then

$$\sqrt{n} S_n^{-1/2}(\bar{X}_n - \mu) \xrightarrow[\mathcal{D}]{} \text{N}(0, I), \tag{1.6}$$

where S_n is the sample variance defined as

$$S_n = \frac{1}{n-1} \sum_{i=1}^{n} (X_i - \bar{X}_n)(X_i - \bar{X}_n)^{\text{T}}.$$

In fact, by direct calculation

$$S_n = \frac{n}{n-1} \left(\frac{1}{n} \sum_{i=1}^{n} X_i X_i^{\text{T}} - \bar{X}_n \bar{X}_n^{\text{T}} \right).$$

By the law of large numbers, $\bar{X}_n \xrightarrow[\mathcal{P}]{} E(X)$ and $\frac{1}{n} \sum_{i=1}^{n} X_i X_i^{\text{T}} \xrightarrow[\mathcal{P}]{} E(XX^{\text{T}})$. Then by continuous mapping, $S_n \xrightarrow[\mathcal{P}]{} E(XX^{\text{T}}) - E(X)E(X)^{\text{T}} = \Sigma$ and $S_n^{1/2} \xrightarrow[\mathcal{P}]{} \Sigma^{1/2}$. Moreover, by the central limit theorem, $\sqrt{n}(\bar{X}_n - \mu) \xrightarrow[\mathcal{D}]{} \text{N}(0, \Sigma)$. Finally, by Slutsky's theorem (Corollary 1.3),

$$\begin{aligned} \sqrt{n} S_n^{-1/2}(\bar{X}_n - \mu) &= (S_n^{1/2})^{-1} \cdot \sqrt{n}(\bar{X}_n - \mu) \\ &\xrightarrow[\mathcal{P}]{} \Sigma^{-1/2} \cdot \text{N}(0, \Sigma) = \text{N}(0, I), \end{aligned}$$

provided that Σ is invertible.

The asymptotic normality (1.6) provides a distributional property needed to justify Wald confidence intervals for components of μ or confidence regions for subvectors of μ. For example, for $0 < c < 1$, the Wald confidence interval for $\mu^{(j)}$, the jth component of μ, is

$$\bar{X}_n^{(j)} \pm z_{c/2} \sqrt{\frac{(S_n)_{jj}}{n}},$$

which achieves asymptotic size $1 - c$, where $\bar{X}_n^{(j)}$ is the jth component of $\bar{X}_n$, $(S_n)_{jj}$ is the jth diagonal component of S_n, and $z_{c/2}$ is the $(1 - c/2)$-quantile of $\text{N}(0, 1)$. To see this, the coverage probability can be shown to satisfy

$$\begin{aligned} &P\left(\mu^{(j)} \in \left[\bar{X}_n^{(j)} - z_{c/2}\sqrt{(S_n)_{jj}/n}, \bar{X}_n^{(j)} + z_{c/2}\sqrt{(S_n)_{jj}/n} \right] \right) \\ &= P\left(\frac{\bar{X}_n^{(j)} - \mu^{(j)}}{\sqrt{(S_n)_{jj}/n}} \in [-z_{c/2}, z_{c/2}] \right) \\ &\to 1 - c, \end{aligned}$$

because $\frac{\bar{X}_n^{(j)}-\mu^{(j)}}{\sqrt{(S_n)_{jj}/n}} \xrightarrow[\mathcal{D}]{} \mathrm{N}(0,1)$ by the univariate version of (1.6) for $\bar{X}_n^{(j)}$. To justify Wald confidence regions for subvectors of μ, the multivariate convergence in (1.6) is required, not just the convergence of scalar components.

The preceding discussion of the t statistic uses only the convergence in probability of the sample variance S_n. Under finite fourth moments, each component of S_n is also asymptotically normal centered at the corresponding component of Σ. The scalar case is illustrated by the following example.

Example 1.3 (Asymptotic distribution of sample standard deviation). Let $X_1, \ldots, X_n$ be IID as X from a distribution with a finite fourth moment. The variance of X is $\sigma^{*2} = m_2^* - m_1^{*2}$, where $m_1^* = E(X)$ and $m_2^* = E(X^2)$. It is of interest to derive the asymptotic distribution of $s_n^2 = \frac{1}{n-1}\sum_{i=1}^n (X_i - \bar{X}_n)^2$. For ease of algebra, we consider $\tilde{s}_n^2 = \frac{1}{n}\sum_{i=1}^n (X_i - \bar{X}_n)^2$ and show that $\sqrt{n}(\tilde{s}_n^2 - \sigma^{*2}) \xrightarrow[\mathcal{D}]{} \mathrm{N}(0, v)$ for some $v > 0$. Then $\sqrt{n}(s_n^2 - \sigma^{*2}) \xrightarrow[\mathcal{D}]{} \mathrm{N}(0, v)$ for the same v [Exercise].

Let $Y_{n1} = \bar{X}_n$ and $Y_{n2} = \frac{1}{n}\sum_{i=1}^n X_i^2$. Then by direct calculation, $\tilde{s}_n^2 = Y_{n2} - Y_{n1}^2$. By applying CLT to the IID pairs (X_i, X_i^2), $i = 1, \ldots, n$, we have

$$\sqrt{n}\begin{pmatrix} Y_{n1} - m_1^* \\ Y_{n2} - m_2^* \end{pmatrix} \xrightarrow[\mathcal{D}]{} \mathrm{N}(0, \Sigma), \quad \Sigma = \begin{pmatrix} \mathrm{var}(X) & \mathrm{cov}(X, X^2) \\ \mathrm{cov}(X^2, X) & \mathrm{var}(X^2) \end{pmatrix}.$$

By applying the delta method (Theorem 1.6) with $g(y_1, y_2) = y_2 - y_1^2$, we obtain

$$\sqrt{n}(\tilde{s}_n^2 - \sigma^{*2}) \xrightarrow[\mathcal{D}]{} \mathrm{N}(0, D\Sigma D^{\mathrm{T}}),$$

where $D = (\frac{\partial g}{\partial y_1}, \frac{\partial g}{\partial y_2})|_{(y_1,y_2)=(m_1^*,m_2^*)} = (-2m_1^*, 1)$. To simplify calculation, we replace X by $X - m_1^*$, which does not affect $\tilde{s}_n^2$. Then

$$\begin{aligned} D &= (0, 1), \\ \Sigma &= \begin{pmatrix} \mathrm{var}(X - m_1^*) & \mathrm{cov}(X - m_1^*, (X - m_1^*)^2) \\ \mathrm{cov}((X - m_1^*)^2, X - m_1^*) & \mathrm{var}((X - m_1^*)^2) \end{pmatrix}, \end{aligned}$$

which lead to

$$\begin{aligned} v = D\Sigma D^{\mathrm{T}} &= \mathrm{var}((X - m_1^*)^2) \\ &= E((X - m_1^*)^4) - \sigma^{*4}. \end{aligned}$$

By another application of the delta method, we also find

$$\sqrt{n}(\tilde{s}_n - \sigma^*) \xrightarrow[\mathcal{D}]{} \mathrm{N}\left(0, \frac{1}{4\sigma^{*2}}\mathrm{var}((X - m_1^*)^2)\right).$$

If $X \sim \mathrm{N}(\mu^*, \sigma^{*2})$, then by direct calculation, $\mathrm{var}((X - \mu^*)^2) = 2\sigma^{*4}$ and $\sqrt{n}(\tilde{s}_n - \sigma^*) \xrightarrow[\mathcal{D}]{} \mathrm{N}(0, \frac{1}{2}\sigma^{*2})$. If $X \sim \mathrm{Exponential}(\theta^*)$ with mean θ^* and variance $\sigma^{*2} = \theta^{*2}$, then by direct calculation, $\mathrm{var}((X - \theta^*)^2) = 8\theta^{*4}$ and $\sqrt{n}(\tilde{s}_n - \sigma^*) \xrightarrow[\mathcal{D}]{} \mathrm{N}(0, 2\theta^{*2})$.

1.4 Appendices

1.4.1 Equivalence of norms

We discuss the equivalence of norms on a fixed-dimensional vector space, for example $\mathbb{R}^d$, as mentioned after Definition 1.2. By definition, a norm $\|\cdot\|$ on $\mathbb{R}^d$ is a function, $\mathbb{R}^d \to [0, \infty)$ satisfying the following conditions:

(i) Linearity: $\|\alpha x| = |\alpha|\|x\|$ for any $x \in \mathbb{R}^d$ and $\alpha \in \mathbb{R}$;

(ii) Positive definiteness: $\|x\| \geq 0$, and $\|x\| = 0$ if and only if $x = 0$;

(iii) Triangle inequality: $\|x + y\| \leq \|x\| + \|y\|$ for any $x, y \in \mathbb{R}^d$.

We show that $\|\cdot\|$ is equivalent to the L_2 norm, $\|\cdot\|_2$, i.e., there exist two constants $c_1, c_2 > 0$ such that for any $x \in \mathbb{R}^d$,

$$c_1\|x\|_2 \leq \|x\| \leq c_2\|x\|_2. \tag{1.7}$$

[Question: How would the proof proceed if the L_1 norm $\|\cdot\|_1$ is used in place of $\|\cdot\|_2$?] Then all the norms on $\mathbb{R}^d$ are equivalent to each other, by the transitivity of the equivalence relationship: if two norms are equivalent to $\|\cdot\|_2$, then they are equivalent to each other.

First, we show the upper bound (or the second inequality) in (1.7). Write $x = (x^{(1)}, \ldots, x^{(d)})^{\mathrm{T}}$ or $x = \sum_{j=1}^d x^{(j)} e_j$, where e_j is the jth basis vector, with the jth component being 1 and the remaining components being 0. Then by the linearity and triangle inequality for $\|\cdot\|$ and by the Cauchy–Schwartz inequality,

$$\|x\| \leq \sum_{j=1}^{d} |x^{(j)}|\|e_j\| \leq c_2\|x\|_2, \tag{1.8}$$

where $c_2 = (\sum_{j=1}^d \|e_j\|^2)^{1/2}$, i.e., the L_2 norm of $(\|e_1\|, \ldots, \|e_d\|)^{\mathrm{T}}$.

Second, we show the lower bound (or the first inequality) in (1.7). It suffices to show that there exists $c_1 > 0$ such that $\|x\| \geq c_1$ for $x \in \mathbb{R}^d$ with $\|x\|_2 = 1$. From (1.8), the function $\|\cdot\|$ is Lipschitz continuous (hence continuous) with respect to the L_2 norm: for any $x, y \in \mathbb{R}^d$,

$$\big|\|x\| - \|y\|\big| \leq \|x - y\| \leq c_2\|x - y\|_2.$$

The first inequality, $|\|x\| - \|y\|| \leq \|x - y\|$, follows because $\|x\| - \|y\| \leq \|x - y\|$ and $\|y\| - \|x\| \leq \|y - x\|$ by the triangle inequality and $\|x - y\| = \|y - x\|$ by the linearity. Note that the unit sphere $\mathbb{S} = \{x \in \mathbb{R}^d : \|x\|_2 = 1\}$ is a bounded and closed set with respect to $\|\cdot\|_2$ in $\mathbb{R}^d$. Hence $\|\cdot\|$ attains a minimum on $\mathbb{S}$. Let $c_1 = \|x^*\| = \min_{x \in \mathbb{S}} \|x\|$, where $x^* \in \mathbb{S}$ satisfying $\|x^*\|_2 = 1$. Then $x^* \neq 0$ by the positivity of $\|\cdot\|_2$ and hence $c_1 = \|x^*\| > 0$ by the positivity of $\|\cdot\|$. This completes the proof.

1.4.2 Additional convergence results

Convergence in mean or in the L_1 norm is another type of convergence of random variables. For simplicity, the scalar case is considered.

Definition 1.7. A sequence of random variables X_n with finite $E(X_n)$, $n = 1, 2, \ldots$, is said to converge in mean or in the L_1 norm, denoted as $X_n \xrightarrow[L_1]{} X$, if $\lim_{n\to\infty} E|X_n - X| = 0$.

Convergence in mean, $X_n \xrightarrow[L_1]{} X$, implies both convergence in probability, $X_n \xrightarrow[P]{} X$, and convergence of means, $E(X_n) \to E(X)$.

Proposition 1.2 (Implication of convergence in mean). *For a sequence of random variables X_n with $E|X_n| < \infty$, $n = 1, 2, \ldots$, if $X_n \xrightarrow[L_1]{} X$, then $X_n \xrightarrow[P]{} X$, $E(X_n) \to E(X)$, and $E|X_n| \to E|X|$ with $E|X| < \infty$.*

Proof. By the Markov inequality, for any $\epsilon > 0$,

$$P(|X_n - X| > \epsilon) \leq \frac{E|X_n - X|}{\epsilon}.$$

Then $P(|X_n - X| > \epsilon) \to 0$ because $E|X_n - X| \to 0$ as $n \to \infty$. Moreover, by the triangle inequality, $E|X| \leq E|X_n| + E|X_n - X| < \infty$, and by properties of expectations,

$$|E(X_n) - E(X)| = |E(X_n - X)| \leq E|X_n - X|,$$

which implies that $E(X_n) - E(X) \to 0$ as $n \to \infty$. Similarly, by properties of expectations and the triangle inequality, $||x| - |y|| \leq |x - y|$,

$$|E|X_n| - E|X|| = |E(|X_n| - |X|)| \leq E\big||X_n| - |X|\big| \leq E|X_n - X|,$$

which implies that $E|X_n| \to E|X|$ as $n \to \infty$. □

The following results are important and frequently used. In particular, the dominated convergence theorem provides a sufficient condition for obtaining convergence in mean from convergence in probability (or almost sure).

Theorem 1.9. *Let X_n, $n \geq 1$, be a sequence of random variables such that $(X_n)_- \leq Y$ for a random variable Y with $E|Y| < \infty$, where $x_- = \max(-x, 0)$. For example, $X_n \geq 0$ for $n \geq 1$.*

(i) [Monotone convergence theorem] If, in addition, X_n is non-decreasing in n, i.e., $X_n \leq X_{n+1}$ for $n \geq 1$, then $\lim_{n\to\infty} E(X_n) = E(\lim_{n\to\infty} X_n)$.

(ii) [Fatou's lemma] $E(\liminf_{n\to\infty} X_n) \leq \liminf_{n\to\infty} E(X_n)$.

Theorem 1.10 (Dominated convergence theorem). *Let X_n, $n \geq 1$, be a sequence of random variables such that $|X_n| \leq Y$ for a random variable Y with $E(Y) < \infty$.*

(i) If $X_n \xrightarrow[\text{a.s.}]{} X$ *for a random variable* X*, then* $E|X_n - X| \to 0$ *and* $E(X_n) \to E(X)$.

(ii) If $X_n \xrightarrow[\mathcal{P}]{} X$ *for a random variable* X*, then* $E|X_n - X| \to 0$ *and* $E(X_n) \to E(X)$.

Theorem 1.11 (Pratt's theorem). *Let* X_n*,* L_n*, and* U_n*,* $n \geq 1$*, be sequences of random variables such that* $L_n \leq X_n \leq U_n$ *for* $n \geq 1$*, and let* L *and* U *be random variables with* $E|L| < \infty$ *and* $E|U| < \infty$.

(i) If $X_n \xrightarrow[\text{a.s.}]{} X$ *for a random variable* X*,* $L_n \xrightarrow[\text{a.s.}]{} L$*,* $U_n \xrightarrow[\text{a.s.}]{} U$*,* $E(L_n) \to E(L)$*, and* $E(U_n) \to E(U)$*, then* $E|X| < \infty$ *and* $E(X_n) \to E(X)$.

(ii) If $X_n \xrightarrow[\mathcal{P}]{} X$ *for a random variable* X*,* $L_n \xrightarrow[\mathcal{P}]{} L$*,* $U_n \xrightarrow[\mathcal{P}]{} U$*,* $E(L_n) \to E(L)$*, and* $E(U_n) \to E(U)$*, then* $E|X| < \infty$ *and* $E(X_n) \to E(X)$.

As an interesting application of Pratt's theorem, Proposition 1.2 can be strengthened to the following equivalence result.

Theorem 1.12 (L_1 convergence theorem). *Let* X_n *be a sequence of random variables with* $E|X_n| < \infty$ *for* $n \geq 1$*. Then* $X_n \xrightarrow[L_1]{} X$ *if and only if* $X_n \xrightarrow[\mathcal{P}]{} X$ *and* $E|X_n| \to E|X|$ *with* $E|X| < \infty$.

Proof. The part of "$\Longrightarrow$" is from Proposition 1.2. To show the part of "$\Longleftarrow$", we use Pratt's theorem. Suppose that $X_n \xrightarrow[\mathcal{P}]{} X$ and $E|X_n| \to E|X|$ with $E|X| < \infty$. Let $L_n = L = 0$, $U_n = |X_n| + |X|$, and $U = 2|X|$ for $n \geq 1$. Then

$$
\begin{aligned}
&L_n = 0 \leq |X_n - X| \leq U_n = |X_n| + |X|,\\
&|X_n - X| \xrightarrow[\mathcal{P}]{} 0 \text{ by } X_n \xrightarrow[\mathcal{P}]{} X,\\
&L_n \xrightarrow[\mathcal{P}]{} L,\ U_n \xrightarrow[\mathcal{P}]{} U \text{ by } X_n \xrightarrow[\mathcal{P}]{} X \text{ and hence } |X_n| \xrightarrow[\mathcal{P}]{} |X|,\\
&E(U_n) = E(|X_n| + |X|)\\
&\qquad \to 2E|X| = E(U) < \infty \text{ by } E|X_n| \to E|X| \text{ with } E|X| < \infty.
\end{aligned}
$$

Therefore, $E|X_n - X| \to 0$ by Pratt's theorem. □

Exercises

Ex 1.1.

(i) For any fixed dimension d, show that there exist constants $c_1 = c_1(d)$ and $c_2 = c_2(d)$, possible depending on d but not on x, such that

$$c_1\|x\|_2 \leq \|x\|_1 \leq c_2\|x\|_2, \quad x \in \mathbb{R}^d.$$

Find the constants $c_1(d)$ and $c_2(d)$ as tight as possible. What are the limits of $c_1(d)$ and $c_2(d)$ as $d \to \infty$?

(ii) Do there exist constants c_1 and c_2, independently of x and d, such that

$$c_1\|x\|_2 \leq \|x\|_1 \leq c_2\|x\|_2, \quad x \in \mathbb{R}^d, \quad d \geq 1?$$

Show your proof.

Ex 1.2. Suppose that $X_n \leq Y_n$ almost surely for $n \geq 1$.

(i) Show that if $X_n \xrightarrow[\text{a.s.}]{} X$ and $Y_n \xrightarrow[\text{a.s.}]{} Y$, then $X \leq Y$ almost surely.

(ii) Show that if $X_n \xrightarrow[\mathcal{P}]{} X$ and $Y_n \xrightarrow[\mathcal{P}]{} Y$, then $X \leq Y$ almost surely.

Ex 1.3. Let $(X_n)_{n\geq 1}$ be a sequence of random variables.

(i) For a constant $c_1 > 0$, show that $X_n = o_p(1)$ if the following condition holds: for any $\epsilon > 0$, there exists $N \geq 1$ such that for all $n \geq N$, $P(|X_n| > \epsilon) \leq c_1\epsilon$.

(ii) For a constant $c_2 > 0$, show that $X_n = o_p(1)$ if the following condition holds: for any $\epsilon > 0$, there exists $N \geq 1$ such that for all $n \geq N$, $P(|X_n| > \epsilon) \leq c_2\epsilon^2$.

(iii) Can you generalize/strengthen results (i) and (ii) as much as possible?

Ex 1.4. Let $(X_n)_{n\geq 1}$ be *extended* random variables which may take values $\pm\infty$ but $P(|X_n| < \infty) = 1$ for each n. Show that if $X_n = O_p(1)$, then for any $\epsilon > 0$, there exist some $M > 0$ such that $P(|X_n| \leq M) > 1 - \epsilon$ for $n \geq 1$.

Ex 1.5.

(i) For a sequence of real vectors $(a_n)_{n\geq 1}$, show that $a_n = o(1)$ if and only if $(a_n)_{n\geq 1}$ is asymptotically bounded by any $\epsilon > 0$.

(ii) For a sequence of random vectors $(X_n)_{n\geq 1}$, show that $X_n = o_p(1)$ if and only if $(X_n)_{n\geq 1}$ is asymptotically bounded in probability by any $\epsilon > 0$.

Ex 1.6.

(i) Let X and $(X_n)_{n\geq 1}$ be random variables. Show that if $X_n \xrightarrow[\mathcal{D}]{} X$, then $X_n = O_p(1)$. In addition, show that if further $|X| \leq M$ almost surely for a constant $M > 0$, then $(X_n)_{n\geq 1}$ is asymptotically bounded in probability by $M + c$ for any constant $c > 0$.

(ii) State an extension of the result from (i) to the case where X and $(X_n)_{n\geq 1}$ are random vectors, and provide a proof.

Ex 1.7. Let $(X_n)_{n\geq 1}$ and Y be (finite) random variables.

(i) If $(X_n)_{n\geq 1}$ is asymptotically bounded almost surely by Y [i.e., there exists $N \geq 1$ such that $|X_n| \leq Y$ almost surely for $n \geq N$], then $X_n = O_p(1)$.

(ii) If $(X_n)_{n\geq 1}$ is asymptotically bounded in probability by Y [i.e., for any $\eta > 0$, there exists $N \geq 1$ such that $P(|X_n| \leq Y) \geq 1-\eta$ for $n \geq N$], then $X_n = O_p(1)$.

Ex 1.8. Show that

$$o_p(1) + O_p(1) = O_p(1),$$
$$o_p(1)O_p(1) = o_p(1).$$

Ex 1.9. Let $X_1, \ldots, X_n$ be IID from Exponential(θ^*) with mean θ^* and variance θ^{*2}. Define $\bar{X}_n = \frac{1}{n}\sum_{i=1}^n X_i$ and $\tilde{s}_n^2 = \frac{1}{n}\sum_{i=1}^n (X_i - \bar{X}_n)^2$. Use the delta method to find Σ such that as $n \to \infty$,

$$\sqrt{n}\begin{pmatrix} \bar{X}_n - \theta^* \\ \tilde{s}_n - \theta^* \end{pmatrix} \xrightarrow[\mathcal{D}]{} \mathrm{N}(0, \Sigma).$$

2

Classical theory for M- and Z-estimation

In this chapter, we study classical large sample theory for M- and Z-estimation through the following topics, where the sample size tends to infinity but the number of parameters is fixed. First, we discuss general, high-level results about consistency of M- and Z-estimation. Then we present a uniform law of large numbers and discuss consistency and asymptotic normality for M- and Z-estimation based on smoothness and dominance. Finally we discuss consistency and asymptotic normality for M-estimation with convex sample objective functions that may not be twice differentiable as in quantile estimation. We present direct proofs as much as possible, assuming basic knowledge of real analysis.

We study large sample theory for M- and Z-estimation in the classical setting where the parameter size is fixed as the sample size tends to infinity. For example, in linear or generalized linear regression, the number of covariate terms (and hence associated coefficients) is fixed as the sample size tends to infinity. These topics are covered by many books in various manners, including Ferguson (1996) and van der Vaart (1998) both of which provide a broad coverage of important topics, and Newey and McFadden (1994) which provides a fairly comprehensive discussion from an econometric perspective. Our discussion draws on, among others, van der Vaart (1998) for general results, Ferguson (1996) for smoothness-based theory, and Pollard (1991), Niemiro (1992), and Hjort and Pollard (1993) for convexity-based theory.

2.1 Introduction

Suppose that $Z_1, \ldots, Z_n$ are IID variables as Z. An M-estimator (with M indicating "minimization") is defined by minimizing a sample objective function.

DOI: 10.1201/9781003680734-2

Definition 2.1. An M-estimator $\hat{\theta}_n$ is defined in the form $\hat{\theta}_n = \text{argmin}_{\theta \in \Theta} K_n(\theta)$ with a sample objective function

$$K_n(\theta) = \frac{1}{n} \sum_{i=1}^{n} \rho(Z_i; \theta), \tag{2.1}$$

where $\Theta \subset \mathbb{R}^d$ is a parameter space, and $\rho(\cdot; \theta)$ is a scalar objective function for an individual datum. The target value $\bar{\theta}$ for $\hat{\theta}_n$ is defined as $\bar{\theta} = \text{argmin}_{\theta \in \Theta} K(\theta)$ with the population objective function

$$K(\theta) = E\{\rho(Z; \theta)\}. \tag{2.2}$$

A Z-estimator (with Z indicating "zero") is defined by solving a sample estimating equation.

Definition 2.2. A Z-estimator $\hat{\theta}_n$ is defined as a solution to $T_n(\theta) = 0$ on Θ with a sample estimating function

$$T_n(\theta) = \frac{1}{n} \sum_{i=1}^{n} \psi(Z_i; \theta), \tag{2.3}$$

where $\Theta \subset \mathbb{R}^d$ is a parameter space, and $\psi(\cdot; \theta)$ is a $d \times 1$ estimating function (of the same dimension as θ) for an individual datum. The target value $\bar{\theta}$ for $\hat{\theta}_n$ is defined as a solution to $T(\theta) = 0$ with the population estimating function

$$T(\theta) = E\{\psi(Z; \theta)\}. \tag{2.4}$$

M-estimation and Z-estimation are closely related. With $\rho(z; \theta)$ differentiable in θ for fixed z, the M-estimator $\hat{\theta}_n$ in Definition 2.1 is also a Z-estimator, with

$$\psi(z; \theta) = \frac{\partial}{\partial \theta} \rho(z; \theta).$$

Conversely, a Z-estimator may not be an M-estimator. Given an estimating function $\psi(z; \theta)$, there may not exist a scalar objective function $\rho(z; \theta)$ satisfying the preceding relationship. [Exercise: Give an example where this happens.]

We present a list of examples to illustrate the broad presence of M- and Z-estimation.

Example 2.1 (Maximum likelihood). Consider a parametric model postulating that $Z_1, \ldots, Z_n$ are IID from a density function $f(\cdot; \theta^*)$. The maximum likelihood estimator (MLE) of θ^* is defined as

$$\hat{\theta}_n = \text{argmax}_{\theta \in \Theta} \underbrace{\sum_{i=1}^{n} \log f(Z_i; \theta)}_{\ell_n(\theta)},$$

where $\ell_n(\theta)$ is called the log-likelihood function. Then $\hat{\theta}_n$ is an M-estimator with $\rho(z;\theta) = -\log f(z;\theta)$. With $\log f(z;\theta)$ differentiable in θ for fixed z, the MLE $\hat{\theta}_n$ is a solution to

$$0 = \frac{\partial}{\partial\theta}\ell_n(\theta) = \underbrace{\sum_{i=1}^{n} \frac{\partial}{\partial\theta} \log f(Z_i;\theta)}_{S_n(\theta)},$$

where $S_n(\theta)$ is called the score function. Hence $\hat{\theta}_n$ is also a Z-estimator with $\psi(z;\theta) = \frac{\partial}{\partial\theta}\log f(z;\theta)$.

Example 2.2 (Robust estimation). For simplicity, assume that $Z_1,\ldots,Z_n$ are IID scalar variables as Z. For estimating the mean $\mu^* = E(Z)$, the sample mean $\hat{\mu}_n = \frac{1}{n}\sum_{i=1}^{n} Z_i$ can be determined as

$$\hat{\mu}_n = \text{argmin}_{\mu\in\mathbb{R}} \sum_{i=1}^{n} (Z_i - \mu)^2,$$

hence an M-estimator with $\rho(z;\mu) = (z-\mu)^2$. The median of Z, denoted as ν^*, is defined any value satisfying $P(Z \geq \nu^*) \geq \frac{1}{2}$ and $P(Z \leq \nu^*) \geq \frac{1}{2}$. Equivalently, ν^* can be determined as $\nu^* = \text{argmin}_{\nu\in\mathbb{R}} E(|Z-\nu|)$ (Exercise 2.1). Note that ν^* may not be unique. For estimating ν^*, the sample median is defined as

$$\hat{\nu}_n = \text{argmin}_{\nu\in\mathbb{R}} \sum_{i=1}^{n} |Z_i - \nu|,$$

hence an M-estimator with $\rho(z;\mu) = |z-\mu|$. As a compromise between the sample mean and median, Huber's location estimator is defined as

$$\hat{\theta}_n = \text{argmin}_{\theta\in\mathbb{R}} \sum_{i=1}^{n} \rho_\kappa(Z_i - \theta),$$

where $\kappa > 0$ is a tuning parameter and

$$\rho_\kappa(x) = \begin{cases} \frac{1}{2}x^2, & \text{if } |x| \leq \kappa, \\ \kappa|x| - \frac{1}{2}\kappa^2, & \text{if } |x| > \kappa. \end{cases} \tag{2.5}$$

The function $\rho_\kappa(\cdot)$ is quadratic in the interval $[-\kappa,\kappa]$ and linear outside the interval, and is differentiable everywhere and twice differentiable in $\mathbb{R}$ except at $\pm\kappa$. See Figure 2.1 for an illustration. For a symmetric distribution, $\hat{\mu}_n$, $\hat{\nu}_n$, and $\hat{\theta}_n$ are estimators of the center $\mu^* = \nu^*$, but exhibit different properties of efficiency and robustness.

Example 2.3 (Quantile estimation). Suppose that $Z_1,\ldots,Z_n$ are IID scalar variables as Z. For $0 < \tau < 1$, the τth quantile of Z, denoted as q_τ^*, is defined

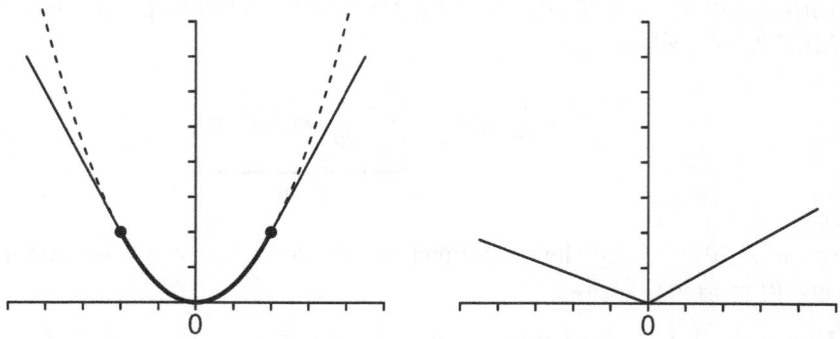

FIGURE 2.1
Left: $\rho_\kappa(\cdot)$ (solid) for Huber's estimator with $\kappa = 2$ and quadratic $x^2/2$ outside $[-\kappa, \kappa]$ (dashed); Right: $\rho_\tau(\cdot)$ for quantile regression with $\tau = 0.6$.

any value satisfying $P(Z \geq q^*) \geq 1 - \tau$ and $P(Z \leq q^*) \geq \tau$. Equivalently, q_τ^* can be determined as $q_\tau^* = \text{argmin}_{q \in \mathbb{R}} E\{\rho_\tau(Z - q)\}$, where $\rho_\tau(x) = \tau x_+ + (1-\tau)(-x)_+$, i.e.,

$$\rho_\tau(x) = \begin{cases} \tau x, & x > 0, \\ -(1-\tau)x, & x \leq 0. \end{cases}$$

See Figure 2.1 for an illustration. For estimating q_τ^*, the sample τth quantile is defined as

$$\hat{q}_{\tau,n} = \text{argmin}_{q \in \mathbb{R}} \sum_{i=1}^n \rho_\tau(Z_i - q),$$

hence an M-estimator with $\rho(z; q) = \rho_\tau(z - q)$. For $\tau = 1/2$, the sample 50% quantile recovers the sample median defined in Example 2.2.

Example 2.4 (Regression models). Let $Z_i = (X_i, Y_i)$, where X_i is a covariate vector of dimension p and Y_i is a (scalar) response variable. Consider a linear regression model

$$E(Y_i | X_i) = X_i^{\mathrm{T}} \beta^*.$$

The least squares estimator $\hat{\beta}_n$ is defined as

$$\hat{\beta}_n = \text{argmin}_{\beta \in \mathbb{R}^p} \sum_{i=1}^n (Y_i - X_i^{\mathrm{T}} \beta)^2,$$

hence an M-estimator with $\rho(x, y; \beta) = (y - x^{\mathrm{T}} \beta)^2$. Alternatively, in the case of binary response Y_i, consider a logistic regression model

$$P(Y_i = 1 | X_i) = \frac{\exp(X_i^{\mathrm{T}} \beta^*)}{1 + \exp(X_i^{\mathrm{T}} \beta^*)}.$$

The maximum likelihood estimator $\hat{\beta}_n$ is defined as

$$\hat{\beta}_n = \text{argmax}_{\beta \in \mathbb{R}^p} \sum_{i=1}^n \left\{ Y_i \log \frac{\exp(X_i^{\mathrm{T}}\beta^*)}{1+\exp(X_i^{\mathrm{T}}\beta^*)} + (1-Y_i) \frac{1}{1+\exp(X_i^{\mathrm{T}}\beta^*)} \right\}$$

$$= \text{argmax}_{\beta \in \mathbb{R}^p} \sum_{i=1}^n \{ Y_i X_i^{\mathrm{T}}\beta - \log(1+\exp(X_i^{\mathrm{T}}\beta)) \},$$

hence an M-estimator with $\rho(x, y; \beta) = -yx^{\mathrm{T}}\beta + \log(1+\exp(x^{\mathrm{T}}\beta))$.

2.2 Consistency: General results

We present general results about consistency of M- and Z-estimation. An M-estimator is defined as $\hat{\theta}_n = \text{argmin}_{\theta \in \Theta} K_n(\theta)$ with a sample objective function $K_n(\theta)$. The target parameter value $\bar{\theta}$ for $\hat{\theta}_n$ is defined as $\bar{\theta} = \text{argmin}_{\theta \in \Theta} K(\theta)$ with the population objective function $K(\theta) = E\{K_n(\theta)\}$. In this section, it is allowed that $K_n(\theta)$ may not be of the sample average form (2.1) and $K(\theta)$ may not be of the form (2.2). In fact, $K_n(\theta)$ is just treated as a random function in θ with the expectation $K(\theta)$, and may not be defined from an IID sample $(Z_1, \ldots, Z_n)$.

We say that $\hat{\theta}_n$ is consistent (for $\bar{\theta}$) if $\hat{\theta}_n \xrightarrow[\mathcal{P}]{} \bar{\theta}$ as $n \to \infty$. The following result gives a sufficient condition for consistency of $\hat{\theta}_n$.

Theorem 2.1 (General consistency of M-estimation). *Suppose that*

(i) [Isolated target parameter] $\inf_{\theta \in \Theta: \|\theta - \bar{\theta}\| > \epsilon} K(\theta) > K(\bar{\theta})$ *for each* $\epsilon > 0$;

(ii) [Uniform convergence] $\sup_{\theta \in \Theta} |K_n(\theta) - K(\theta)| \xrightarrow[\mathcal{P}]{} 0$;

(iii) [Existence of estimator] $\hat{\theta}_n$ *exists as a minimizer of* $K_n(\theta)$ *on* Θ *with probability tending to one.*

Then $\hat{\theta}_n \xrightarrow[\mathcal{P}]{} \bar{\theta}$, *i.e., for any* $\epsilon > 0$, $P(\hat{\theta}_n$ *exists and* $\|\hat{\theta}_n - \bar{\theta}\| \le \epsilon) \to 1$ *as* $n \to \infty$.

We make some comments about the conditions. Condition (i) implies that $\bar{\theta}$ is a unique minimizer of $K(\theta)$ on Θ. But the converse may not be true. See Figure 2.2 for an example to illustrate this. Condition (ii) says that $K_n(\theta) \xrightarrow[\mathcal{P}]{} K(\theta)$ uniformly in $\theta \in \Theta$, and is a more stringent condition than saying that $K_n(\theta) \xrightarrow[\mathcal{P}]{} K(\theta)$ pointwise for each $\theta \in \Theta$. Condition (iii) is introduced because $K_n(\theta)$ due to randomness may not attain a minimum on Θ.

Proof. First, we give a proof in the case where $\hat{\theta}_n$ exists almost surely for $n \ge 1$. We show that for any $\epsilon > 0$, $P(\|\hat{\theta}_n - \bar{\theta}\| > \epsilon) \to 0$ as $n \to \infty$. By condition (i), for each $\epsilon > 0$, there exists $\delta > 0$ such that if $\|\theta - \bar{\theta}\| > \epsilon$, then

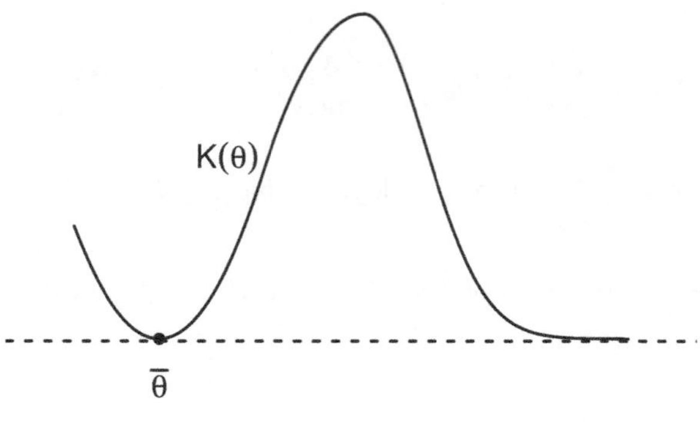

FIGURE 2.2
An example where condition (i) in Theorem 2.1 is violated, even though $\bar{\theta}$ is a unique minimizer of $K_n(\theta)$.

$K(\theta) - K(\bar{\theta}) \geq \delta$. In fact, we may take $\delta = \inf_{\theta \in \Theta : \|\theta - \bar{\theta}\| > \epsilon} K(\theta) - K(\bar{\theta}) > 0$. Then

$$P(\|\hat{\theta}_n - \bar{\theta}\| > \epsilon) \leq P(K(\hat{\theta}_n) - K(\bar{\theta}) \geq \delta). \tag{2.6}$$

It suffices to show that $P(K(\hat{\theta}_n) - K(\bar{\theta}) \geq \delta) \to 0$. Consider the decomposition

$$K(\hat{\theta}_n) - K(\bar{\theta}) = \underbrace{K(\hat{\theta}_n) - K_n(\hat{\theta}_n)}_{\xrightarrow{P} 0} + \underbrace{K_n(\hat{\theta}_n) - K_n(\bar{\theta})}_{\leq 0 \text{ by definition}} + \underbrace{K_n(\bar{\theta}) - K(\bar{\theta})}_{\xrightarrow{P} 0},$$

where the two convergences in probability follow from condition (ii). Intuitively, it may be reasoned that $K(\hat{\theta}_n) - K(\bar{\theta})$ converges in probability to a limit no greater than 0 and hence $P(K(\hat{\theta}_n) - K(\bar{\theta}) \geq \delta) \to 0$. However, it is unknown *at this moment* whether $K(\hat{\theta}_n) - K(\bar{\theta})$ converges in probability at all. To circumvent this issue, we directly deduce an upper bound of $K(\hat{\theta}_n) - K(\bar{\theta})$ from the above decomposition:

$$\begin{aligned} K(\hat{\theta}_n) - K(\bar{\theta}) &\leq \sup_{\theta \in \Theta} |K_n(\theta) - K(\theta)| + |K_n(\bar{\theta}) - K(\bar{\theta})| \\ &\leq 2 \sup_{\theta \in \Theta} |K_n(\theta) - K(\theta)|. \end{aligned}$$

Then

$$P(K(\hat{\theta}_n) - K(\bar{\theta}) \geq \delta) \leq P(\sup_{\theta \in \Theta} |K_n(\theta) - K(\theta)| \geq \delta/2),$$

which tends to 0 by condition (ii). Hence $P(K(\hat{\theta}_n) - K(\bar{\theta}) \geq \delta) \to 0$.

Next, consider the case where $\hat{\theta}_n$ may exist with probability smaller than one. Let A_n be the event that $\hat{\theta}_n$ exists. The preceding proof shows that for any $\epsilon > 0$, $P(A_n \cap \{\|\hat{\theta}_n - \bar{\theta}\| > \epsilon\}) \to 0$. Then $P(A_n \cap \{\|\hat{\theta}_n - \bar{\theta}\| \leq \epsilon\}) \to 1$ because $P(A_n) \to 1$ by condition (iii). □

A Z-estimator $\hat{\theta}_n$ is defined as a solution to $T_n(\theta) = 0$ on Θ with a sample estimating function $T_n(\theta)$. The target parameter value $\bar{\theta}$ is defined as a solution to $T(\theta) = 0$, with the population estimating function $T(\theta) = E\{T_n(\theta)\}$. It is allowed that $T_n(\theta)$ may not be of the sample average form (2.3) and $T(\theta)$ may not be of the form (2.4).

Theorem 2.2 (General consistency of Z-estimation). *Suppose that*

(i) [Isolated target parameter] $\inf_{\theta\in\Theta:\|\theta-\bar{\theta}\|>\epsilon} \|T(\theta)\| > \|T(\bar{\theta})\| = 0$ *for each* $\epsilon > 0$;

(ii) [Uniform convergence] $\sup_{\theta\in\Theta} \|T_n(\theta) - T(\theta)\| \xrightarrow[\mathcal{P}]{} 0$;

(iii) [Existence of estimator] $\hat{\theta}_n$ *exists as a solution to* $T_n(\theta) = 0$ *on* Θ *with probability tending to one.*

Then $\hat{\theta}_n \xrightarrow[\mathcal{P}]{} \bar{\theta}$. *The result also holds if* $\hat{\theta}_n$ *is defined as a minimizer of* $\|T_n(\theta)\|$, *instead of a solution to* $T_n(\theta) = 0$, *and assumed to exist with probability tending to one.*

Proof. Take $K_n(\theta) = \|T_n(\theta)\|$ and $K(\theta) = \|T(\theta)\|$, and apply Theorem 2.1. Condition (ii) here implies condition (ii) in Theorem 2.1, $\sup_{\theta\in\Theta} |\|T_n(\theta)\| - \|T(\theta)\|| \xrightarrow[\mathcal{P}]{} 0$, by the triangle inequality $|\|T_n(\theta)\| - \|T(\theta)\|| \leq \|T_n(\theta) - T(\theta)\|$. [Note that $K_n(\theta)$ is *not* of the sample-average form (2.1) even when $T_n(\theta)$ is of the form (2.3). Hence Z-estimator $\hat{\theta}_n$ is treated as an M-estimator, but in a more general sense than in Definition 2.1.] □

In the scalar case, the following two results give consistency of a Z-estimator under more primitive conditions than in Theorem 2.2.

Theorem 2.3 (Consistency of Z-estimation under monotonicity). *Suppose that* θ *is scalar and*

(i) [Isolated target parameter] $T(\bar{\theta} - \epsilon) > 0 = T(\bar{\theta}) > T(\bar{\theta} + \epsilon)$ *for each* $\epsilon > 0$;

(ii) [Pointwise convergence] $T_n(\theta) \xrightarrow[\mathcal{P}]{} T(\theta)$ *for each* θ;

(iii) [Monotonicity of estimating function] $T_n(\theta)$ *is non-increasing in* θ;

(iv) [Existence of estimator] $\hat{\theta}_n$ *exists as a solution to* $T_n(\theta) = 0$ *on* Θ *with probability tending to one.*

Then $\hat{\theta}_n \xrightarrow[\mathcal{P}]{} \bar{\theta}$.

Before presenting a proof, we make some comments about Theorem 2.3, compared with Theorem 2.2. Condition (ii) above indicates pointwise convergence in probability, whereas condition (ii) in Theorem 2.2 indicates uniform convergence in probability. Moreover, given conditions (ii) and (iii), $T(\theta)$ is

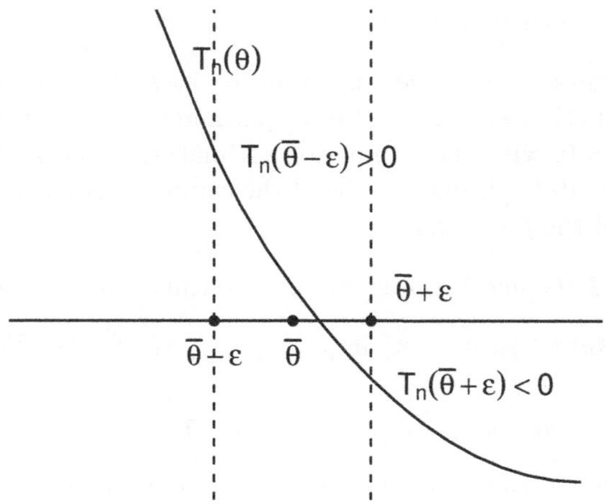

FIGURE 2.3
Relating the event $\hat{\theta}_n \in [\bar{\theta} - \epsilon, \bar{\theta} + \epsilon]$ to the values $T_n(\bar{\theta} \pm \epsilon)$.

non-increasing in θ and condition (i) above implies condition (i) in Theorem 2.2. [Exercise: Prove these statements.] If $T_n(\theta) = \frac{\mathrm{d}}{\mathrm{d}\theta} K_n(\theta)$ for some objective function $K_n(\theta)$, then the non-increasingness of $T_n(\theta)$ corresponds to the concavity of $K_n(\theta)$. A similar result holds if $T_n(\theta)$ is non-decreasing in θ, or equivalently $K_n(\theta)$ is convex, and condition (i) is modified properly. Hence Theorem 2.3 can be seen to establish consistency of M-estimation based on concavity (or convexity).

Proof. We only consider the case where $\hat{\theta}_n$ exists almost surely for $n \geq 1$; the case where $\hat{\theta}_n$ may not exist can be handled similarly as in the proof of Theorem 2.1. For any solution $\hat{\theta}_n$ to $T_n(\theta) = 0$, we show that $P(|\hat{\theta}_n - \bar{\theta}| > \epsilon) \to 0$ or $P(|\hat{\theta}_n - \bar{\theta}| \leq \epsilon) \to 1$, for each $\epsilon > 0$. Note that

$$\begin{aligned} P(|\hat{\theta}_n - \bar{\theta}| \leq \epsilon) &= P(\hat{\theta}_n \in [\bar{\theta} - \epsilon, \bar{\theta} + \epsilon]) \\ &\geq P(T_n(\bar{\theta} - \epsilon) > 0 > T_n(\bar{\theta} + \epsilon)), \end{aligned} \tag{2.7}$$

because by condition (iii), if $T_n(\bar{\theta} - \epsilon) > 0 > T_n(\bar{\theta} + \epsilon)$, then any solution $\hat{\theta}_n$ necessarily falls in the interval $[\bar{\theta} - \epsilon, \bar{\theta} + \epsilon]$. See Figure 2.3. It suffices to show that $P(T_n(\bar{\theta} - \epsilon) > 0 > T_n(\bar{\theta} + \epsilon)) \to 1$. By the union bound,

$$\begin{aligned} & P(T_n(\bar{\theta} - \epsilon) > 0 > T_n(\bar{\theta} + \epsilon)) \\ &= 1 - P(T_n(\bar{\theta} - \epsilon) \leq 0 \text{ or } T_n(\bar{\theta} + \epsilon) \geq 0) \\ &\geq 1 - P(T_n(\bar{\theta} - \epsilon) \leq 0) - P(T_n(\bar{\theta} + \epsilon) \geq 0). \end{aligned} \tag{2.8}$$

Moreover,

$$P(T_n(\bar{\theta} - \epsilon) \leq 0) = P(T_n(\bar{\theta} - \epsilon) - T(\bar{\theta} - \epsilon) \leq -T(\bar{\theta} - \epsilon)) \to 0,$$

because $-T(\bar{\theta}-\epsilon) < 0$ by condition (i), and $T_n(\bar{\theta}-\epsilon) \xrightarrow[\mathcal{P}]{} T(\bar{\theta}-\epsilon)$ by condition (ii). Similarly, $P(T_n(\bar{\theta}+\epsilon) \geq 0) \to 0$. Hence $P(T_n(\bar{\theta}-\epsilon) > 0 > T_n(\bar{\theta}+\epsilon)) \to 1$.

Note that the proof can also be presented such that it uses the following inequality, instead of (2.7) and (2.8):

$$P(|\hat{\theta}_n - \bar{\theta}| > \epsilon) \leq P(T_n(\bar{\theta}-\epsilon) \leq 0 \text{ or } T_n(\bar{\theta}+\epsilon) \geq 0).$$

This gives an upper bound on $P(|\hat{\theta}_n - \bar{\theta}| > \epsilon)$ in a different manner from the related inequality (2.6) in the proof of Theorem 2.1 (General consistency). Nevertheless, the essence of inequality (2.6) or (2.7) is to upper or lower bound the probability of an event about $\hat{\theta}_n$ in terms of the probability of an event about $T_n(\bar{\theta}-\epsilon)$ and $T_n(\bar{\theta}+\epsilon)$. □

Theorem 2.4 (Consistency of Z-estimation under continuity). *Suppose that θ is scalar and*

(i) [Isolated target parameter] $T(\bar{\theta}-\epsilon) > 0 = T(\bar{\theta}) > T(\bar{\theta}+\epsilon)$ for each $\epsilon > 0$;

(ii) [Pointwise convergence] $T_n(\theta) \xrightarrow[\mathcal{P}]{} T(\theta)$ for each θ;

(iii) [Continuity of estimating function] $T_n(\theta)$ is continuous in θ;

(iv) [Existence of estimator] $\hat{\theta}_n$ exists as a unique solution to $T_n(\theta) = 0$ on Θ with probability tending to one.

Then $\hat{\theta}_n \xrightarrow[\mathcal{P}]{} \bar{\theta}$.

We note that the conditions in Theorem 2.4 are the same as those in Theorem 2.3, except for the difference in condition (iii). In addition, Theorem 2.4 depends on the assumption that $\hat{\theta}_n$ is a unique solution to $T_n(\theta) = 0$, where Theorem 2.3 holds for $\hat{\theta}_n$ which is any solution to $T_n(\theta) = 0$ and may not be unique. If the uniqueness of $\hat{\theta}_n$ as a solution to $T_n(\theta) = 0$ were removed from condition (iii), then the conclusion needs to be modified as follows: A solution $\hat{\theta}_n$ to $T_n(\theta) = 0$ can be chosen for $n \geq 1$ such that $\hat{\theta}_n \xrightarrow[\mathcal{P}]{} \bar{\theta}$.

Proof. We only consider the case where $\hat{\theta}_n$ exists almost surely for $n \geq 1$. We show that $P(|\hat{\theta}_n - \bar{\theta}| \leq \epsilon) \to 1$, for each $\epsilon > 0$. Note that

$$\begin{aligned} P(|\hat{\theta}_n - \bar{\theta}| \leq \epsilon) &= P(\hat{\theta}_n \in [\bar{\theta}-\epsilon, \bar{\theta}+\epsilon]) \\ &\geq P(T_n(\bar{\theta}-\epsilon) > 0 > T_n(\bar{\theta}+\epsilon)), \end{aligned}$$

because if $T_n(\bar{\theta}-\epsilon) > 0 > T_n(\bar{\theta}+\epsilon)$, then there exists a solution to $T_n(\theta) = 0$ in the interval $[\bar{\theta}-\epsilon, \bar{\theta}+\epsilon]$ by the continuity of $T_n(\theta)$, and $\hat{\theta}_n$ as the unique solution to $T_n(\theta) = 0$ necessarily falls in this interval. The above inequality is similar to (2.7), but is justified in a different manner. The remaining proof is the same as the proof of Theorem 2.3 after (2.7). □

If $T_n(\theta) = \frac{\mathrm{d}}{\mathrm{d}\theta} K_n(\theta)$ for some objective function $K_n(\theta)$, then the continuity of $T_n(\theta)$ indicates that $K_n(\theta)$ is continuously differentiable. Hence

Theorem 2.4 can be seen to establish consistency of M-estimation based on smoothness. In fact, smoothness (more precisely, smoothness and dominance) and convexity represent two types of primitive conditions for obtaining uniform convergence of sample objective functions and hence consistency of M-estimation, as discussed in the subsequent sections.

2.3 Consistency based on smoothness

2.3.1 A uniform law of large numbers

We present a uniform law of large numbers (ULLN), which gives the uniform convergence of sample objective functions in probability as required in Theorem 2.1. See Ferguson (1996), Theorem 16(b), for a one-sided version of ULLN and van de Geer (2000), Lemmas 3.1 and 3.10, for a treatment using metric entropies, where almost surely convergence is considered in both references. Suppose that $Z_1, \ldots, Z_n$ are IID variables as Z. Let

$$K_n(\theta) = \frac{1}{n}\sum_{i=1}^{n} \rho(Z_i;\theta),$$
$$K(\theta) = E\{\rho(Z;\theta)\},$$

where Θ is a parameter space, and $\rho(z;\theta)$ is a scalar function of z, parameterized by θ.

Theorem 2.5 (ULLN). *Assume that the following conditions are satisfied:*

(i) [Compactness] $\Theta \subset \mathbb{R}^d$ *is compact, i.e.,* $\Theta \subset \mathbb{R}^d$ *is closed and bounded;*

(ii) [Continuity] $\rho(z;\theta)$ *is continuous in* θ *for each* z*;*

(iii) [Dominance] There exists a nonnegative function $M(z)$ *such that* $|\rho(z;\theta)| \le M(z)$ *for each* z *and* $\theta \in \Theta$ *and* $E\{M(Z)\} < \infty$.

Then $K(\theta)$ *is continuous in* θ*, and* $\sup_{\theta\in\Theta} |K_n(\theta) - K(\theta)| \xrightarrow{\mathcal{P}} 0$*, i.e.,*

$$\sup_{\theta\in\Theta} \left| \frac{1}{n}\sum_{i=1}^{n} \rho(Z_i;\theta) - E\{\rho(Z;\theta)\} \right| \xrightarrow{\mathcal{P}} 0. \tag{2.9}$$

Before handling the general case, it is helpful to consider the special case where Θ is of a finite size, say $\Theta = \{\theta_1, \ldots, \theta_J\}$ for a finite J (independent of n). In this case, by Theorem 1.7 (Law of large numbers), for each $j = 1, \ldots, J$,

$$\frac{1}{n}\sum_{i=1}^{n} \rho(Z_i;\theta_j) \xrightarrow{\mathcal{P}} E\{\rho(Z;\theta_j)\}.$$

Then (2.9) follows from the union bound: for any $\epsilon > 0$,

$$\begin{aligned}
&P\left(\sup_{\theta\in\Theta}\left|\frac{1}{n}\sum_{i=1}^{n}\rho(Z_i;\theta) - E\{\rho(Z;\theta)\}\right| > \epsilon\right)\\
&= P\left(\bigcup_{j=1}^{J}\left\{\left|\frac{1}{n}\sum_{i=1}^{n}\rho(Z_i;\theta_j) - E\{\rho(Z;\theta_j)\}\right| > \epsilon\right\}\right)\\
&\leq \sum_{j=1}^{J} P\left(\left|\frac{1}{n}\sum_{i=1}^{n}\rho(Z_i;\theta_j) - E\{\rho(Z;\theta_j)\}\right| > \epsilon\right) \to 0.
\end{aligned}$$

For the general case of Θ, a central task is to tackle the difficulty that the union bound argument does not *directly* work.

Proof. First, we show that $K(\theta)$ is continuous in θ, i.e., for any $\theta_0 \in \Theta$,

$$\lim_{\theta\to\theta_0} E\{\rho(Z;\theta)\} = E\{\rho(Z;\theta_0)\},$$

or equivalently, for any $\theta_0 \in \Theta$ and any sequence $\theta_k \in \Theta$, $k \geq 1$, with $\lim_{k\to\infty}\theta_k = \theta_0$,

$$\lim_{k\to\infty} E\{\rho(Z;\theta_k)\} = E\{\rho(Z;\theta_0)\}.$$

By continuity condition (ii), $\rho(z;\theta_k) \to \rho(z;\theta_0)$ for each z and hence $\rho(Z;\theta_k)\xrightarrow[\text{a.s.}]{}\rho(Z;\theta_0)$. With dominance condition (iii), the above convergence follows from Theorem 1.10 (Dominated convergence).

Next, we show the uniform convergence (2.9) in the general case of Θ. For any $\theta \in \Theta$ and $\delta > 0$, denote

$$B(\theta,\delta) = \{\tilde{\theta} \in \mathbb{R}^d : \|\tilde{\theta} - \theta\| \leq \delta\}.$$

By the compactness (more precisely, the boundedness) of Θ, let $\{B(\theta_j,\delta) : j = 1,\ldots,J\}$ be a finite collection of balls such that $\Theta \subset \cup_{j=1}^{J} B(\theta_j,\delta)$. Note that both J and $\{\theta_j : j = 1,\ldots,J\}$ depend on δ. [This construction is direct and does not require the Heine–Borel theorem that a collection of open sets whose union contains a compact set S has a finite collection whose union contains S.] For $\theta \in \Theta$, we pick j such that $\theta \in B(\theta_j,\delta)$ and consider the bound

$$\begin{aligned}
&\left|\frac{1}{n}\sum_{i=1}^{n}\rho(Z_i;\theta) - E\{\rho(Z;\theta)\}\right|\\
&\leq \left|\frac{1}{n}\sum_{i=1}^{n}\rho(Z_i;\theta) - \frac{1}{n}\sum_{i=1}^{n}\rho(Z_i;\theta_j)\right| + \left|\frac{1}{n}\sum_{i=1}^{n}\rho(Z_i;\theta_j) - E\{\rho(Z;\theta_j)\}\right| \qquad (2.10)\\
&\quad + \left|E\{\rho(Z;\theta_j)\} - E\{\rho(Z;\theta)\}\right|.
\end{aligned}$$

Then

$$\sup_{\theta\in\Theta}\left|\frac{1}{n}\sum_{i=1}^{n}\rho(Z_i;\theta)-E\{\rho(Z;\theta)\}\right|\leq\max_{j=1,\ldots,J}(\Delta_{1j}+\Delta_{2j}+\Delta_{3j}),$$

where

$$\Delta_{1j}=\sup_{\theta\in\Theta\cap B(\theta_j,\delta)}\left|\frac{1}{n}\sum_{i=1}^{n}\rho(Z_i;\theta)-\frac{1}{n}\sum_{i=1}^{n}\rho(Z_i;\theta_j)\right|$$

$$\Delta_{2j}=\left|\frac{1}{n}\sum_{i=1}^{n}\rho(Z_i;\theta_j)-E\{\rho(Z;\theta_j)\}\right|,$$

$$\Delta_{3j}=\sup_{\theta\in\Theta\cap B(\theta_j,\delta)}\left|E\{\rho(Z;\theta_j)\}-E\{\rho(Z;\theta)\}\right|.$$

From the preceding bound, for any $\epsilon>0$,

$$P\left(\sup_{\theta\in\Theta}\left|\frac{1}{n}\sum_{i=1}^{n}\rho(Z_i;\theta)-E\{\rho(Z;\theta)\}\right|>\epsilon\right)$$
$$\leq P\left(\max_{j=1,\ldots,J}\Delta_{1j}>\frac{\epsilon}{2}\right)+P\left(\max_{j=1,\ldots,J}\Delta_{2j}>\frac{\epsilon}{4}\right)+P\left(\max_{j=1,\ldots,J}\Delta_{3j}>\frac{\epsilon}{4}\right).$$

Then it suffices to show that $P\left(\max_{j=1,\ldots,J}\Delta_{1j}>\frac{\epsilon}{2}\right)\to0$, $P\left(\max_{j=1,\ldots,J}\Delta_{2j}>\frac{\epsilon}{4}\right)\to0$, and $P\left(\max_{j=1,\ldots,J}\Delta_{3j}>\frac{\epsilon}{4}\right)\to0$, for a suitable choice of δ. While other ways of splitting ϵ are allowed (for example, three equal parts), the particular split above is made for the convenience of a remark at the end of the proof.

Proof of $P(\max_{j=1,\ldots,J}\Delta_{2j}>\frac{\epsilon}{4})\to0$. Similarly as in the case of finite Θ,

$$P\left(\max_{j=1,\ldots,J}\Delta_{2j}>\frac{\epsilon}{4}\right)\leq\sum_{j=1}^{J}P\left(\Delta_{2j}>\frac{\epsilon}{4}\right)\to0.$$

Proof of $P(\max_{j=1,\ldots,J}\Delta_{3j}>\frac{\epsilon}{4})\to0$. For pedagogical purposes, we present a seemingly reasonable but incorrect proof. By the continuity of $E\{\rho(Z;\theta)\}$ in θ, for each θ_j, there exists $\delta_{1j}>0$ such that for $\|\theta-\theta_j\|\leq\delta_{1j}$, $|E\{\rho(Z;\theta)\}-E\{\rho(Z;\theta_j)\}|\leq\frac{\epsilon}{4}$. If we choose $\delta\leq\min_{j=1,\ldots,J}\delta_{1j}$, then $\max_{j=1,\ldots,J}\Delta_{3j}\leq\frac{\epsilon}{4}$, that is, $P(\max_{j=1,\ldots,J}\Delta_{3j}>\frac{\epsilon}{4})=0$. This proof is incorrect because δ_{1j} depends on θ_j, which in turn depends on the choice of δ. Therefore, δ may not be chosen to satisfy $\delta\leq\min_{j=1,\ldots,J}\delta_{1j}$.

To address this issue, we use the Heine–Cantor theorem that a continuous function on a compact set is uniformly continuous. [A function f is said to be uniformly continuous on a set $\mathcal{S}$ if for any $\epsilon>0$, there exists $\nu>0$ such that $|f(x)-f(\tilde{x})|\leq\epsilon$ for any $x,\tilde{x}\in\mathcal{S}$ satisfying $\|x-\tilde{x}\|\leq\nu$.] As proved earlier, the function $K(\theta)=E\{\rho(Z;\theta)\}$ is continuous on the compact set Θ, and hence is uniformly continuous, i.e., there exists $\delta_1 > 0$ such that for $\|\theta - \tilde{\theta}\| \leq \delta_1$, $|E\{\rho(Z;\theta)\} - E\{\rho(Z;\tilde{\theta})\}| \leq \frac{\epsilon}{4}$. Then for any $\delta \leq \delta_1$,

$$\max_{j=1,\ldots,J} \Delta_{3j} \leq \frac{\epsilon}{4}, \tag{2.11}$$

that is, $P(\max_{j=1,\ldots,J} \Delta_{3j} > \frac{\epsilon}{4}) = 0$.

Proof of $P(\max_{j=1,\ldots,J} \Delta_{1j} > \frac{\epsilon}{2}) \to 0$. For each j, we consider the bound:

$$\begin{aligned}\Delta_{1j} &\leq \frac{1}{n}\sum_{i=1}^{n} \underbrace{\sup_{\theta\in\Theta\cap B(\theta_j,\delta)} |\rho(Z_i;\theta) - \rho(Z_i;\theta_j)|}_{r_j(Z_i;\delta)} \\ &\leq \left|\frac{1}{n}\sum_{i=1}^{n} r_j(Z_i;\delta) - E\{r_j(Z;\delta)\}\right| + E\{r_j(Z;\delta)\}.\end{aligned}$$

The second inequality is reminiscent of the two steps (controlling the deviation and the mean of supremum) in Section 3.4. Then

$$\begin{aligned}&P(\max_{j=1,\ldots,J} \Delta_{1j} > \frac{\epsilon}{2}) \\ &\leq P\left(\max_{j=1,\ldots,J} \left|\frac{1}{n}\sum_{i=1}^{n} r_j(Z_i;\delta) - E\{r_j(Z;\delta)\}\right| > \frac{\epsilon}{4}\right) \\ &\quad + P(\max_{j=1,\ldots,J} E\{r_j(Z;\delta)\} > \frac{\epsilon}{4}).\end{aligned} \tag{2.12}$$

By condition (iii), $0 \leq r_j(z;\delta) \leq 2M(z)$ and $E\{r_j(Z;\delta)\} < \infty$. By the law of large numbers similarly as in the case of finite Θ,

$$\begin{aligned}&P\left(\max_{j=1,\ldots,J} \left|\frac{1}{n}\sum_{i=1}^{n} r_j(Z_i;\delta) - E\{r_j(Z;\delta)\}\right| > \frac{\epsilon}{4}\right) \\ &\leq \sum_{j=1}^{J} P\left(\left|\frac{1}{n}\sum_{i=1}^{n} r_j(Z_i;\delta) - E\{r_j(Z;\delta)\}\right| > \frac{\epsilon}{4}\right) \to 0.\end{aligned} \tag{2.13}$$

To control $\max_j E\{r_j(Z;\delta)\}$, let

$$r_*(z;\nu) = \sup_{\theta,\tilde{\theta}\in\Theta, \|\theta-\tilde{\theta}\|\leq\nu} |\rho(z;\theta) - \rho(z;\tilde{\theta})|,$$

as suggested by the argument to control $\max_j \Delta_{3j}$. By condition (ii), for each z, $\rho(z;\theta)$ is continuous in θ and hence uniformly continuous on the compact set Θ, which implies that $r_*(z;\nu) \to 0$ as $\nu \to 0$. By condition (iii), $0 \leq r_*(z;\nu) \leq 2M(z)$ and $E\{M(Z)\} < \infty$. Then by the dominated convergence theorem, $\lim_{\nu\to 0} E\{r_*(Z;\nu)\} = E(0) = 0$. From this property, there exists $\delta_2 > 0$ such that for $\nu \leq \delta_2$, $E\{r_*(Z;\nu)\} \leq \frac{\epsilon}{4}$. Hence for any $\delta \leq \delta_2$,

$$\max_{j=1,\ldots,J} E\{r_j(Z;\delta)\} \leq E\{r_*(Z;\delta)\} \leq \frac{\epsilon}{4}, \tag{2.14}$$

that is, $P(\max_{j=1,\ldots,J} E\{r_j(Z_i;\delta)\} > \frac{\epsilon}{4}) = 0$. For $\delta \le \delta_2$, combining (2.12)–(2.14) gives $P(\max_{j=1,\ldots,J} \Delta_{1j} > \frac{\epsilon}{2}) \to 0$.

Finally, the proof is completed by choosing $\delta = \min(\delta_1, \delta_2)$. More carefully, note that for $\|\theta - \tilde{\theta}\| \le \delta$, $|E\{\rho(Z;\theta)\} - E\{\rho(Z;\tilde{\theta})\}| \le E\{r_*(Z;\delta)\}$. Then with the split of ϵ, (2.11) is implied by (2.14) for $\delta \le \delta_2$. Therefore, the proof is also completed by picking δ_1 the same as δ_2 and setting $\delta = \delta_2$. □

By an examination of the arguments, a slightly shorter proof can be obtained for the property $P(\max_{j=1,\ldots,J} \Delta_{1j} > \frac{\epsilon}{2}) \to 0$. In fact, by the definition of $r_*(z,\nu)$, we have for each j,

$$\begin{aligned}\Delta_{1j} &\le \frac{1}{n}\sum_{i=1}^{n} r_*(Z_i;\delta)\\ &\le \left|\frac{1}{n}\sum_{i=1}^{n} r_*(Z_i;\delta) - E\{r_*(Z;\delta)\}\right| + E\{r_*(Z;\delta)\}.\end{aligned}$$

Then for $\delta \le \delta_2$, $E\{r_*(Z;\delta)\} \le \frac{\epsilon}{4}$ from (2.14), and

$$\begin{aligned}P(\max_{j=1,\ldots,J} \Delta_{1j} > \frac{\epsilon}{2}) &\le P\left(\frac{1}{n}\sum_{i=1}^{n} r_*(Z_i;\delta) > \frac{\epsilon}{2}\right)\\ &\le P\left(\left|\frac{1}{n}\sum_{i=1}^{n} r_*(Z_i;\delta) - E\{r_*(Z;\delta)\}\right| > \frac{\epsilon}{4}\right),\end{aligned}$$

which tends to 0 by the law of large numbers.

The preceding discussion also shows that for any $\epsilon > 0$, there exists $\delta > 0$ such that

$$P\left(\sup_{\theta,\tilde{\theta}\in\Theta, \|\theta-\tilde{\theta}\|\le\delta} \left|\frac{1}{n}\sum_{i=1}^{n}\rho(Z_i;\theta) - \frac{1}{n}\sum_{i=1}^{n}\rho(Z_i;\tilde{\theta})\right| > \epsilon\right) \to 0. \qquad (2.15)$$

This property, defined as *strong* stochastic equi-continuity in Section 2.6.1, can be seen as a key to the uniform convergence (2.9). See Section 2.6.1 for further discussion on uniform convergence and stochastic equi-continuity.

To help understanding, we provide an alternative proof of (2.9), adapted from the proof of Theorem 16(b) in Ferguson (1996), which uses the Heine–Borel theorem about covering, but not directly the Heine–Cantor theorem about uniform continuity. The first proof uses the Heine–Cantor theorem, but not directly the Heine–Borel theorem. [Nevertheless, the proof of the Heine–Cantor theorem relies on the Heine–Borel theorem.]

Second proof of uniform convergence (2.9). Fix any $\epsilon > 0$. For each $\theta \in \Theta$, let

$$r_\theta(z;\nu) = \sup_{\tilde{\theta}\in\Theta\cap B(\theta,\nu)} |\rho(z;\tilde{\theta}) - \rho(z;\theta)|.$$

By condition (ii), $r_\theta(z;\nu) \to 0$ as $\nu \to 0$. By condition (iii), $0 \le r_\theta(z;\nu) \le 2M(z)$ and $E\{M(Z)\} < 0$. Then by the dominated convergence theorem, $E\{r_\theta(Z;\theta)\} \to 0$ as $\nu \to 0$. For $\theta \in \Theta$, pick $\nu_\theta > 0$ such that $E\{r_\theta(Z;\theta)\} \le \frac{\epsilon}{4}$. The collection of balls $\{B(\theta,\nu_\theta) : \theta \in \Theta\}$ cover Θ, a compact subset of $\mathbb{R}^d$. By the Heine–Borel theorem, there exists a finite sub-collection of balls with centers $\theta_1, \ldots, \theta_J$ such that $\Theta \subset \cup_{j=1}^J B(\theta_j, \nu_{\theta_j})$. Then similarly to (2.10), we have

$$\sup_{\theta\in\Theta}\left|\frac{1}{n}\sum_{i=1}^n \rho(Z_i;\theta) - E\{\rho(Z;\theta)\}\right| \le \max_{j=1,\ldots,J}(\tilde{\Delta}_{1j} + \tilde{\Delta}_{2j} + \tilde{\Delta}_{3j}),$$

where

$$\begin{aligned}
\tilde{\Delta}_{1j} &= \sup_{\theta\in\Theta\cap B(\theta_j,\nu_{\theta_j})}\left|\frac{1}{n}\sum_{i=1}^n \rho(Z_i;\theta) - \frac{1}{n}\sum_{i=1}^n \rho(Z_i;\theta_j)\right| \\
\tilde{\Delta}_{2j} &= \left|\frac{1}{n}\sum_{i=1}^n \rho(Z_i;\theta_j) - E\{\rho(Z;\theta_j)\}\right|, \\
\tilde{\Delta}_{3j} &= \sup_{\theta\in\Theta\cap B(\theta_j,\nu_{\theta_j})} \left|E\{\rho(Z;\theta_j)\} - E\{\rho(Z;\theta)\}\right|.
\end{aligned}$$

For each j, by the construction of the balls $B(\theta_j, \nu_{\theta_j})$,

$$\begin{aligned}
\tilde{\Delta}_{3j} &\le E\{r_{\theta_j}(Z;\nu_{\theta_j})\} \le \frac{\epsilon}{4}, \\
\tilde{\Delta}_{1j} &\le \left|\frac{1}{n}\sum_{i=1}^n r_{\theta_j}(Z;\nu_{\theta_j}) - E\{r_{\theta_j}(Z;\nu_{\theta_j})\}\right| + E\{r_{\theta_j}(Z;\nu_{\theta_j})\} \\
&\le \left|\frac{1}{n}\sum_{i=1}^n r_{\theta_j}(Z;\nu_{\theta_j}) - E\{r_{\theta_j}(Z;\nu_{\theta_j})\}\right| + \frac{\epsilon}{4}.
\end{aligned}$$

Then the uniform convergence (2.9) is obtained:

$$\begin{aligned}
&P\left(\sup_{\theta\in\Theta}\left|\frac{1}{n}\sum_{i=1}^n \rho(Z_i;\theta) - E\{\rho(Z;\theta)\}\right| > \epsilon\right) \\
&\le P(\max_{j=1,\ldots,J} \Delta_{2j} > \frac{\epsilon}{4}) \\
&\quad + P\left(\max_{j=1,\ldots,J}\left|\frac{1}{n}\sum_{i=1}^n r_{\theta_j}(Z_i;\nu_{\theta_j}) - E\{r_{\theta_j}(Z;\nu_{\theta_j})\}\right| > \frac{\epsilon}{4}\right),
\end{aligned}$$

which tends to 0 by the law of large numbers. □

Another interesting aspect of the second proof is that the dominance condition (iii) is only used locally, i.e., in a non-empty neighborhood of each $\theta \in \Theta$. Therefore, Theorem 2.5 remains valid if condition (iii) is replaced by the following condition (which is seemingly weaker):

(iii†) [Local dominance] For each $\theta \in \Theta$, there exists a nonnegative function $M(z;\theta)$ such that $|\rho(z;\tilde{\theta})| \leq M(z;\theta)$ for each z and $\tilde{\theta} \in \Theta$ in a neighborhood of θ and $E\{M(Z;\theta)\} < \infty$.

By further reasoning, conditions (iii) and (iii†) are in fact equivalent to each other with a compact set Θ. The proof is left as Exercise 2.3.

2.3.2 Consistency of M-estimation

Combining Theorem 2.1 and the uniform law of large numbers (Theorem 2.5) gives the following consistency result.

Theorem 2.6 (Consistency of M-estimation under smoothness and dominance). *Consider the setting of Definition 2.1, where $K_n(\theta)$ and $K(\theta)$ are defined as in Theorem 2.5, $\hat{\theta}_n = \operatorname{argmin}_{\theta\in\Theta} K_n(\theta)$ and $\bar{\theta} = \operatorname{argmin}_{\theta\in\Theta} K(\theta)$. Assume that*

(i) [Same as in Theorem 2.5] $\Theta \subset \mathbb{R}^d$ is compact;

(ii) [Same as in Theorem 2.5] $\rho(z;\theta)$ is continuous in θ for each z;

(iii) [Same as in Theorem 2.5] There exists a nonnegative function $M(z)$ such that $|\rho(z;\theta)| \leq M(z)$ for each z and $\theta \in \Theta$ and $E\{M(Z)\} < \infty$;

(iv) [Unique target parameter] $\bar{\theta}$ is a unique minimizer of $K(\theta)$ on Θ.

Then $\hat{\theta}_n \underset{P}{\to} \bar{\theta}$.

Proof. We apply Theorem 2.1, where condition (ii) is satisfied by Theorem 2.5. The estimator $\hat{\theta}_n$ always exists because $K_n(\theta)$ is continuous on Θ (a compact set) and hence attains a minimum on Θ. It remains to verify condition (i) in Theorem 2.1, i.e., for any $\epsilon > 0$,

$$\inf_{\theta\in\Theta:\|\theta-\bar{\theta}\|>\epsilon} K(\theta) > K(\bar{\theta}).$$

In fact, by the compactness of Θ, the set $\Theta_\epsilon = \{\theta \in \Theta : \|\theta - \bar{\theta}\| \geq \epsilon\}$ is compact [Exercise]. By Theorem 2.5, the function $K(\theta) = E\{\rho(Z;\theta)\}$ is continuous in θ, and hence attains a minimum on Θ_ϵ, i.e., there exists $\tilde{\theta} \in \Theta_\epsilon$ such that $K(\tilde{\theta}) = \inf_{\theta\in\Theta_\epsilon} K(\theta)$. Then by condition (iv),

$$K(\bar{\theta}) < K(\tilde{\theta}) = \inf_{\theta\in\Theta_\epsilon} K(\theta),$$

which, together with $\inf_{\theta\in\Theta_\epsilon} K(\theta) \leq \inf_{\theta\in\Theta:\|\theta-\bar{\theta}\|>\epsilon} K(\theta)$, implies that $K(\bar{\theta}) < \inf_{\theta\in\Theta:\|\theta-\bar{\theta}\|>\epsilon} K(\theta)$. □

We discuss application of Theorem 2.6 to maximum likelihood estimation, as described in Example 2.1. For a parametric model $f(\cdot;\theta)$, the MLE

$\bar{\theta}_n$ is an M-estimator with $\rho(z;\theta) = -\log f(z;\theta)$. The target value is $\bar{\theta} = \text{argmin}_{\theta\in\Theta} E\{-\log f(Z;\theta)\}$, which can be equivalently stated as

$$\begin{aligned}\bar{\theta} &= \text{argmin}_{\theta\in\Theta} E\{-\log \frac{f(Z;\theta)}{f^*(Z)}\}\\ &= \text{argmin}_{\theta\in\Theta} D_{\text{KL}}(f^* \| f(\cdot;\theta)),\end{aligned} \tag{2.16}$$

where $f^*(\cdot)$ denotes the probability density function of Z, and $D_{\text{KL}}(\cdot\|\cdot)$ denotes the Kullback–Leibler (KL) divergence. For convenience, $f^*(\cdot)$ is called the data density (or data distribution) and $f(\cdot;\theta)$ is called the model density (or model distribution). Then $D_{\text{KL}}(f^*\|f(\cdot;\theta))$ is the KL divergence from the model distribution to the data distribution.

For two probability distributions P and Q with density functions $p(\cdot)$ and $q(\cdot)$, the KL divergence from Q to P is defined as

$$D_{\text{KL}}(p\|q) = E_p(\log \frac{p}{q}) = \int p(z) \log \frac{p(z)}{q(z)} \, \mathrm{d}z,$$

where

$$p(z)\log\frac{p(z)}{q(z)} = \begin{cases} 0, & \text{if } p(z) = 0, \\ \infty, & \text{if } p(z) > 0,\ q(z) = 0. \end{cases}$$

Hence $D_{\text{KL}}(p\|q) = \infty$ if $P(\{z : q(z) = 0\}) > 0$, i.e., P is not absolutely continuous with respect to Q. In other words, for $D_{\text{KL}}(p\|q)$ to be finite, P needs to be absolutely continuous with respect to Q. [We say that P is absolutely continuous with respect to Q if $P(\{z : q(z) = 0\}) = 0$ or $Q(A) = 0$ implies $P(A) = 0$ for any event A.] By Jensen's inequality,

$$D_{\text{KL}}(p\|q) \geq 0, \tag{2.17}$$

where the equality holds if and only if $P = Q$ or $p(z) = q(z)$ almost surely under P or Q.

We consider two distinct scenarios. In the first scenario, the parametric model $f(\cdot;\theta)$ is assumed to be correctly specified, i.e., $f^*(\cdot) = f(\cdot;\theta^*)$ for some true value θ^*. From the property (2.17), any choice of θ^* is a minimizer of $K(\theta) = E\{-\log f(Z;\theta)\}$ and hence a valid choice of $\bar{\theta}$. Conversely, any choice of $\bar{\theta}$ is a valid choice of θ^* such that $f^*(\cdot) = f(\cdot;\theta^*)$. Therefore, $\bar{\theta}$ is unique and unequivocally $\bar{\theta} = \theta^*$ if and only if θ^* is unique, i.e., the following identification condition holds:

$$f(z;\theta) = f(z;\theta^*) \text{ almost surely under } f(\cdot;\theta^*) \text{ if and only if } \theta = \theta^*. \tag{2.18}$$

Typically, the parameter space Θ can be properly restricted such that (2.18) is satisfied. Then application of Theorem 2.6 to the MLE $\hat{\theta}_n$ gives the following result.

Corollary 2.1 (Consistency of MLE with correct model specification). *Suppose that $Z_1, \ldots, Z_n$ are IID variables from $f(\cdot;\theta^*)$ for some (unknown) true value θ^*. If conditions (i)–(iii) in Theorem 2.5 with $\rho(z;\theta) = -\log f(z;\theta)$ and the identification condition* (2.18) *are satisfied, then $\hat{\theta}_n$ is consistent for θ^*, i.e., $\hat{\theta}_n \xrightarrow[P]{} \theta^*$. The result also holds with $\rho(z;\theta) = -\log \frac{f(z;\theta)}{f(z;\theta^*)}$ or other choice of $\rho(z;\theta)$ which differs from $-\log f(z;\theta)$ by a fixed function of z, independently of θ.*

In the second scenario, the parametric model $f(\cdot;\theta)$ may be misspecified, i.e., there exists no value θ^* such that $f^*(\cdot) = f(\cdot;\theta^*)$. The target value $\bar{\theta}$ remains well-defined by (2.16). Application of Theorem 2.6 to the MLE $\hat{\theta}_n$ gives the following result. [It is possible that $D_{\mathrm{KL}}(f^* \| f(\cdot;\theta))$ admits two different minimizers, $\theta_1 \neq \theta_2$, and the two densities $f(\cdot;\theta_1)$ and $f(\cdot;\theta_2)$ represent different distributions. Hence uniqueness in the identification of $f(\cdot;\theta)$ by θ may not imply uniqueness of $\bar{\theta}$.]

Corollary 2.2 (Consistency of MLE with model misspecification). *Suppose that $Z_1, \ldots, Z_n$ are IID variables from $f^*(\cdot)$ which may differ from $f(\cdot;\theta)$ for each $\theta \in \Theta$. If conditions (i)–(iii) in Theorem 2.5 with $\rho(z;\theta) = -\log f(z;\theta)$ are satisfied and $\bar{\theta}$ is a unique minimizer of $D_{\mathrm{KL}}(f^* \| f(\cdot;\theta))$, then $\hat{\theta}_n$ is consistent for $\bar{\theta}$, i.e., $\hat{\theta}_n \xrightarrow[P]{} \bar{\theta}$. The result also holds with $\rho(z;\theta) = -\log \frac{f(z;\theta)}{f(z;\bar{\theta})}$ or other choice of $\rho(z;\theta)$ which differs from $-\log f(z;\theta)$ by a fixed function of z, independently of θ.*

From Corollaries 2.1 and 2.2, we make the following observations. An M-estimator (for example, the MLE) is usually consistent for its target value, subject to suitable regularity conditions, but independently of how the M-estimator is constructed (i.e., the choice of $\rho(z;\theta)$) or whether the related modeling assumptions are satisfied. The choice of $\rho(z;\theta)$, however, determines the meaning of the target value, in particular, whether the target value recovers the true parameter value when the modeling assumptions are satisfied. For maximum likelihood estimation, it is remarkable that the choice of the negative log-likelihood, $-\log f(z;\theta)$, not only admits an intuitively simple interpretation but also theoretically leads to consistency for the true parameter value under correct model specification.

2.3.3 Consistency of Z-estimation

Suppose that $Z_1, \ldots, Z_n$ are IID variables as Z. As in Definition 2.2, let

$$T_n(\theta) = \frac{1}{n} \sum_{i=1}^{n} \psi(Z_i;\theta),$$
$$T(\theta) = E\{\psi(Z;\theta)\},$$

where Θ is a parameter space, and $\psi(z;\theta)$ is a vector function of z, parameterized by θ, and assumed to be of the same dimension as θ.

A consistency result for Z-estimation can also be obtained by combining Theorem 2.2 and a multivariate version of Theorem 2.5. The proof is left as Exercise 2.7.

Theorem 2.7 (Consistency of Z-estimation under smoothness and dominance). *Let $\hat{\theta}_n$ be a solution to $T_n(\theta) = 0$, and $\bar{\theta}$ be a solution to $T(\theta) = 0$ on Θ. Assume that the following conditions are satisfied:*

(i) [Compactness] $\Theta \subset \mathbb{R}^d$ is compact;

(ii) [Continuity] $\psi(z; \theta)$ is continuous in θ for each z;

(iii) [Dominance] There exists a nonnegative function $M(z)$ such that $\|\psi(z; \theta)\| \leq M(z)$ for each z and $\theta \in \Theta$ and $E\{M(Z)\} < \infty$;

(iv) [Unique target parameter] $\bar{\theta}$ is a unique solution to $T(\theta) = 0$ on Θ;

(v) [Existence of estimator] $\hat{\theta}_n$ exists as a solution to $T_n(\theta) = 0$ on Θ with probability tending to one.

Then $\hat{\theta}_n \underset{P}{\to} \bar{\theta}$.

In contrast with Theorem 2.6, condition (v) is introduced because $T_n(\theta) = 0$ may not admit a solution on Θ. Alternatively, if $\hat{\theta}_n$ is redefined as a minimizer of $\|T_n(\theta)\|$ on Θ, then Theorem 2.7 holds with condition (v) removed, because $\hat{\theta}_n$ always exists (by the continuity of $\|T_n(\theta)\|$ and compactness of Θ), although $\hat{\theta}_n$ may not solve $T_n(\theta) = 0$.

2.4 Asymptotic normality based on smoothness

2.4.1 Asymptotic normality of Z-estimation

For pedagogical purposes, we provide two results, depending on slightly different sets of regularity conditions, to establish asymptotic normality of Z-estimation, provided that consistency is achieved. As a slight generalization, we allow that $\hat{\theta}_n$ satisfies $T_n(\hat{\theta}_n) = o_p(n^{-1/2})$ instead of $T_n(\hat{\theta}_n) = 0$ exactly.

Theorem 2.8 (Asymptotic normality of Z-estimation under smoothness and dominance). *In the setting of Definition 2.2, let $\hat{\theta}_n$ be defined such that $T_n(\hat{\theta}_n) = o_p(n^{-1/2})$. Assume that the following conditions are satisfied, with $\theta = (\theta^{(1)}, \ldots, \theta^{(d)})^{\mathrm{T}}$ and $\psi = (\psi^{(1)}, \ldots, \psi^{(d)})^{\mathrm{T}}$:*

(i) [Interior target parameter] $\bar{\theta}$ is in the interior of Θ with $T(\bar{\theta}) = 0$;

(ii) [Smoothness] $\psi(z; \theta)$ is twice differentiable in θ for each z;

(iii) [Dominance] There exists a nonnegative function $M(z)$ such that $E\{M(Z)\} < \infty$ and for each z and θ in a neighborhood of $\bar{\theta}$,

$$\max_{j=1,\ldots,d} \left\| \frac{\partial^2 \psi^{(j)}}{\partial\theta\partial\theta^{\mathrm{T}}}(z;\theta) \right\| \le M(z);$$

(iv) [Finite moments and invertibility] $G = \operatorname{var}\{\psi(Z;\bar{\theta})\} = E\{\psi(Z;\bar{\theta})\psi(Z;\bar{\theta})^{\mathrm{T}}\}$ exists with finite elements, and $H = E\{\frac{\partial\psi}{\partial\theta^{\mathrm{T}}}(Z;\bar{\theta})\}$ exists and is invertible, where $\frac{\partial\psi}{\partial\theta^{\mathrm{T}}}$ is a $d \times d$ matrix with (j,k) element $\frac{\partial\psi^{(j)}}{\partial\theta^{(k)}}$.

If $\hat{\theta}_n \xrightarrow[\mathcal{P}]{} \bar{\theta}$, then

$$\sqrt{n}(\hat{\theta}_n - \bar{\theta}) = -H^{-1}\frac{1}{\sqrt{n}}\sum_{i=1}^{n}\psi(Z_i;\bar{\theta}) + o_p(1), \tag{2.19}$$

and hence, with $H^{-\mathrm{T}} = (H^{-1})^{\mathrm{T}}$,

$$\sqrt{n}(\hat{\theta}_n - \bar{\theta}) \xrightarrow[\mathcal{D}]{} N(0, H^{-1}GH^{-\mathrm{T}}). \tag{2.20}$$

Proof. Suppose $\hat{\theta}_n \xrightarrow[\mathcal{P}]{} \bar{\theta}$, i.e., $\hat{\theta}_n - \bar{\theta} = o_p(1)$. We use a matrix extension of the o_p and O_p notation (Section 1.2). A sequence of random matrices is $o_p(1)$ or $O_p(1)$ if and only if each component is $o_p(1)$ or, respectively, $O_p(1)$. For $j = 1, \ldots, d$, by a second-order Taylor expansion with conditions (i) and (ii),

$$\begin{aligned} o_p(n^{-1/2}) &= T_n^{(j)}(\hat{\theta}_n) \\ &= T_n^{(j)}(\bar{\theta}) + \frac{1}{n}\sum_{i=1}^{n}\frac{\partial\psi^{(j)}}{\partial\theta^{\mathrm{T}}}(Z_i;\bar{\theta})(\hat{\theta}_n - \bar{\theta}) \\ &\quad + \frac{1}{2}(\hat{\theta}_n - \bar{\theta})^{\mathrm{T}}\left\{\frac{1}{n}\sum_{i=1}^{n}\frac{\partial^2\psi^{(j)}}{\partial\theta\partial\theta^{\mathrm{T}}}(Z_i;\theta)\Big|_{\theta=\tilde{\theta}_{nj}}\right\}(\hat{\theta}_n - \bar{\theta}), \end{aligned}$$

where $\tilde{\theta}_{nj}$ lies between $\hat{\theta}_n$ and $\bar{\theta}$, and $T_n^{(j)}(\theta) = \frac{1}{n}\sum_{i=1}^{n}\psi^{(j)}(Z_i;\theta)$. [Question: Can we use a second-order Taylor expansion of $T_n(\hat{\theta}_n)$ around $\bar{\theta}$ and complete the proof?] By condition (iii), it can be shown that (Exercise 2.8)

$$\frac{1}{n}\sum_{i=1}^{n}\frac{\partial^2\psi^{(j)}}{\partial\theta\partial\theta^{\mathrm{T}}}(Z_i;\theta)\Big|_{\theta=\tilde{\theta}_{nj}} = O_p(1),$$

which indicates that each component of the random matrix on the left-hand side is $O_p(1)$. Then by a multivariate extension of Proposition 1.1,

$$(\hat{\theta}_n - \bar{\theta})^{\mathrm{T}}\left\{\frac{1}{n}\sum_{i=1}^{n}\frac{\partial^2\psi^{(j)}}{\partial\theta\partial\theta^{\mathrm{T}}}(Z_i;\theta)\Big|_{\theta=\tilde{\theta}_{nj}}\right\} = o_p(1)O_p(1) = o_p(1).$$

Substituting this into the Taylor expansion shows that for $j = 1, \ldots, d$,

$$o_p(n^{-1/2}) = T_n^{(j)}(\bar{\theta}) + \frac{1}{n}\sum_{i=1}^{n} \frac{\partial \psi^{(j)}}{\partial \theta^{\mathrm{T}}}(Z_i; \bar{\theta})(\hat{\theta}_n - \bar{\theta}) + o_p(1)(\hat{\theta}_n - \bar{\theta}).$$

By Corollary 1.1, stacking components which are $o_p(1)$ into a vector (or a matrix) remains $o_p(1)$. Then stacking the d components from the preceding display leads to

$$o_p(n^{-1/2}) = T_n(\bar{\theta}) + \frac{1}{n}\sum_{i=1}^{n} \frac{\partial \psi}{\partial \theta^{\mathrm{T}}}(Z_i; \bar{\theta})(\hat{\theta}_n - \bar{\theta}) + o_p(1)(\hat{\theta}_n - \bar{\theta}),$$

where $o_p(1)$ represents a random matrix such that each component is $o_p(1)$. By condition (iv) and the law of large numbers, $\frac{1}{n}\sum_{i=1}^{n} \frac{\partial \psi}{\partial \theta^{\mathrm{T}}}(Z_i; \bar{\theta}) \xrightarrow[P]{} H$. Then

$$o_p(n^{-1/2}) = T_n(\bar{\theta}) + (H + o_p(1))(\hat{\theta}_n - \bar{\theta}),$$

and after multiplying both sides by $\sqrt{n}$ and rearranging yields

$$(H + o_p(1))\sqrt{n}(\hat{\theta}_n - \bar{\theta}) = -\sqrt{n}T_n(\bar{\theta}) + o_p(1).$$

By condition (iv) and the central limit theorem, $\sqrt{n}T_n(\bar{\theta}) \xrightarrow[\mathcal{D}]{} N(0, G)$ and hence $\sqrt{n}T_n(\bar{\theta}) = O_p(1)$. Further multiplying both sides above by $(H + o_p(1))^{-1}$ yields

$$\begin{aligned}
\sqrt{n}(\hat{\theta}_n - \bar{\theta}) &= -(H + o_p(1))^{-1}(\sqrt{n}T_n(\bar{\theta}) + o_p(1)) \\
&= -(H^{-1} + o_p(1))(\sqrt{n}T_n(\bar{\theta}) + o_p(1)) \\
&= -H^{-1}\sqrt{n}T_n(\bar{\theta}) + o_p(1),
\end{aligned}$$

which is the asymptotic expansion (2.19). The second equality above uses $(H + o_p(1))^{-1} = H^{-1} + o_p(1)$ (Exercise 2.9), and the third equality uses Proposition 1.1. Finally, the asymptotic normality (2.20) follows from the asymptotic expansion (2.19) in combination with the cental limit theorem, $\sqrt{n}T_n(\bar{\theta}) \xrightarrow[\mathcal{D}]{} N(0, G)$ and Slutsky's theorem. □

The asymptotic expansion (2.19) can be equivalently written as

$$\hat{\theta}_n - \bar{\theta} = -H^{-1}\frac{1}{n}\sum_{i=1}^{n} \psi(Z_i; \bar{\theta}) + o_p(n^{-1/2}), \tag{2.21}$$

which indicates that $\hat{\theta}_n - \bar{\theta}$ behaves asymptotically as the sample average $-H^{-1}\frac{1}{n}\sum_{i=1}^{n}\psi(Z_i; \bar{\theta})$ (which is $O_p(n^{-1/2})$), up to a remainder of order $o_p(n^{-1/2})$. The asymptotic expansion (2.21) is called a first-order or $n^{-1/2}$ asymptotic expansion for $\hat{\theta}_n$, where the leading term, of order $O_p(n^{-1/2})$, is a sample average of IID variables and the remainder is $o_p(n^{-1/2})$. The function

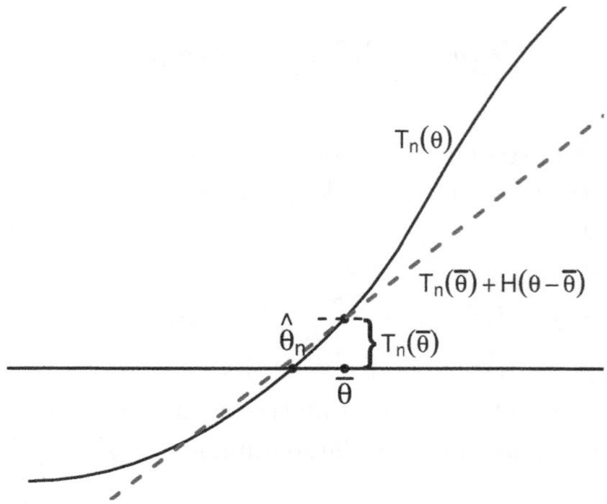

FIGURE 2.4
Linear approximation of $T_n(\theta)$ near $\bar{\theta}$. Then $\hat{\theta}_n - \bar{\theta}$ is approximated as $-H^{-1}T_n(\bar{\theta})$.

$-H^{-1}\psi(Z;\bar{\theta})$ is called the influence function of $\hat{\theta}_n$. The rescaled asymptotic variance for $\hat{\theta}_n$, $\Sigma = H^{-1}GH^{-\mathrm{T}}$, is called the sandwich variance formula. [The unscaled asymptotic variance for $\hat{\theta}_n$ is Σ/n.]

Heuristically, the main idea in the preceding proof can be summarized as follows. For θ in a neighborhood of $\bar{\theta}$, $T_n(\theta)$ behaves approximately as a linear function:

$$T_n(\theta) \approx T_n(\bar{\theta}) + H(\theta - \bar{\theta}).$$

Then $\hat{\theta}_n$ as a solution of $T_n(\theta) = 0$ behaves like the solution of $T_n(\bar{\theta}) + H(\theta - \bar{\theta}) = 0$, that is, $\hat{\theta}_n \approx \bar{\theta} - H^{-1}T_n(\bar{\theta})$. See Figure 2.4 for an illustration. The technical essence of the proof is then to make precise the orders of remainders in such approximations.

The following result establishes the asymptotic expansion (2.19) under a weaker set of conditions, depending only on the first-order derivatives of components of $\psi(z;\theta)$, but not second-order derivatives. This is achieved by carefully using a first-order Taylor expansion and the uniform law of large numbers (Theorem 2.5).

Theorem 2.9 (Asymptotic normality of Z-estimation under smoothness and dominance). *In the setting of Definition 2.2, let $\hat{\theta}_n$ be defined such that $T_n(\hat{\theta}_n) = o_p(n^{-1/2})$. Assume that the following conditions are satisfied:*

(i) [Same as in Theorem 2.8] $\bar{\theta}$ is in the interior of Θ with $T(\bar{\theta}) = 0$;

(ii) [Smoothness] $\psi(z;\theta)$ is continuously differentiable in θ for each z;

(*iii*) *[Local dominance near $\bar{\theta}$] There exists a nonnegative function $M(z)$ such that $E\{M(Z)\} < \infty$ and for each z and θ in a neighborhood of $\bar{\theta}$,*

$$\left\| \frac{\partial \psi}{\partial \theta^{\mathrm{T}}}(z;\theta) \right\| \leq M(z); \tag{2.22}$$

(*iv*) *[Same as in Theorem 2.8] $G = \mathrm{var}\{\psi(Z;\bar{\theta})\} = E\{\psi(Z;\bar{\theta})\psi(Z;\bar{\theta})^{\mathrm{T}}\}$ exists with finite elements, and $H = E\{\frac{\partial \psi}{\partial \theta^{\mathrm{T}}}(Z;\bar{\theta})\}$ exists and is invertible.*

If $\hat{\theta}_n \xrightarrow[\mathcal{P}]{} \bar{\theta}$, then (2.19) *and* (2.20) *hold.*

Before the proof, we discuss how additional conditions can be incorporated to ensure consistency $\hat{\theta}_n \xrightarrow[\mathcal{P}]{} \bar{\theta}$. Suppose that condition (iii) is strengthened to local dominance everywhere in Θ:

(iii†) [Local dominance] For each $\theta \in \Theta$, there exists a nonnegative function $M(z;\theta)$ such that $E\{M(Z;\theta)\} < \infty$ and for each z and $\tilde{\theta} \in \Theta$ in a neighborhood of θ,

$$\left\| \frac{\partial \psi}{\partial \theta^{\mathrm{T}}}(z;\tilde{\theta}) \right\| \leq M(z;\theta).$$

Condition (iii†) implies condition (iii) with $M(z) = M(z;\bar{\theta})$. Similarly to the equivalence between dominance condition (iii) and local dominance condition (iii†) for ULLN in Section 2.3.1, condition (iii†) above is equivalent to the following dominance condition provided Θ is compact:

(iii‡) [Dominance] There exists a nonnegative function $M(z)$ such that $E\{M(Z)\} < \infty$ and for each z and $\theta \in \Theta$ (not just in a neighborhood of $\bar{\theta}$), (2.22) holds.

Moreover, given the boundedness of Θ, conditions (ii) and (iii‡) above imply conditions (ii) and (iii) in Theorem 2.7 (Consistency of Z-estimation). [Exercise: Prove these statements.] Therefore, if conditions (i), (iv), and (v) in Theorem 2.7 and conditions (i), (ii), (iii†) or (iii‡), and (iv) above are satisfied, then $\hat{\theta}_n \xrightarrow[\mathcal{P}]{} \bar{\theta}$ and (2.19) and (2.20) hold.

Proof. Suppose $\hat{\theta}_n \xrightarrow[\mathcal{P}]{} \bar{\theta}$. By a first-order Taylor expansion with conditions (i) and (ii),

$$\begin{aligned} o_p(n^{-1/2}) &= T_n(\hat{\theta}_n) \\ &= T_n(\bar{\theta}) + \frac{1}{n}\sum_{i=1}^{n} \frac{\partial \psi}{\partial \theta^{\mathrm{T}}}(Z_i;\theta)\Big|_{\theta=\tilde{\theta}_n} (\hat{\theta}_n - \bar{\theta}), \end{aligned}$$

where $\tilde{\theta}_n$ lies between $\hat{\theta}_n$ and $\bar{\theta}$. Then

$$\sqrt{n}(\hat{\theta}_n - \bar{\theta}) = -\left\{ \frac{1}{n}\sum_{i=1}^{n} \frac{\partial \psi}{\partial \theta^{\mathrm{T}}}(Z_i;\theta)\Big|_{\theta=\tilde{\theta}_n} \right\}^{-1} (\sqrt{n}T_n(\bar{\theta}) + o_p(1)).$$

From the proof of Theorem 2.8, it suffices to show that

$$\underbrace{\frac{1}{n}\sum_{i=1}^{n}\frac{\partial\psi}{\partial\theta^{\mathrm{T}}}(Z_i;\theta)\Big|_{\theta=\tilde{\theta}_n}}_{H_n(\tilde{\theta}_n)} \xrightarrow[\mathcal{P}]{} H,$$

where $H_n(\theta) = \frac{1}{n}\sum_{i=1}^{n}\frac{\partial\psi}{\partial\theta^{\mathrm{T}}}(Z_i;\theta)$.

For $\delta > 0$, denote $B(\bar{\theta},\delta) = \{\theta \in \mathbb{R}^d : \|\theta - \bar{\theta}\| \leq \delta\}$. For any $\epsilon > 0$ and $\delta > 0$ satisfying $B(\bar{\theta},\delta) \subset \Theta$, consider the bound

$$\begin{aligned}
&P(\|H_n(\tilde{\theta}_n) - H\| > \epsilon) \\
&\leq P(\|\hat{\theta}_n - \bar{\theta}\| > \delta) + P(\|\hat{\theta}_n - \bar{\theta}\| \leq \delta, \|H_n(\tilde{\theta}_n) - H\| > \epsilon) \\
&\leq P(\|\hat{\theta}_n - \bar{\theta}\| > \delta) + P(\sup_{\|\theta-\bar{\theta}\|\leq\delta} \|H_n(\theta) - H\| > \epsilon).
\end{aligned}$$

By assumption $\hat{\theta}_n \xrightarrow[\mathcal{P}]{} \bar{\theta}$, the first probability tends to 0: $P(\|\hat{\theta}_n - \bar{\theta}\| > \delta) \to 0$. It remains to show that the second probability also tends to 0 for a suitable choice of δ. In fact, this probability can be bounded as

$$\begin{aligned}
&P(\sup_{\|\theta-\bar{\theta}\|\leq\delta} \|H_n(\theta) - H\| > \epsilon) \\
&\leq P\left(\sup_{\|\theta-\bar{\theta}\|\leq\delta} \|H_n(\theta) - H(\theta)\| > \frac{\epsilon}{2}\right) + P\left(\sup_{\|\theta-\bar{\theta}\|\leq\delta} \|H(\theta) - H\| > \frac{\epsilon}{2}\right),
\end{aligned} \tag{2.23}$$

where $H(\theta) = E\{\frac{\partial\psi}{\partial\theta^{\mathrm{T}}}(Z;\theta)\}$ and hence $H = H(\bar{\theta})$. By condition (iii), there exists $\delta_1 > 0$ such that $B(\bar{\theta},\delta_1) \subset \Theta$ and (2.22) holds for $\theta \in B(\bar{\theta},\delta_1)$. Then a matrix extension of Theorem 2.5 applies to $\rho(z;\theta) = \frac{\partial\psi}{\partial\theta^{\mathrm{T}}}(z;\theta)$ on the compact set $B(\bar{\theta},\delta_1)$, indicating that $H(\theta)$ is continuous on $B(\bar{\theta},\delta_1)$ and

$$P(\sup_{\|\theta-\bar{\theta}\|\leq\delta_1} \|H_n(\theta) - H(\theta)\| > \frac{\epsilon}{2}) \to 0. \tag{2.24}$$

By the continuity of $H(\theta)$ on $B(\bar{\theta},\delta_1)$, there exists $0 < \delta_2 \leq \delta_1$ such that $\|H(\theta) - H\| \leq \frac{\epsilon}{2}$ for $\theta \in B(\bar{\theta},\delta_2)$, and hence $P(\sup_{\|\theta-\bar{\theta}\|\leq\delta_2} \|H(\theta) - H\| > \frac{\epsilon}{2}) = 0$. Combining this with (2.23) and (2.24) shows that for $\delta = \delta_2$, $P(\sup_{\|\theta-\bar{\theta}\|\leq\delta} \|H_n(\theta) - H\| > \epsilon) \to 0$. □

Example 2.5 (Joint asymptotic distribution of sample mean and standard deviation). Let $X_1, \ldots, X_n$ be IID as X from a distribution with mean μ^*, variance σ^{*2}, and a finite fourth moment. Define $\bar{X}_n = \frac{1}{n}\sum_{i=1}^{n} X_i$ and $\tilde{s}_n^2 = \frac{1}{n}\sum_{i=1}^{n}(X_i - \bar{X}_n)^2$. In Example 1.3, the asymptotic variance of $\tilde{s}_n$ is obtained using the delta method. The joint asymptotic variance of $(\bar{X}_n, \tilde{s}_n)$ can also be derived using the delta method (Exercise 1.9). Alternatively, we apply the sandwich variance formula to find the asymptotic variance matrix.

We treat the pair $(\hat{\theta}_{n1}, \hat{\theta}_{n2}) = (\bar{X}_n, \tilde{s}_n)$ as jointly a solution to

$$\frac{1}{n}\sum_{i=1}^{n}(X_i - \theta_1) = 0,$$
$$\frac{1}{n}\sum_{i=1}^{n}(X_i - \theta_1)^2 - \theta_2^2 = 0.$$

In other words, $\hat{\theta}_n = (\hat{\theta}_{n1}, \hat{\theta}_{n2})^{\mathrm{T}}$ is a Z-estimator defined by the estimating function in $\theta = (\theta_1, \theta_2)^{\mathrm{T}}$:

$$\psi(X;\theta) = \begin{pmatrix} X - \theta_1 \\ (X-\theta_1)^2 - \theta_2^2 \end{pmatrix}.$$

The target value, determined by solving $E\{\psi(X;\theta)\} = 0$, is $\bar{\theta} = (\bar{\theta}_1, \bar{\theta}_2)^{\mathrm{T}} = (\mu^*, \sigma^*)^{\mathrm{T}}$. Then $\sqrt{n}(\hat{\theta}_n - \bar{\theta}) \underset{\mathcal{D}}{\to} \mathrm{N}(0, \Sigma)$, where

$$G = \mathrm{var}\{\psi(X;\bar{\theta})\} = \begin{pmatrix} \sigma^{*2} & \mathrm{cov}(X,(X-\mu^*)^2) \\ \mathrm{cov}((X-\mu^*)^2, X) & \mathrm{var}((X-\mu^*)^2) \end{pmatrix},$$
$$H = E\left\{\frac{\partial\psi}{\partial\theta^{\mathrm{T}}}(X;\bar{\theta})\right\} = E\begin{pmatrix} -1 & 0 \\ -2(X-\mu^*) & -2\sigma^* \end{pmatrix} = \begin{pmatrix} -1 & 0 \\ 0 & -2\sigma^* \end{pmatrix}.$$

Therefore, the asymptotic variance matrix is

$$\Sigma = H^{-1}GH^{-\mathrm{T}} = \begin{pmatrix} \sigma^{*2} & \frac{1}{2\sigma^*}\mathrm{cov}(X,(X-\mu^*)^2) \\ \frac{1}{2\sigma^*}\mathrm{cov}((X-\mu^*)^2, X) & \frac{1}{4\sigma^{*2}}\mathrm{var}((X-\mu^*)^2) \end{pmatrix}.$$

If $X \sim \mathrm{N}(\mu^*, \sigma^{*2})$, then by direct calculation, $\Sigma = \begin{pmatrix} \sigma^{*2} & 0 \\ 0 & \frac{\sigma^{*2}}{2} \end{pmatrix}$. If $X \sim$ Exponential(θ^*) with $(\mu^*, \sigma^{*2}) = (\theta^*, \theta^{*2})$, then $\Sigma = \begin{pmatrix} \theta^{*2} & \theta^{*2} \\ \theta^{*2} & 2\theta^{*2} \end{pmatrix}$.

To conclude this section, we briefly discuss variance estimation and Wald confidence intervals. The rescaled asymptotic variance for $\hat{\theta}_n$ is $\Sigma = H^{-1}GH^{-\mathrm{T}}$. By substituting sample averages for expectations and $\hat{\theta}_n$ for $\bar{\theta}$ in the definitions of H and G, a natural estimator of Σ is $\Sigma_n = \Sigma_n(\hat{\theta}_n)$, called the sandwich variance estimator, where

$$\Sigma_n(\theta) = H_n(\theta)^{-1}G_n(\theta)H_n(\theta)^{-\mathrm{T}},$$
$$G_n(\theta) = \frac{1}{n}\sum_{i=1}^{n}\psi(Z_i;\theta)\psi(Z_i;\theta)^{\mathrm{T}}, \quad H_n(\theta) = \frac{1}{n}\sum_{i=1}^{n}\frac{\partial\psi}{\partial\theta^{\mathrm{T}}}(Z_i;\theta).$$

Under suitable regularity conditions as $n \to \infty$ and d is fixed, it can be shown that Σ_n is a consistent estimator of Σ (Exercise 2.10), i.e.,

$$\Sigma_n \underset{\mathcal{P}}{\to} \Sigma. \tag{2.25}$$

By Slutsky's theorem, combining asymptotic normality (2.20) and consistent variance estimation (2.25) leads to valid Wald confidence intervals for components of $\bar{\theta}$ or confidence regions for subvectors of $\bar{\theta}$. This is similar to the justification of Wald confidence intervals (or regions) for population mean μ in Section 1.3. For example, for $0 < c < 1$, the Wald confidence interval for $\bar{\theta}^{(j)}$, the jth component of $\bar{\theta}$, is

$$\hat{\theta}_n^{(j)} \pm z_{c/2}\sqrt{\frac{(\Sigma_n)_{jj}}{n}},$$

which achieves asymptotic size $1 - c$ in covering $\bar{\theta}^{(j)}$, i.e.,

$$P\left(\bar{\theta}^{(j)} \in \left[\hat{\theta}_n^{(j)} - z_{c/2}\sqrt{(\Sigma_n)_{jj}/n},\ \hat{\theta}_n^{(j)} + z_{c/2}\sqrt{(\Sigma_n)_{jj}/n}\right]\right) \to 1 - c,$$

where $\hat{\theta}_n^{(j)}$ is the jth component of $\hat{\theta}_n$ and $(\Sigma_n)_{jj}$ is the jth diagonal component of Σ_n.

2.4.2 Asymptotic normality of M-estimation

As mentioned earlier, an M-estimator with $\rho(\cdot;\theta)$ differentiable in θ is also a Z-estimator. In this case, the asymptotic normality for M-estimation can be readily transferred from that for Z-estimation in Theorem 2.9.

Theorem 2.10 (Asymptotic normality of M-estimation under smoothness and dominance). *In the setting of Definition 2.1, assume that conditions (i)–(iv) in Theorem 2.9 are satisfied with* $\psi(z;\theta) = \frac{\partial}{\partial\theta}\rho(z;\theta)$*:*

(i) [Interior target parameter] $\bar{\theta}$ *is in the interior of* Θ*;*

(ii) [Smoothness] $\rho(z;\theta)$ *is continuously twice differentiable in* θ *for each* z*;*

(iii) [Local dominance near $\bar{\theta}$*] There exists a nonnegative function* $M(z)$ *such that* $E\{M(Z)\} < \infty$ *and for each* z *and* θ *in a neighborhood of* $\bar{\theta}$*,*

$$\left\|\frac{\partial^2\rho}{\partial\theta\partial\theta^{\mathrm{T}}}(z;\theta)\right\| \leq M(z); \tag{2.26}$$

(iv) [Finite moments and invertibility] $G = \mathrm{var}\{\frac{\partial\rho}{\partial\theta}(Z;\bar{\theta})\}$ *exists with finite elements, and* $H = E\{\frac{\partial^2\rho}{\partial\theta\partial\theta^{\mathrm{T}}}(Z;\bar{\theta})\}$ *exists and is invertible. [Note that* $H = H^{\mathrm{T}}$ *is symmetric.]*

If $\hat{\theta}_n \xrightarrow[P]{} \bar{\theta}$*, then* (2.19) *and* (2.20) *hold:*

$$\sqrt{n}(\hat{\theta}_n - \bar{\theta}) = -H^{-1}\frac{1}{\sqrt{n}}\sum_{i=1}^{n}\frac{\partial\rho}{\partial\theta}(Z_i;\bar{\theta}) + o_p(1), \tag{2.27}$$

$$\sqrt{n}(\hat{\theta}_n - \bar{\theta}) \xrightarrow[D]{} N(0, H^{-1}GH^{-1}). \tag{2.28}$$

Proof. By the differentiability of $\rho(z;\theta)$ in θ, the M-estimator $\hat{\theta}_n$, defined as a minimizer of $K_n(\theta) = \frac{1}{n}\sum_{i=1}^n \rho(Z_i;\theta)$, satisfies $0 = \frac{1}{n}\sum_{i=1}^n \frac{\partial}{\partial\theta}\rho(Z_i;\bar{\theta})$. Moreover, condition (iii) implies that $\|\frac{\partial}{\partial\theta}\rho(Z;\theta)\|$ is bounded uniformly in a neighborhood of $\bar{\theta}$ by a nonnegative function of Z with a finite expectation, and by Proposition 2.1 (Exchanging differentiation and expectation), ensures that $\frac{\partial}{\partial\theta}E\{\rho(Z;\theta)\} = E\{\frac{\partial}{\partial\theta}\rho(Z;\theta)\}$. The target value $\bar{\theta}$, defined as a minimizer of $E\{\rho(Z;\theta)\}$, satisfies $0 = E\{\frac{\partial}{\partial\theta}\rho(Z;\bar{\theta})\}$, as required for Z-estimation. The result then follows from Theorem 2.9 by treating $\hat{\theta}_n$ as a Z-estimator. □

We discuss how additional conditions can be incorporated to ensure consistency $\hat{\theta}_n \xrightarrow[P]{} \bar{\theta}$. Similarly to the discussion of condition (iii) in Theorem 2.9, suppose that condition (iii) is strengthened to

(iii†) [Local dominance] For each $\theta \in \Theta$, there exists a nonnegative function $M(z;\theta)$ such that $E\{M(Z;\theta)\} < \infty$ and for each z and $\tilde{\theta} \in \Theta$ in a neighborhood of θ,

$$\left\| \frac{\partial^2 \rho}{\partial\theta\partial\theta^{\mathrm{T}}}(z;\tilde{\theta}) \right\| \le M(z;\theta),$$

which is equivalent to the following condition provided Θ is compact:

(iii‡) [Dominance] There exists a nonnegative function $M(z)$ such that $E\{M(Z)\} < \infty$ and for each z and each $\theta \in \Theta$, (2.26) holds.

Moreover, given the boundedness of Θ, conditions (ii) and (iii‡) above imply conditions (ii) and (iii) in Theorem 2.6 (Consistency of M-estimation). Therefore, if conditions (i) and (iv) in Theorem 2.6 and conditions (i), (ii), (iii†) or (iii‡), and (iv) above are satisfied, then $\hat{\theta}_n \xrightarrow[P]{} \bar{\theta}$ and (2.27) and (2.28) hold.

Consistent variance estimation and valid Wald confidence intervals can also be obtained as in Section 2.4.1 (Z-estimation). Under suitable regularity conditions, a consistent estimator of $\Sigma = H^{-1}GH^{-1}$ is $\Sigma_n = \Sigma_n(\hat{\theta}_n)$, where

$$\Sigma_n(\theta) = H_n(\theta)^{-1}G_n(\theta)H_n(\theta)^{-1},$$
$$G_n(\theta) = \frac{1}{n}\sum_{i=1}^n \frac{\partial\rho}{\partial\theta}(Z_i;\theta)\frac{\partial\rho}{\partial\theta^{\mathrm{T}}}(Z_i;\theta), \quad H_n(\theta) = \frac{1}{n}\sum_{i=1}^n \frac{\partial^2\rho}{\partial\theta\partial\theta^{\mathrm{T}}}(Z_i;\theta).$$

The Wald confidence interval, $\hat{\theta}_n^{(j)} \pm z_{c/2}\sqrt{\frac{(\Sigma_n)_{jj}}{n}}$, can be shown to achieve asymptotic size $1-c$ in covering $\bar{\theta}^{(j)}$, the jth component of $\bar{\theta}$.

Next, we discuss application of Theorem 2.10 to maximum likelihood estimation, as described in Example 2.1. For a parametric model $f(\cdot;\theta)$, the MLE $\hat{\theta}_n$ is an M-estimator with $\rho(z;\theta) = -\log f(z;\theta)$. The target value is $\bar{\theta} = \min_{\theta\in\Theta} E\{-\log f(Z;\theta)\}$ or equivalently stated as (2.16). We consider two distinct scenarios as in Section 2.3.2.

In the first scenario, the parametric model $f(\cdot;\theta)$ is assumed to be correctly specified, i.e., $f^*(\cdot) = f(\cdot;\theta^*)$ for some true value θ^*, where $f^*(\cdot)$ is the true data density. As discussed in Section 2.3.2, $\bar{\theta} = \theta^*$ (i.e., $\bar{\theta}$ can be any choice of θ^*, and vice versa). Then G and H are determined as

$$G = \text{var}_{\theta^*}\left\{\frac{\partial}{\partial\theta}\log f(Z;\theta^*)\right\}, \quad H = -E_{\theta^*}\left\{\frac{\partial^2}{\partial\theta\partial\theta^{\mathrm{T}}}\log f(Z;\theta^*)\right\},$$

where $E_\theta(\cdot)$ and $\text{var}_\theta(\cdot)$ denote the expectation and variance with $Z \sim f(\cdot;\theta)$. Under the regularity conditions from Theorems 2.6 and 2.10 with $\rho(z;\theta) = -\log f(z;\theta)$, the MLE $\hat{\theta}_n$ satisfies $\hat{\theta}_n \xrightarrow{P} \theta^*$, (2.27), and (2.28). [Condition (iv) in Theorem 2.6 can be replaced by the identification condition (2.18), so that θ^* is unique and $\bar{\theta}\,(=\theta^*)$ is also unique.] From the proof of Theorem 2.10, the first *Bartlett identity* holds:

$$E_{\theta^*}\left\{\frac{\partial}{\partial\theta}\log f(Z;\theta^*)\right\} = 0.$$

This is equivalent to $\int \frac{\partial}{\partial\theta} f(z;\theta^*)\,\mathrm{d}z = \frac{\partial}{\partial\theta}\int f(z;\theta)\,\mathrm{d}z|_{\theta^*}\ (=0)$, and can also be deduced provided that $\|\frac{\partial}{\partial\theta} f(z;\theta)\|$ is uniformly bounded in a neighborhood of θ^* by a nonnegative, integrable function of z. Justification for exchanging differentiation and integration with respect to a non-probability measure is similar to that for exchanging differentiation and expectation (with respect to a probability measure) as stated in Proposition 2.1 (Section 2.6.3). In addition, assume that the second *Bartlett identity* holds: $G = H$, that is,

$$\text{var}_{\theta^*}\left\{\frac{\partial}{\partial\theta}\log f(Z;\theta^*)\right\} = -E_{\theta^*}\left\{\frac{\partial^2}{\partial\theta\partial\theta^{\mathrm{T}}}\log f(Z;\theta^*)\right\}. \tag{2.29}$$

This is equivalent to $\int \frac{\partial^2}{\partial\theta\partial\theta^{\mathrm{T}}} f(z;\theta^*)\,\mathrm{d}z = \frac{\partial^2}{\partial\theta\partial\theta^{\mathrm{T}}}\int f(z;\theta)\,\mathrm{d}z|_{\theta^*}\ (=0)$, and can be deduced provided that $\|\frac{\partial^2}{\partial\theta\partial\theta^{\mathrm{T}}} f(z;\theta)\|$ is uniformly bounded in a neighborhood of θ^* by a nonnegative, integrable function of z. [This condition is distinct from the associated condition (iii) in Theorem 2.10.] Then the asymptotic variance in (2.28), $\Sigma = H^{-1}GH^{-1}$, reduces to $\Sigma = G^{-1} = H^{-1}$, and (2.27) and (2.28) reduce to

$$\sqrt{n}(\hat{\theta}_n - \theta^*) = I^{-1}\frac{1}{\sqrt{n}}\sum_{i=1}^n \frac{\partial \log f}{\partial\theta}(Z_i;\theta^*) + o_p(1), \tag{2.30}$$

$$\sqrt{n}(\hat{\theta}_n - \theta^*) \xrightarrow{D} N(0, I^{-1}),$$

where $I = G = H$ is called the *Fisher information* matrix. See Section 2.6.4 for a discussion of asymptotic efficiency of MLE.

From the preceding discussion, given the correct specification of model $f(\cdot;\theta)$, the rescaled asymptotic variance for $\hat{\theta}_n$, $\Sigma = H^{-1}GH^{-1} = I^{-1}$, can be consistently estimated by both the sandwich estimator $\Sigma_n(\hat{\theta}_n)$ and the simpler estimator $H_n(\hat{\theta}_n)^{-1}$ or $G_n(\hat{\theta}_n)^{-1}$, where

$$\Sigma_n(\theta) = H_n(\theta)^{-1} G_n(\theta) H_n(\theta)^{-1},$$

$$G_n(\theta) = \frac{1}{n}\sum_{i=1}^{n} \frac{\partial \log f}{\partial \theta}(Z_i;\theta)\frac{\partial \log f}{\partial \theta^{\mathrm{T}}}(Z_i;\theta),$$

$$H_n(\theta) = -\frac{1}{n}\sum_{i=1}^{n} \frac{\partial^2 \log f}{\partial \theta \partial \theta^{\mathrm{T}}}(Z_i;\theta).$$

For Σ_n defined as any of these variance estimators, $\Sigma_n(\hat{\theta}_n)$, $H_n(\hat{\theta}_n)^{-1}$, or $G_n(\hat{\theta}_n)^{-1}$, the Wald confidence interval, $\hat{\theta}_n^{(j)} \pm z_{c/2}\sqrt{\frac{(\Sigma_n)_{jj}}{n}}$, can be shown to achieve asymptotic size $1-c$ in covering $\theta^{*(j)}$, the jth component of true value θ^*. There is, however, a substantive difference between these variance estimators when model $f(\cdot;\theta)$ may be misspecified, as discussed below.

In the second scenario, the parametric model $f(\cdot;\theta)$ may be misspecified, i.e., there exists no value θ^* such that $f^*(\cdot) = f(\cdot;\theta^*)$. The target value $\bar{\theta}$ remains well-defined by (2.16). Then G and H are determined as

$$G = \text{var}\left\{\frac{\partial}{\partial \theta}\log f(Z;\bar{\theta})\right\}, \quad H = -E\left\{\frac{\partial^2}{\partial \theta \partial \theta^{\mathrm{T}}}\log f(Z;\bar{\theta})\right\},$$

where $E(\cdot)$ and var$(\cdot)$ denotes the expectation and variance with $Z \sim f^*(\cdot)$. Under suitable regularity conditions, the MLE $\hat{\theta}_n$ satisfies $\hat{\theta}_n \xrightarrow[P]{} \bar{\theta}$, (2.27), and (2.28), with $\rho(z;\theta) = -\log f(z;\theta)$. In general, the (second) Bartlett identity does not hold, $G \neq H$, and hence the asymptotic variance in (2.28), $\Sigma = H^{-1}GH^{-1}$, no longer reduces to H^{-1} or G^{-1}.

When the parametric model $f(\cdot;\theta)$ may be misspecified, $H_n(\hat{\theta}_n)^{-1}$ and $G_n(\hat{\theta}_n)^{-1}$ are in general inconsistent for V, but the variance estimator $\Sigma_n(\hat{\theta}_n)$ remains consistent for V under regularity conditions. Then the Wald confidence interval, $\hat{\theta}_n^{(j)} \pm z_{c/2}\sqrt{\frac{(\Sigma_n(\hat{\theta}_n))_{jj}}{n}}$, can be shown to achieve asymptotic size $1-c$ in covering $\bar{\theta}^{(j)}$, the jth component of target value $\bar{\theta}$. In this sense, $\Sigma_n(\hat{\theta}_n)$ is a *model-robust* variance estimator (being consistent for V in the presence of possible model misspecification), whereas $H_n(\hat{\theta}_n)^{-1}$ or $G_n(\hat{\theta}_n)^{-1}$ is a *model-based* variance estimator (being consistent for V only with correct model specification).

As a concrete example, we discuss maximum likelihood estimation in logistic regression described in Example 2.4. Formally, logistic regression is a semi-parametric model (because the conditional distribution of Y given X is parametrically specified, but the marginal distribution of X is unspecified), and maximum likelihood estimation involves maximizing the conditional likelihood of β defined from the conditional distribution of Y given X. Nevertheless, maximum conditional likelihood estimation in logistic regression, considered as M-estimation with an IID sample, satisfies similar properties as maximum likelihood estimation in parametric models discussed above.

Example 2.6 (Logistic regression). For an IID sample, $Z_i = (X_i, Y_i)$, $i = 1, \ldots, n$, consider a logistic regression model as in Example 2.4:

$$P(Y_i = 1|X_i) = \frac{\exp(X_i^{\mathrm{T}}\beta^*)}{1 + \exp(X_i^{\mathrm{T}}\beta^*)}. \tag{2.31}$$

The MLE $\hat{\beta}$ is an M-estimator, $\hat{\beta} = \operatorname{argmin}_{\beta \in B} n^{-1} \sum_{i=1}^n \rho(X_i, Y_i; \beta)$, with

$$\rho(X, Y; \beta) = -\left\{ Y X^{\mathrm{T}}\beta - \log(1 + \mathrm{e}^{X^{\mathrm{T}}\beta}) \right\}.$$

The gradient and Hessian of $\rho(X, Y; \beta)$ in β are

$$\frac{\partial \rho}{\partial \beta}(X, Y; \beta) = (Y - \mu(\beta))X,$$
$$\frac{\partial^2 \rho}{\partial \beta \partial \beta^{\mathrm{T}}}(X, Y; \beta) = \mu(\beta)(1 - \mu(\beta))XX^{\mathrm{T}},$$

where $\mu(\beta) = \mathrm{e}^{X^{\mathrm{T}}\beta}/(1 + \mathrm{e}^{X^{\mathrm{T}}\beta})$, with the dependency on X suppressed. To illustrate asymptotic theory based on smoothness, assume that (i) $\mathcal{B}$ is compact, for example, $\{\beta \in \mathbb{R}^p : \|\beta\|_2 \leq b\}$ for a constant $b > 0$, (ii) the target value, $\bar{\beta} = \operatorname{argmin}_{\beta \in \mathcal{B}} E\{\rho(X, Y; \beta)\}$, lies in the interior of $\mathcal{B}$, (iii) $E(\|X\|_2^2) < \infty$, and (iv) $H = \frac{\partial^2 \rho}{\partial \beta \partial \beta^{\mathrm{T}}}(X, Y; \bar{\beta})$ exists and is invertible. Then it can be verified [Exercise] that the conditions in Theorems 2.6 and 2.10 are satisfied, and hence $\hat{\beta}_n \underset{P}{\to} \bar{\beta}$ and $\sqrt{n}(\hat{\beta}_n - \bar{\beta}) \underset{\mathcal{D}}{\to} \mathrm{N}(0, \Sigma(\bar{\beta}))$, where

$$\Sigma(\beta) = H(\beta)^{-1} G(\beta) H(\beta)^{-1},$$
$$G(\beta) = E\left\{(Y - \mu(\beta))^2 XX^{\mathrm{T}}\right\},$$
$$H(\beta) = E\left\{\mu(\beta)(1 - \mu(\beta))XX^{\mathrm{T}}\right\}.$$

[Note that the parameter space $\mathcal{B}$ is usually unrestricted as $\mathbb{R}^p$. In that case, asymptotic theory based on convexity in Section 2.5 can be applied (Exercise 2.18).] A consistent estimator of $\Sigma(\bar{\beta})$ is the sandwich variance estimator (or model-robust variance estimator) $\Sigma_n(\hat{\beta}_n) = H_n(\hat{\beta}_n)^{-1} G_n(\hat{\beta}_n) H_n(\hat{\beta}_n)^{-1}$, where

$$\Sigma_n(\beta) = H_n(\beta)^{-1} G_n(\beta) H_n(\beta)^{-1},$$
$$G_n(\beta) = \frac{1}{n} \sum_{i=1}^n (Y_i - \mu_i(\beta))^2 X_i X_i^{\mathrm{T}},$$
$$H_n(\beta) = \frac{1}{n} \sum_{i=1}^n \mu_i(\beta)(1 - \mu_i(\beta)) X_i X_i^{\mathrm{T}},$$

with $\mu_i(\beta) = \mathrm{e}^{X_i^{\mathrm{T}}\beta}/(1 + \mathrm{e}^{X_i^{\mathrm{T}}\beta})$. The preceding properties hold while logistic regression (2.31) may be misspecified and β^* may not exist but $\bar{\beta}$ is well-defined.

If logistic regression (2.31) is correctly specified with true value β^*, then $\bar{\beta}$ reduces to β^*, and $\sqrt{n}(\hat{\beta}_n - \beta^*) \xrightarrow{\mathcal{D}} \mathrm{N}(0, H(\beta^*)^{-1})$, because it can be directly shown that $H(\beta^*) = G(\beta^*)$ and hence $\Sigma(\beta^*) = H(\theta^*)^{-1}$. The asymptotic variance $\Sigma(\beta^*)$ can be consistently estimated by the model-based estimator $H_n(\hat{\beta}_n)^{-1}$ as well as the model-robust estimator $\Sigma_n(\hat{\beta}_n)$.

2.5 Asymptotic theory based on convexity

There are at least two motivations for studying asymptotic theory of M-estimation based on convexity of the associated objective function, in addition to that based on smoothness in Sections 2.3.2 and 2.4.2. First, the assumption of a compact parameter set Θ may be satisfied by placing large bounds on possible parameter values, for example, possible coefficient vectors in logistic regression. But such bounds are typically not used in numerical implementation of M-estimation. Second, there are important applications, including Huber's robust estimation (Example 2.2) and quantile estimation (Example 2.3), where the objective function $\rho(z;\theta)$ is not twice differentiable in θ and hence Theorem 2.10 is not applicable to establish asymptotic normality. The convexity of $\rho(z;\theta)$ in θ can be exploited to establish asymptotic theory (including consistency and asymptotic normality) for M-estimation, without requiring the compactness of Θ or twice differentiability of $\rho(z;\theta)$ in θ.

2.5.1 Consistency of convex M-estimation

The following lemma, which is fundamental to convexity-based theory, shows that pointwise convergence of random convex functions on an open set $\mathcal{T}$ implies uniform convergence on any compact subset $\mathcal{S}$. This is slightly generalized from the convexity lemma in Pollard (1991) and Niemiro (1992), where $\mathcal{T}$ is assumed to be an open *convex* set in $\mathbb{R}^d$. A function $f : \mathcal{T} \to \mathbb{R}$ is said to be convex on a possibly non-convex set $\mathcal{T} \subset \mathbb{R}^d$ if for any $s, t \in \mathcal{T}$ and any $0 \le \lambda \le 1$ satisfying $\lambda s + (1-\lambda)t \in \mathcal{T}$, we have

$$f(\lambda s + (1-\lambda)t) \le \lambda f(s) + (1-\lambda) f(t).$$

Proofs of Lemma 2.1 and associated material about convex functions are provided in Section 2.6.2. [Note: The conclusion in Lemma 2.1 may fail if $\mathcal{T}$ is not an open set. Consider the deterministic example where $\mathcal{T} = \mathcal{S} = [0,1]$ and $F_n(s) = s^n$, which is convex on $\mathcal{T}$.]

Lemma 2.1 (Uniform convergence of random convex functions). *Let $F_n : \mathcal{T} \to \mathbb{R}$, $n \ge 1$, be a sequence of random convex functions on an open set $\mathcal{T} \subset \mathbb{R}^d$, and $F : \mathcal{T} \to \mathbb{R}$ be a non-random function such that $F_n(s) \xrightarrow{\mathcal{P}} F(s)$*

for each $s \in \mathcal{T}$. Then $F(\cdot)$ is convex on $\mathcal{T}$, and for each compact set $\mathcal{S} \subset \mathcal{T}$,

$$\sup_{s \in \mathcal{S}} |F_n(s) - F(s)| \underset{\mathcal{P}}{\to} 0.$$

From Lemma 2.1, a general consistency result can be deduced as follows, whose proof is left as Exercise 2.12.

Lemma 2.2 (Convergence of random minimizers under convexity). *In the setting of Lemma 2.1, let $\alpha_n = \text{argmin}_{s \in \mathcal{T}} F_n(s)$ and $\alpha = \text{argmin}_{s \in \mathcal{T}} F(s)$. If α is a unique minimizer of $F(\cdot)$ on $\mathcal{T}$, then $\alpha_n \underset{\mathcal{P}}{\to} \alpha$.*

Instead of proving and applying the general result in Lemma 2.2, we directly prove the consistency of M-estimation with a convex objective function. Compared with Theorem 2.6 (Consistency of M-estimation based on smoothness and dominance), it is not required here that Θ is compact or $\rho(\cdot;\theta)$ is dominated by an integrable function $M(\cdot)$.

Theorem 2.11 (Consistency of M-estimation under convexity). *In the setting of Definition 2.1 for M-estimator $\hat{\theta}_n$, assume that the following conditions are satisfied:*

(i) [Open parameter set] $\Theta \subset \mathbb{R}^d$ is an open set;

(ii) [Convexity] $\rho(z;\theta)$ is convex in θ for each z;

(iii) [Finite population objective] $K(\theta) = E\{\rho(Z;\theta)\}$ is finite for θ in a neighborhood of $\bar{\theta}$;

(iv) [Unique target parameter] $\bar{\theta}$ is a unique minimizer of $K(\theta)$ on Θ.

Then $\hat{\theta}_n \underset{\mathcal{P}}{\to} \bar{\theta}$. [Note: Under conditions (i)–(iii), it can be shown that $K(\theta) = E\{\rho(Z;\theta)\}$ exists on Θ, being either finite or ∞ (Exercise 2.13).]

Proof. For any $\epsilon > 0$, let $B_\epsilon = \{\theta \in \mathbb{R}^d : \|\theta - \bar{\theta}\| \leq \epsilon\}$. Without loss of generality, assume that $B_{2\epsilon} \subset \Theta$ by condition (i) and $K(\theta)$ is finite for $\theta \in B_{2\epsilon}$ by condition (iii). It is desired to show that $P(\hat{\theta}_n$ exists and $\hat{\theta}_n \in B_\epsilon) \to 1$.

First, by condition (ii), $K_n(\theta)$ is convex on Θ and hence is continuous on Θ because convexity implies continuity on an open set (Lemma 2.8). Then

$$\begin{aligned} P(\hat{\theta}_n \text{ exists and } \hat{\theta}_n \in B_\epsilon) &\geq P(K_n(\theta) > K_n(\bar{\theta}) \text{ for } \theta \notin B_\epsilon) \\ &\geq P(\inf_{\theta \in \partial B_\epsilon} K_n(\theta) > K_n(\bar{\theta})), \end{aligned} \tag{2.32}$$

where $\partial B_\epsilon = \{\theta \in \mathbb{R}^d : \|\theta - \bar{\theta}\| = \epsilon\}$. The first inequality follows because if $K_n(\theta) > K_n(\bar{\theta})$ for $\theta \notin B_\epsilon$, then a minimizer of $K_n(\theta)$ on Θ is obtained as a minimizer of $K_n(\theta)$ on B_ϵ (which exists by continuity of K_n). The second inequality holds because if $\inf_{\theta \in \partial B_\epsilon} K_n(\theta) > K_n(\bar{\theta})$, then $K_n(\theta) > K_n(\bar{\theta})$ for $\theta \notin B_\epsilon$, by convexity of $K_n(\theta)$ on Θ (Exercise 2.14). See Figure 2.5 for a 1-dimensional illustration. Conceptually, inequality (2.32) with convex objective function $K_n(\cdot)$ is the M-estimation counterpart of inequality (2.7) with monotone estimating function $T_n(\cdot)$ in Z-estimation.

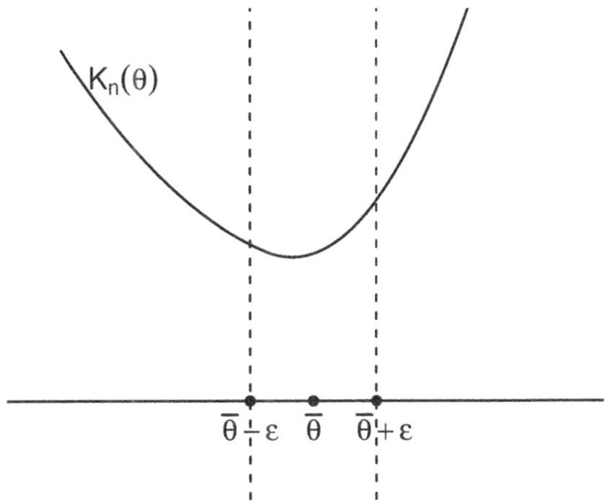

FIGURE 2.5
Relating the event $\hat{\theta}_n \in [\bar{\theta} - \epsilon, \bar{\theta} + \epsilon]$ to the event that $K_n(\theta) > K_n(\bar{\theta})$ for $\theta = \bar{\theta} \pm \epsilon$ in the case of 1-dimensional Θ.

Next, we show that $P(\inf_{\theta \in \partial B_\epsilon} K_n(\theta) > K_n(\bar{\theta})) \to 1$. By condition (ii) and the choice of ϵ, $K(\theta)$ is convex and hence continuous on the open set $B_{2\epsilon} \setminus \partial E_{2\epsilon}$, which contains the compact set ∂B_ϵ. By condition (iv) and continuity of $K(\theta)$ on ∂B_ϵ, $\inf_{\theta \in \partial B_\epsilon} K(\theta) = K(\tilde{\theta}) > K(\bar{\theta})$ for some $\tilde{\theta} \in \partial B_\epsilon$. [Note: By convexity of $K(\theta)$ on Θ, we also have $\inf_{\theta \notin B_\epsilon} K(\theta) > K(\bar{\theta})$ (Exercise 2.15). Hence condition (i) in Theorem 2.1 (General consistency of M-estimation) is satisfied, although this property is not used in the subsequent proof.] Take $\delta = \inf_{\theta \in \partial B_\epsilon} K(\theta) - K(\bar{\theta}) > 0$, depending on ϵ. Then

$$
\begin{aligned}
& P(\inf_{\theta \in \partial B_\epsilon} K_n(\theta) > K_n(\bar{\theta})) \\
& \geq P\left(\inf_{\theta \in \partial B_\epsilon} K_n(\theta) > K(\bar{\theta}) + \frac{\delta}{2} > K_n(\bar{\theta})\right) \\
& = P\left(\inf_{\theta \in \partial B_\epsilon} K_n(\theta) > \inf_{\theta \in \partial B_\epsilon} K(\theta) - \frac{\delta}{2},\ K_n(\bar{\theta}) < K(\bar{\theta}) + \frac{\delta}{2}\right), \qquad (2.33)
\end{aligned}
$$

Therefore, it suffices to show that

$$
P\left(K_n(\bar{\theta}) < K(\bar{\theta}) + \frac{\delta}{2}\right) \to 1,
$$

$$
P\left(\inf_{\theta \in \partial B_\epsilon} K_n(\theta) > \inf_{\theta \in \partial B_\epsilon} K(\theta) - \frac{\delta}{2}\right) \to 1.
$$

By condition (iii) and the law of large numbers, $K_n(\theta) \xrightarrow[\mathcal{P}]{} K(\theta)$ for each θ. The first convergence above can be obtained from

$$P(K_n(\bar{\theta}) < K(\bar{\theta}) + \frac{\delta}{2}) \geq P(|K_n(\bar{\theta}) - K(\bar{\theta})| < \frac{\delta}{2}), \tag{2.34}$$

which tends to 1 because $K_n(\bar{\theta}) \xrightarrow[\mathcal{P}]{} K(\bar{\theta})$. The second convergence can be shown as follows:

$$\begin{aligned} & P\left(\inf_{\theta \in \partial B_\epsilon} K_n(\theta) > \inf_{\theta \in \partial B_\epsilon} K(\theta) - \frac{\delta}{2}\right) \\ \geq\ & P\left(|\inf_{\theta \in \partial B_\epsilon} K_n(\theta) - \inf_{\theta \in \partial B_\epsilon} K(\theta)| < \frac{\delta}{2}\right) \\ \geq\ & P\left(\sup_{\theta \in \partial B_\epsilon} |K_n(\theta) - K(\theta)| < \frac{\delta}{2}\right), \end{aligned} \tag{2.35}$$

which tends to 1 because $\sup_{\theta \in \partial B_\epsilon} |K_n(\theta) - K(\theta)| \xrightarrow[\mathcal{P}]{} 0$ by applying Lemma 2.1 with $F_n = K_n$ (which is convex on an open set $\mathcal{T} = \Theta$), $F = K$, and $\mathcal{S} = \partial B_\epsilon$ (which is a compact subset of Θ). The second step above uses $|\inf_{\theta \in \partial B_\epsilon} K_n(\theta) - \inf_{\theta \in \partial B_\epsilon} K(\theta)| \leq \sup_{\theta \in \partial B_\epsilon} |K_n(\theta) - K(\theta)|$. □

2.5.2 Asymptotic normality of convex M-estimation

The following lemma is similar to Lemma 2.2 and indicates that if a random convex function $F_n(s)$ is arbitrarily close in probability to $Q_n(s)$, a quadratic random function with a non-random square term and a unique minimizer that is $O_p(1)$, then any minimizer of $F_n(s)$ converges in probability to the minimizer of $Q_n(s)$. [Question: Can Lemma 2.3 be extended to allow that $F_n(s) = Q_n(s) + r_n(s) + c_n$, where $Q_n(s)$ may be a non-quadratic, convex random function with a unique minimizer that is $O_p(1)$?]

Lemma 2.3 (Closeness of random minimizers under convexity). *For $n \geq 1$, let $F_n : \mathcal{T}_n \to \mathbb{R}$ be a random convex function on $\mathcal{T}_n \subset \mathbb{R}^d$. Suppose that $\mathcal{T}_n$ is asymptotically unrestricted, i.e., for any $M > 0$ there exists $N \geq 1$ such that $\{s \in \mathbb{R}^d : \|s\| \leq M\} \subset \mathcal{T}_n$ for $n \geq N$, and that $F_n(s)$ can be represented in the form*

$$F_n(s) = \underbrace{\frac{1}{2} s^{\mathrm{T}} V s + u_n^{\mathrm{T}} s}_{Q_n(s)} + r_n(s) + c_n,$$

where V is a positive-definite matrix, $u_n = O_p(1)$, $r_n(s) = o_p(1)$ for each (fixed) s, and c_n is an arbitrary constant depending on n. Let $\alpha_n = \mathrm{argmin}_{s \in \mathcal{T}_n} F_n(s)$ and $\beta_n = \mathrm{argmin}_{s \in \mathbb{R}^d} Q_n(s)$ with $Q_n(s) = \frac{1}{2} s^{\mathrm{T}} V s + u_n^{\mathrm{T}} s$. Then $\beta_n = -V^{-1} u_n$ and $\alpha_n = \beta_n + o_p(1)$.

Proof. Without loss of generality, assume that $c_n = 0$. The proof is, at a high level, similar to that of Theorem 2.11, but involves additional reasoning to handle the complication that the "approximation" $Q_n(s)$, $n \geq 1$, is also a sequence of random functions. [If u_n is replaced by a non-random vector u_0, then the result follows directly from Lemma 2.1.] In addition, to focus on main ideas, we assume that $\mathcal{T}_n = \mathbb{R}^d$ for each $n \geq 1$. The proof in the general case of $\mathcal{T}_n$ is left as Exercise 2.16.

For any $0 < \epsilon \leq 1$, consider the bound

$$P(\alpha_n \text{ exists and } \|\alpha_n - \beta_n\| \leq \epsilon) \geq P(\inf_{s \in \partial B_{n,\epsilon}} F_n(s) > F_n(\beta_n)),$$

by convexity of $F_n(s)$ similarly as in (2.32), where $\partial B_{n,\epsilon} = \{s \in \mathbb{R}^d : \|s - \beta_n\| = \epsilon\}$. If $\|s - \beta_n\| = \epsilon$, then

$$Q_n(s) - Q_n(\beta_n) = \frac{1}{2}(s - \beta_n)^{\mathrm{T}} V (s - \beta_n) = \frac{1}{2}\epsilon^2 (t^{\mathrm{T}} V t),$$

where $t = (s - \beta_n)/\epsilon$ with $\|t\| = 1$. Take $\delta = \inf_{\|t\|=1} \frac{1}{2}\epsilon^2 (t^{\mathrm{T}} V t) > 0$. [For the L_2 norm $\|\cdot\|$, $\delta = \frac{1}{2}\epsilon^2 \kappa$, where $\kappa > 0$ is the smallest eigenvalue of V.] Then similarly as in (2.33),

$$\begin{aligned}
&P\left(\inf_{s \in \partial B_{n,\epsilon}} F_n(s) > F_n(\beta_n)\right) \\
&\geq P\left(\inf_{s \in \partial B_{n,\epsilon}} F_n(s) > \inf_{s \in \partial B_{n,\epsilon}} Q_n(s) - \frac{\delta}{2},\ F_n(\beta_n) < Q_n(\beta_n) + \frac{\delta}{2}\right) \\
&\geq P\left(\inf_{s \in \partial B_{n,\epsilon}} F_n(s) > \inf_{s \in \partial B_{n,\epsilon}} Q_n(s) - \frac{\delta}{2}\right) + P\left(F_n(\beta_n) < Q_n(\beta_n) + \frac{\delta}{2}\right) - 1.
\end{aligned}$$

Moreover, similarly as in (2.34) and (2.35),

$$P(F_n(\beta_n) < Q_n(\beta_n) + \frac{\delta}{2}) \geq P(|F_n(\beta_n) - Q_n(\beta_n)| \leq \frac{\delta}{2}),$$

and

$$\begin{aligned}
&P\left(\inf_{s \in \partial B_{n,\epsilon}} F_n(s) > \inf_{s \in \partial B_{n,\epsilon}} Q_n(s) - \frac{\delta}{2}\right) \\
&\geq P\left(|\inf_{s \in \partial B_{n,\epsilon}} F_n(s) - \inf_{s \in \partial B_{n,\epsilon}} Q_n(s)| < \frac{\delta}{2}\right) \\
&\geq P\left(\sup_{\|s - \beta_n\| = \epsilon} |F_n(s) - Q_n(s)| < \frac{\delta}{2}\right).
\end{aligned}$$

Hence it suffices to show that $P(\sup_{\|s-\beta_n\| \leq \epsilon} |F_n(s) - Q_n(s)| < \frac{\delta}{2}) \to 1$ as $n \to \infty$, i.e., to show that for any $0 < \eta < 1$, there exists $N \geq 1$ such that for $n \geq N$, $P(\sup_{\|s-\beta_n\| \leq \epsilon} |F_n(s) - Q_n(s)| < \frac{\delta}{2}) > 1 - \eta$.

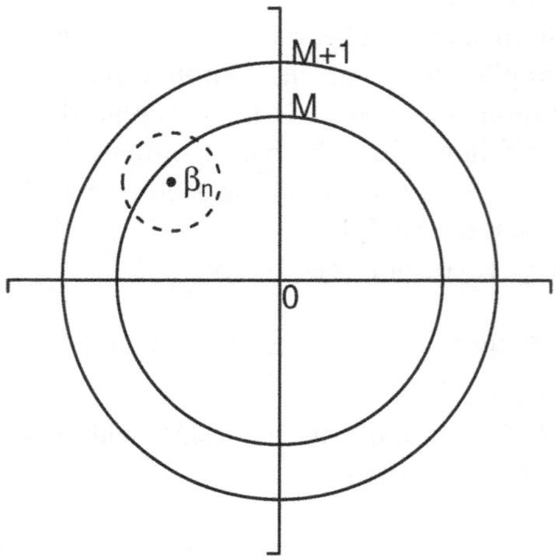

FIGURE 2.6
Two-dimensional illustration of $B_{n,\epsilon} = \{s : \|s - \beta_n\| \leq \epsilon\}$, which is contained in $\{s : \|s\| \leq M + 1\}$ for $\|\beta_n\| \leq M$ and $0 < \epsilon \leq 1$.

First, by direct minimization of quadratic $Q_n(s)$ over $\mathbb{R}^d$, we have $\beta_n = -V^{-1}u_n$, which is $O_p(1)$ because $u_n = O_p(1)$ by assumption. Then there exist $M > 0$ and $N_1 \geq 1$ such that for $n \geq N_1$, $P(\|\beta_n\| \leq M) > 1 - \frac{\eta}{2}$. If $\|s - \beta_n\| \leq \epsilon (\leq 1)$ and $\|\beta_n\| \leq M$, then $\|s\| \leq M + 1$, as illustrated in Figure 2.6. Moreover, for each s,

$$F_n(s) - Q_n(s) = \underbrace{F_n(s) - u_n^{\mathrm{T}} s}_{\tilde{F}_n(s)} - \underbrace{\frac{1}{2} s^{\mathrm{T}} V s}_{\tilde{F}(s)} = o_p(1),$$

where $\tilde{F}_n(s)$ is convex in s and $\tilde{F}(s)$ is non-random. By applying Lemma 2.1 to $\tilde{F}_n(s)$ and $\tilde{F}(s)$ on the set $\{s : \|s\| \leq M + 1\}$, we have $\sup_{\|s\| \leq M+1} |F_n(s) - Q_n(s)| \xrightarrow{P} 0$. Then there exists $N_2 \geq 1$ such that for $n \geq N_2$, $P(\sup_{\|s\| \leq M+1} |F_n(s) - Q_n(s)| < \frac{\delta}{2}) > 1 - \frac{\eta}{2}$. Hence for $n \geq \max(N_1, N_2)$,

$$\begin{aligned}
&P\left(\sup_{\|s-\beta_n\| \leq \epsilon} |F_n(s) - Q_n(s)| < \frac{\delta}{2}\right) \\
&\geq P\left(\|\beta_n\| \leq M, \quad \sup_{\|s-\beta_n\| \leq \epsilon} |F_n(s) - Q_n(s)| < \frac{\delta}{2}\right) \\
&\geq P\left(\|\beta_n\| \leq M, \quad \sup_{\|s\| \leq M+1} |F_n(s) - Q_n(s)| < \frac{\delta}{2}\right) \\
&> 1 - \eta,
\end{aligned}$$

where the last step uses the union bound. □

For each z, let $h(z;\theta)$ be a sub-gradient of convex function $\rho(z;\theta)$ in θ, such that for any $\theta, \tilde{\theta} \in \Theta$,

$$\rho(z;\theta) + (\tilde{\theta} - \theta)^{\mathrm{T}} h(z;\theta) \leq \rho(z;\tilde{\theta}).$$

Equivalently, this indicates that for any $\theta \in \Theta$ and $s \in \mathbb{R}^d$ with $\theta + s \in \Theta$,

$$s^{\mathrm{T}} h(z;\theta) \leq \rho(z;\theta + s) - \rho(z;\theta) \leq s^{\mathrm{T}} h(z;\theta + s). \tag{2.36}$$

See Hiriart-Urruty and Lemaréchal (2001), Chapter D, and Tan (2025), Section 1.10.6, for additional information. The following result establishes asymptotic normality of M-estimation with a convex objective function.

Theorem 2.12 (Asymptotic normality of M-estimation under convexity). *In the setting of Definition 2.1 for M-estimator $\hat{\theta}_n$, assume that the following conditions are satisfied:*

(i) [Open parameter set] $\bar{\theta}$ lies in the interior of $\Theta\ (\subset \mathbb{R}^d)$;

(ii) [Convexity] $\rho(z;\theta)$ is convex in θ for each z;

(iii) [Finite moments] $E\{\|h(Z;\theta)\|^2\} < \infty$ for θ in a neighborhood of $\bar{\theta}$;

(iv) [Smooth population objective and invertibility] $K(\theta) = E\{\rho(Z;\theta)\}$ is finite and differentiable in a neighborhood of $\bar{\theta}$, and $K(\theta)$ is twice differentiable at $\bar{\theta}$, with $H = \frac{\partial^2}{\partial\theta\partial\theta^{\mathrm{T}}} K(\bar{\theta})$ positive definite.

Then $\frac{\partial}{\partial\theta} K(\bar{\theta}) = E\{h(Z;\bar{\theta})\} = 0$ and

$$\sqrt{n}(\hat{\theta}_n - \bar{\theta}) = -H^{-1} \frac{1}{\sqrt{n}} \sum_{i=1}^{n} h(Z_i;\bar{\theta}) + o_p(1), \tag{2.37}$$

$$\sqrt{n}(\hat{\theta}_n - \bar{\theta}) \underset{\mathcal{D}}{\to} N(0, H^{-1} G H^{-1}), \tag{2.38}$$

where $G = \mathrm{var}\{h(Z;\theta)\} = E\{h(Z;\bar{\theta}) h(Z;\bar{\theta})^{\mathrm{T}}\}$.

Note that (2.37) and (2.38) are established directly without first showing $\hat{\theta}_n \underset{\mathcal{P}}{\to} \bar{\theta}$. Nevertheless, condition (iii) in Theorem 2.12 implies condition (iii) in Theorem 2.11, and positive definiteness of H implies the uniqueness of $\bar{\theta}$ as a minimizer of $K(\theta)$. Hence Theorem 2.11 can also be called to obtain $\hat{\theta}_n \underset{\mathcal{P}}{\to} \bar{\theta}$. Compared with Theorem 2.10 (Asymptotic normality based on smoothness and dominance), the main difference is that $\rho(z;\theta)$ may not be twice differentiable in θ and $\frac{\partial^2}{\partial\theta\partial\theta^{\mathrm{T}}} \rho(z;\theta)$ may not exist. The matrix H is defined as $\frac{\partial^2}{\partial\theta\partial\theta^{\mathrm{T}}} K(\bar{\theta})$ instead of $E\{\frac{\partial^2}{\partial\theta\partial\theta^{\mathrm{T}}} \rho(Z;\bar{\theta})\}$.

Proof. By applying Proposition 2.3 (Exchanging differentiation and expectation under convexity) to $\rho(z;\theta)$, if $K(\theta) = E\{\rho(Z;\theta)\}$ is differentiable at θ, then $\frac{\partial}{\partial\theta}K(\theta) = E\{h(Z;\theta)\}$. Hence by condition (iii), the target value $\bar{\theta}$, as a minimizer of $K(\theta)$, satisfies $\frac{\partial}{\partial\theta}K(\bar{\theta}) = E\{h(Z;\bar{\theta})\} = 0$.

Next, we use Lemma 2.3 to show that (2.37) holds. This then implies (2.38) by the central limit theorem and Slutsky's theorem. To relate to the setup in Lemma 2.3, consider the second-order Taylor expansion informally:

$$K_n(\theta) \approx K_n(\bar{\theta}) + \frac{\partial K_n}{\partial\theta^{\mathrm{T}}}(\bar{\theta})(\theta-\bar{\theta}) + \frac{1}{2}(\theta-\bar{\theta})^{\mathrm{T}}\frac{\partial^2 K_n}{\partial\theta\partial\theta^{\mathrm{T}}}(\bar{\theta})(\theta-\bar{\theta}).$$

Letting $s = \sqrt{n}(\theta-\bar{\theta})$ or $\theta = \bar{\theta} + \frac{s}{\sqrt{n}}$ gives

$$K_n(\bar{\theta} + \frac{s}{\sqrt{n}}) \approx K_n(\bar{\theta}) + \frac{\partial K_n}{\partial\theta^{\mathrm{T}}}(\bar{\theta})\frac{s}{\sqrt{n}} + \frac{1}{2n}s^{\mathrm{T}}\frac{\partial^2 K_n}{\partial\theta\partial\theta^{\mathrm{T}}}(\bar{\theta})s.$$

Multiplying both sides by n and replacing $\frac{\partial^2 K_n}{\partial\theta\partial\theta^{\mathrm{T}}}(\bar{\theta})$ by $\frac{\partial^2 K}{\partial\theta\partial\theta^{\mathrm{T}}}(\bar{\theta})$ gives

$$\underbrace{nK_n\left(\bar{\theta} + \frac{s}{\sqrt{n}}\right)}_{F_n(s)} \approx nK_n(\bar{\theta}) + \sqrt{n}\frac{\partial K_n}{\partial\theta^{\mathrm{T}}}(\bar{\theta})s + \frac{1}{2}s^{\mathrm{T}}\frac{\partial^2 K_n}{\partial\theta\partial\theta^{\mathrm{T}}}(\bar{\theta})s$$

$$\approx \underbrace{nK_n(\bar{\theta})}_{c_n} + \underbrace{\frac{s^{\mathrm{T}}}{\sqrt{n}}\sum_{i=1}^{n} h(Z_i;\bar{\theta})}_{u_n^{\mathrm{T}}s} + \underbrace{\frac{1}{2}s^{\mathrm{T}}\frac{\partial^2 K}{\partial\theta\partial\theta^{\mathrm{T}}}(\bar{\theta})s}_{s^{\mathrm{T}}Hs} \stackrel{\mathrm{def}}{=} Q_n(s).$$

Let $\alpha_n = \sqrt{n}(\hat{\theta}_n - \bar{\theta})$ and $\beta_n = -H^{-1}u_n = -H^{-1}\frac{1}{\sqrt{n}}\sum_{i=1}^n h(Z_i;\bar{\theta})$. Then $\alpha_n = \mathrm{argmin}_{s\in\mathcal{T}_n}F_n(s)$ and $\beta_n = \mathrm{argmin}_{s\in\mathbb{R}^d}Q_n(s)$, where $\mathcal{T}_n = \{s \in \mathbb{R}^d : \bar{\theta} + \frac{s}{\sqrt{n}} \in \Theta\}$, which is asymptotically unrestricted by condition (i) [Exercise]. The asymptotic expansion (2.37) can be rewritten as $\alpha_n = \beta_n + o_p(1)$. By Lemma 2.3, it suffices to show that for each $s \in \mathcal{T}_n$,

$$nK_n\left(\bar{\theta} + \frac{s}{\sqrt{n}}\right) - nK_n(\bar{\theta}) - \frac{s^{\mathrm{T}}}{\sqrt{n}}\sum_{i=1}^{n} h(Z_i;\bar{\theta}) - \frac{1}{2}s^{\mathrm{T}}Hs \xrightarrow{\mathcal{P}} 0.$$

This can be obtained by combining the following two results: for each $s \in \mathcal{T}_n$,

$$nK_n\left(\bar{\theta} + \frac{s}{\sqrt{n}}\right) - nK_n(\bar{\theta}) - \frac{s^{\mathrm{T}}}{\sqrt{n}}\sum_{i=1}^{n} h(Z_i;\bar{\theta}) - \left\{nK\left(\bar{\theta} + \frac{s}{\sqrt{n}}\right) - nK(\bar{\theta})\right\} \xrightarrow{\mathcal{P}} 0, \tag{2.39}$$

$$nK\left(\bar{\theta} + \frac{s}{\sqrt{n}}\right) - nK(\bar{\theta}) \to \frac{1}{2}s^{\mathrm{T}}Hs. \tag{2.40}$$

We give a proof of (2.39); the proof of (2.40) is left as Exercise 2.17.

To show (2.39), denote for $i = 1, \ldots, n$,

$$X_{ni} = \rho\left(Z_i; \bar{\theta} + \frac{s}{\sqrt{n}}\right) - \rho(Z_i;\bar{\theta}) - \frac{s^{\mathrm{T}}}{\sqrt{n}}h(Z_i;\bar{\theta}),$$

where the dependency on s is suppressed in the notation. Then

$$\sum_{i=1}^{n} X_{ni} = nK_n\left(\bar{\theta} + \frac{s}{\sqrt{n}}\right) - nK_n(\bar{\theta}) - \frac{s^{\mathrm{T}}}{\sqrt{n}}\sum_{i=1}^{n} h(Z_i;\bar{\theta}),$$

$$\sum_{i=1}^{n} E(X_{ni}) = nK\left(\bar{\theta} + \frac{s}{\sqrt{n}}\right) - nK(\bar{\theta}),$$

because $E\{h(Z;\bar{\theta})\} = 0$. Hence (2.39) amounts to $\sum_{i=1}^{n}(X_{ni} - E(X_{ni})) \underset{\mathcal{P}}{\to} 0$. Note that $E\{(\sum_{i=1}^{n}(X_{ni} - E(X_{ni})))^2\} = \mathrm{var}(\sum_{i=1}^{n} X_{ni}) = \sum_{i=1}^{n} \mathrm{var}(X_{ni}) \leq \sum_{i=1}^{n} E(X_{ni}^2) = nE(X_n^2)$, where

$$X_n = \rho\left(Z;\bar{\theta} + \frac{s}{\sqrt{n}}\right) - \rho(Z;\bar{\theta}) - \frac{s^{\mathrm{T}}}{\sqrt{n}} h(Z;\bar{\theta}).$$

By Proposition 1.2, for a sequence of random variables W_n, if $E(W_n^2) \to 0$ then $W_n \underset{\mathcal{P}}{\to} 0$. It suffices to show that $nE(X_n^2) \to 0$.

By the inequalities in (2.36),

$$0 \leq X_n \leq \frac{s^{\mathrm{T}}}{\sqrt{n}}\left\{h\left(Z;\bar{\theta} + \frac{s}{\sqrt{n}}\right) - h(Z;\bar{\theta})\right\},$$

and hence $nE(X_n^2) \leq E(Y_n^2)$, where

$$Y_n = s^{\mathrm{T}}\left\{h(Z;\bar{\theta} + \frac{s}{\sqrt{n}}) - h(Z;\bar{\theta})\right\}.$$

From (2.36), it can also be shown that Y_n is non-increasing in n, i.e., $Y_n \geq Y_{n-1}$ for $n \geq 1$ [Exercise]. Let $Y_* = \lim_{n\to\infty} Y_n$ (≥ 0). By condition (ii), $E(Y_n^2) < \infty$ for large enough n. Then by the dominated convergence theorem, $\lim_{n\to\infty} E(Y_n) = E(Y_*)$. But as $n \to \infty$,

$$E(Y_n) = s^{\mathrm{T}}\left\{\frac{\partial}{\partial\theta}K(\theta)\Big|_{\bar{\theta}+\frac{s}{\sqrt{n}}} - \frac{\partial}{\partial\theta}K(\theta)\Big|_{\bar{\theta}}\right\} \to 0,$$

where the first step follows by Proposition 2.3 and the assumption that $K(\theta)$ is differentiable in a neighborhood of $\bar{\theta}$, and the second step follows because $\frac{\partial}{\partial\theta}K(\theta)$ is differentiable and hence continuous at $\bar{\theta}$. Therefore $E(Y_*) = 0$, which implies that $Y_* = 0$ almost surely because $Y_* \geq 0$. By the dominated convergence theorem again, $\lim_{n\to\infty} E(Y_n^2) = E(Y_*^2) = 0$, which gives the desired convergence, $\lim_{n\to\infty} nE(X_n^2) = 0$, due to $nE(X_n^2) \leq E(Y_n^2)$. □

Example 2.7 (Quantile estimation). Suppose that $Z_1, \ldots, Z_n$ are IID scalar variables as Z. From Example 2.3, for estimating the τ-quantile q_τ^*, the sample τ-quantile is an M-estimator defined as

$$\hat{q}_{\tau,n} = \mathrm{argmin}_{q\in\mathbb{R}} \sum_{i=1}^{n} \rho_\tau(Z_i - q),$$

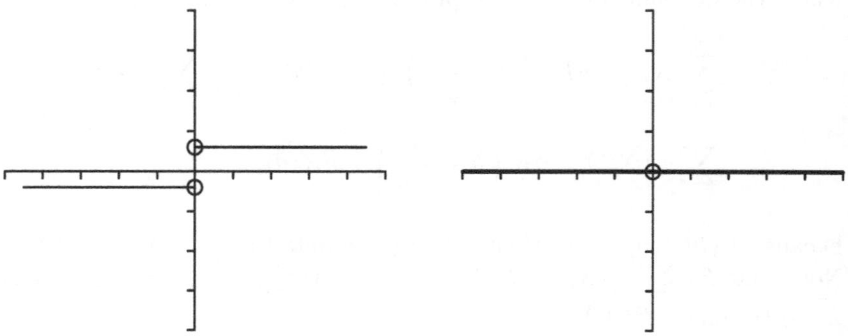

FIGURE 2.7
The first derivative (left) and second derivative of $\rho_\tau(\cdot)$ for quantile regression with $\tau = 0.6$. The first or the second derivative of $\rho_\tau(\cdot)$ does not exist at 0. The set of sub-gradients of $\rho_\tau(\cdot)$ at 0 is $[-(1-\tau), \tau]$.

where $\rho_\tau(x) = \tau x_+ + (1-\tau)(-x)_+$. We apply Theorem 2.12 to derive asymptotic properties of $\hat{q}_{\tau,n}$.

The function $\rho(z;q) = \rho_\tau(z-q)$ is convex in q for each z. A sub-gradient of $\rho(z;q)$ in q is $h(z;q) = -(\tau 1\{z > q\} - (1-\tau)1\{z < q\} + c1\{z = q\})$, with $c \in [-(1-\tau), \tau]$. See Figure 2.7 for an illustration. The expectation of $h(Z;q)$ is

$$E\{h(Z;q)\} = -\{\tau(1 - F(q)) - (1-\tau)F(q-) + c(F(q) - F(q-))\},$$

where $F(\cdot)$ denotes the CDF of Z. If $F(\cdot)$ is continuous in a neighborhood of q^*_τ, then by Proposition 2.3, $K(q) = E\{\rho(Z;q)\}$ is differentiable in the neighborhood with

$$\frac{\partial}{\partial q} K(q) = -\{\tau(1 - F(q)) - (1-\tau)F(q)\}. \tag{2.41}$$

[Question: Why is it relevant to have that $K(q)$ is differentiable in a neighborhood of q^*_τ, instead of just at q^*_τ?] In this case, $F(q^*_\tau) = \tau$ for any choice of q^*_τ as a minimizer of $K(q)$, and taking $h(z;q) = -(\tau 1\{z > q\} - (1-\tau)1\{z \leq q\}) = -\tau + 1\{z \leq q\}$ shows that $h(Z;q^*_\tau)$ is Bernoulli with mean τ and hence

$$G = \text{var}\{h(Z;q^*_\tau)\} = \tau(1-\tau).$$

If further $F(\cdot)$ is differentiable at q^*_τ, with derivative $f(q^*_\tau)$, then $K(q)$ is twice differentiable at q^*_τ and

$$H = \frac{\partial^2}{\partial q^2} K(q^*_\tau) = \tau f(q^*_\tau) + (1-\tau) f(q^*_\tau) = f(q^*_\tau).$$

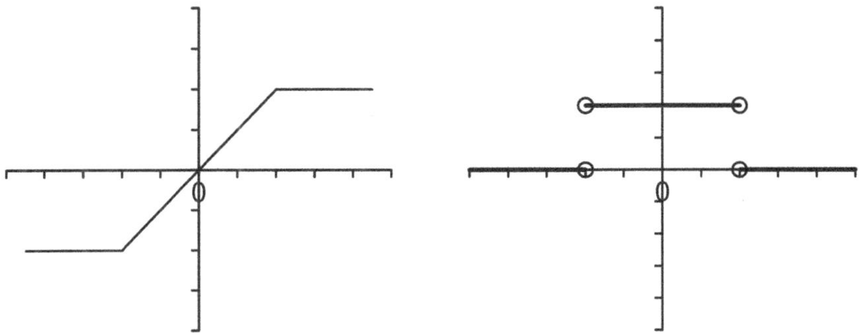

FIGURE 2.8
The first derivative (left) and second derivative of $\rho_\kappa(\cdot)$ for Huber's estimator with $\kappa = 2$. The second derivative of $\rho_\kappa(\cdot)$ does not exist at $\pm\kappa$.

Hence if $f(q_\tau^*) > 0$, then $H > 0$ and q_τ^* is a unique minimizer of $K(q)$. In summary, if $F(\cdot)$ is continuous in a neighborhood of q_τ^* and is differentiable at q_τ^* with derivative $f(q_\tau^*) > 0$, then $\hat{q}_{\tau,n} \underset{P}{\to} q_\tau^*$ and

$$\sqrt{n}(\hat{q}_{\tau,n} - q_\tau^*) \underset{D}{\to} N\left(0, \frac{\tau(1-\tau)}{f(q_\tau^*)^2}\right).$$

This result cannot be deduced from Theorem 2.10 (Asymptotic normality based on smoothness). See Ferguson (1996), Section 13, for a derivation using elementary techniques.

Example 2.8 (Huber's estimation). Suppose that $Z_1, \ldots, Z_n$ are IID scalar variables as Z. From Example 2.2, Huber's location estimator is an M-estimator defined as

$$\hat{\theta}_n = \mathrm{argmin}_{\theta \in \mathbb{R}} \sum_{i=1}^{n} \rho_\kappa(Z_i - \theta),$$

where $\rho_\kappa(\cdot)$ is defined in (2.5). We apply Theorem 2.12 to derive asymptotic properties of $\hat{\theta}_n$. Let $\bar{\theta}$ be a minimizer of $K(\theta) = E\{\rho_\kappa(Z - \theta)\}$.

The function $\rho(z;\theta) = \rho_\kappa(z - \theta)$ is convex and differentiable in θ, with the derivative

$$\begin{aligned} &\frac{\partial}{\partial\theta}\rho(z;\theta) = -\rho_\kappa'(z-\theta) \\ &= \kappa 1\{z-\theta < -\kappa\} - (z-\theta)1\{|z-\theta| \le \kappa\} - \kappa 1\{z - \theta > \kappa\}, \qquad (2.42) \end{aligned}$$

where $\rho_\kappa'(\cdot)$ is the derivative of $\rho_\kappa(\cdot)$. See Figure 2.8 for an illustration. Then by Proposition 2.3, $K(\theta)$ is differentiable with the derivative

$$\frac{\partial}{\partial\theta}K(\theta) = E\left\{\frac{\partial}{\partial\theta}\rho(Z;\theta)\right\}$$

$$= E\{-\rho'_\kappa(Z - \theta)\}.$$

Any choice of $\bar{\theta}$ satisfies $E\{\rho'_\kappa(Z - \bar{\theta})\} = 0$, i.e.,

$$0 = \kappa(F(\bar{\theta} - \kappa) - (1 - F(\bar{\theta} + \kappa))) - E[(Z - \bar{\theta})1\{|Z - \bar{\theta}| \leq \kappa\}]. \tag{2.43}$$

By direct calculation using the expression of $\frac{\partial}{\partial\theta}\rho(z;\theta)$ in (2.42), we have

$$\begin{aligned} G &= \text{var}\left\{\frac{\partial}{\partial\theta}\rho(Z;\bar{\theta})\right\} \\ &= \kappa^2(F(\bar{\theta} - \kappa) + 1 - F(\bar{\theta} + \kappa)) + E[(Z - \bar{\theta})^2 1\{|Z - \bar{\theta}| \leq \kappa\}], \end{aligned}$$

where $F(\cdot)$ denotes the CDF of Z. Moreover, it can be shown (Exercise 2.19) that if $F(\cdot)$ is continuous at $\bar{\theta} \pm \kappa$, then $K(\cdot)$ is twice differentiable at $\bar{\theta}$, i.e., $\frac{\partial}{\partial\theta}K(\cdot)$ is differentiable at $\bar{\theta}$, with

$$H = \frac{\partial^2}{\partial\theta^2}K(\bar{\theta}) = E1\{|Z - \bar{\theta}| \leq \kappa\} = F(\bar{\theta} + \kappa) - F(\bar{\theta} - \kappa).$$

Hence if $F(\bar{\theta} + \kappa) - F(\bar{\theta} - \kappa) > 0$, then $H > 0$ and $\bar{\theta}$ is a unique minimizer of $K(\theta)$. In summary, if $F(\cdot)$ is continuous at $\bar{\theta} \pm \kappa$ and $F(\bar{\theta} + \kappa) - F(\bar{\theta} - \kappa) > 0$, then $\hat{\theta}_n \xrightarrow[\mathcal{P}]{} \bar{\theta}$ and

$$\sqrt{n}(\hat{\theta}_n - \bar{\theta}) \xrightarrow[\mathcal{D}]{} \left(0, \frac{G}{H^2}\right). \tag{2.44}$$

The preceding discussion applies to any distribution for Z, including a non-symmetric distribution. As a special case, suppose Z has a symmetric distribution about θ^* such that $F(\theta^* - z) = 1 - F(\theta^* + z)$ for any $z \geq 0$. Then, by direct verification, θ^* qualifies as a choice of $\bar{\theta}$ satisfying (2.43). Therefore, if $F(\cdot)$ is continuous at $\theta^* \pm \kappa$ and $F(\theta^* + \kappa) - F(\theta^* - \kappa) > 0$, then θ^* is a unique choice for $\bar{\theta}$, and $\hat{\theta}_n$ satisfies $\hat{\theta}_n \xrightarrow[\mathcal{P}]{} \theta^*$ and

$$\sqrt{n}(\hat{\theta}_n - \theta^*) \xrightarrow[\mathcal{D}]{} \left(0, \frac{G}{H^2}\right),$$

where G and H are evaluated with $\bar{\theta} = \theta^*$. See Casella and Berger (2001), Example 10.2.6, for a discussion of Huber's estimator which assumes that Z has a symmetric density about θ^*.

It is of interest to examine the asymptotic variance $\frac{G}{H^2}$ in the two limiting cases of κ. For simplicity, assume that Z has a symmetric distribution about θ^*; otherwise $\bar{\theta}$ may vary as κ increases. Then θ^* coincides with both the mean $\mu^* = E(Z)$ and the median ν^*.

First, as $\kappa \to \infty$, it can be directly shown that

$$\begin{aligned} G &\to \lim_{\kappa\to\infty} \kappa^2 P(|Z - \theta^*| > \kappa) + E\{(Z - \theta^*)^2\}, \\ H &\to 1. \end{aligned}$$

If $E\{(Z-\theta^*)^2\} < \infty$, then $\theta^* = \mu^*$ and, from Exercise 3.1, a finite $\mathrm{var}(Z)$ implies that $\lim_{\kappa\to\infty} \kappa^2 P(|Z-\mu^*| > \kappa) = 0$. Hence in this case, $\frac{G}{H^2} \to \mathrm{var}(Z)$, which is consistent with the observation that $\hat{\theta}_n$ tends to the sample mean $\hat{\mu}_n = n^{-1}\sum_{i=1}^n Z_i$ as $\kappa \to \infty$, and $\sqrt{n}(\hat{\mu}_n - \mu^*) \xrightarrow[D]{} \mathrm{N}(0, \mathrm{var}(Z))$.

Second, as $\kappa \to 0$,

$$\begin{aligned} G/\kappa^2 &= (F(\theta^*-\kappa) + 1 - F(\theta^*+\kappa)) + E[(Z-\theta^*)^2 1\{|Z-\theta^*| \le \kappa\}]/\kappa^2 \\ &\to 1, \\ H/\kappa &= (F(\theta^*+\kappa) - F(\theta^*-\kappa))/\kappa \\ &\to 2f(\theta^*), \end{aligned}$$

where the first limit follows from $F(\theta^*-\kappa) + 1 - F(\theta^*+\kappa) \to 1$ and $E[(Z-\theta^*)^2 1\{|Z-\theta^*| \le \kappa\}]/\kappa^2 \le E[1\{|Z-\theta^*| \le \kappa\}] \to 0$ as $\kappa \to 0$ provided that $F(\cdot)$ is continuous at θ^*, and the second limit holds provided that $F(\cdot)$ is differentiable at θ^* with derivative $f(\theta^*)$. Therefore, in this case, $\frac{G}{H^2} \to 1/(2f(\theta^*))^2$, which is consistent with the observation that $\hat{\theta}_n$ tends to the sample median $\hat{\nu}_n$ as $\kappa \to \infty$, and $\sqrt{n}(\hat{\nu}_n - \nu^*) \xrightarrow[D]{} \mathrm{N}(0, 1/(2f(\theta^*))^2)$ from Example 2.7 with $\tau = 1/2$.

2.6 Appendices

2.6.1 Stochastic equi-continuity

The property (2.15) underlying the uniform law of large numbers (Theorem 2.5) is a strong version of stochastic equi-continuity, which implies the standard definition of stochastic equi-continuity.

Definition 2.3. A sequence of random functions $F_n(\cdot)$, $n \ge 1$, on a set $\mathcal{S}$ is said to be *strongly* stochastically equi-continuous if for any $\epsilon > 0$, there exists $\delta > 0$ such that as $n \to \infty$,

$$P\left(\sup_{s,\tilde{s}\in\mathcal{S}, \|s-\tilde{s}\|\le\delta} |F_n(s) - F_n(\tilde{s})| > \epsilon \right) \to 0. \tag{2.45}$$

A sequence of random functions $F_n(\cdot)$, $n \ge 1$, on a set $\mathcal{S}$ is said to be stochastically equi-continuous if for any $\epsilon > 0$ and $\eta > 0$, there exist $\delta > 0$ and $N \ge 1$ such that for $n \ge N$,

$$P\left(\sup_{s,\tilde{s}\in\mathcal{S}, \|s-\tilde{s}\|\le\delta} |F_n(s) - F_n(\tilde{s})| > \epsilon \right) \le \eta. \tag{2.46}$$

A subtle difference between strong stochastic equi-continuity and the standard version lies in the fact that δ depends only on ϵ in (2.45), whereas δ

depends on ϵ and η in (2.46). Such a difference is reminiscent of that between $O_p(1)$ [asymptotically boundedness in probability] and strong asymptotic boundedness in probability in Definitions 1.4 and 1.5. For example, the sequence $F_n(s) = sX$, $n \geq 1$, with $X \sim \mathrm{N}(0,1)$ is stochastically equi-continuous, but not strongly stochastically equi-continuous. This is related to the fact that $X_n = X$, $n \geq 1$, with $X \sim \mathrm{N}(0,1)$ is $O_p(1)$, but not strongly asymptotically bounded in probability.

The strong stochastic continuity, as well as the standard stochastic continuity, generalizes the concept of equi-continuity for non-random functions. A sequence of non-random functions $f_n(\cdot)$ on a set $\mathcal{S}$ is said to be (asymptotically) equi-continuous if for any $\epsilon > 0$, there exist $\delta > 0$ and $N \geq 1$ such that $|f_n(s) - f_n(\tilde{s})| \leq \epsilon$ for any $n \geq N$ and $s, \tilde{s} \in \mathcal{S}$ satisfying $\|s - \tilde{s}\| \leq \delta$. The concept of equi-continuity, in turn, generalizes uniform continuity for a single non-random function. As a precursor to Theorem 2.5 later, the following result shows that on a compact set, uniform convergence of non-random functions to a continuous function is equivalent to pointwise convergence and equi-continuity. The result can be generalized by relaxing $\mathcal{S}$ to be a bounded, possibly non-closed set and strengthening $f(\cdot)$ to be a uniformly continuous function on $\mathcal{S}$ (Exercise 2.4).

Lemma 2.4 (Uniform convergence and equi-continuity). *Let $f_n : \mathcal{S} \to \mathbb{R}$, $n \geq 1$, be a sequence of functions on a compact set $\mathcal{S} \subset \mathbb{R}^d$, and $f : \mathcal{S} \to \mathbb{R}$ be a function (not assumed continuous) on $\mathcal{S}$. Then the following two statements are equivalent:*

(i) $f(\cdot)$ is continuous on $\mathcal{S}$ and

$$\sup_{s \in \mathcal{S}} |f_n(s) - f(s)| \to 0.$$

(ii) $f_n(s) \to f(s)$ for each $s \in \mathcal{S}$, and the sequence $f_n(\cdot)$, $n \geq 1$, is equi-continuous on $\mathcal{S}$.

By a stochastic extension from Theorem 2.4, a general result (Lemma 2.5) can be established, relating both versions of stochastic equi-continuity to uniform convergence in probability. This is slightly strengthened from Newey (1991), Theorem 2.1, which shows that the uniform convergence (2.47) is equivalent to pointwise convergence and stochastic equi-continuity in (iii), provided that $F(\cdot)$ is continuous on a compact set $\mathcal{S}$.

Lemma 2.5 (Uniform convergence and stochastic equi-continuity). *Let $F_n : \mathcal{S} \to \mathbb{R}$, $n \geq 1$, be a sequence of random functions on a compact set $\mathcal{S} \subset \mathbb{R}^d$, and $F : \mathcal{S} \to \mathbb{R}$ be a non-random function (not assumed continuous) on $\mathcal{S}$. Then the following three statements are equivalent:*

(i) $F(\cdot)$ is continuous on $\mathcal{S}$ and

$$\sup_{s \in \mathcal{S}} |F_n(s) - F(s)| \xrightarrow{\mathcal{P}} 0, \tag{2.47}$$

(ii) $F_n(s) \underset{\mathcal{P}}{\to} F(s)$ *for each* $s \in \mathcal{S}$*, and the sequence* $F_n(\cdot)$*,* $n \geq 1$*, is strongly stochastically equi-continuous on* $\mathcal{S}$*.*

(iii) $F_n(s) \underset{\mathcal{P}}{\to} F(s)$ *for each* $s \in \mathcal{S}$*, and the sequence* $F_n(\cdot)$*,* $n \geq 1$*, is stochastically equi-continuous on* $\mathcal{S}$*.*

Proof.

(ii) $\Rightarrow$ *(iii)*. This is immediate.

(iii) $\Rightarrow$ *(i)*. Suppose that $F_n(s) \underset{\mathcal{P}}{\to} F(s)$ pointwise on $\mathcal{S}$ and the sequence $F_n(\cdot)$ is stochastically equi-continuous on $\mathcal{S}$. First, we show that $F(\cdot)$ is uniformly continuous on $\mathcal{S}$. In fact, for any $\epsilon > 0$ and $\eta = 1/2$, there exist $\delta > 0$ and $N \geq 1$ such that $P\left(\sup_{\|s-\tilde{s}\| \leq \delta} |F_n(s) - F_n(\tilde{s})| > \frac{\epsilon}{3}\right) \leq 1/2$ for $n \geq N$. For any $s, \tilde{s} \in \mathcal{S}$ satisfying $\|s - \tilde{s}\| \leq \delta$,

$$\begin{aligned} &P(|F(s) - F(\tilde{s})| > \epsilon) \\ &\leq P\left(|F(s) - F_n(s)| > \frac{\epsilon}{3}\right) + P\left((|F(\tilde{s}) - F_n(\tilde{s})| > \frac{\epsilon}{3}\right) \\ &\quad + P\left(|F_n(s) - F_n(\tilde{s})| > \frac{\epsilon}{3}\right). \end{aligned}$$

The $\limsup$ of the right-hand side as $n \to \infty$ is no greater than 1/2. Then $P(F(s) - F(\tilde{s})| > \epsilon)$, which is either 0 or 1 due to non-randomness of $F(\cdot)$, is necessarily 0, i.e., $|F(s) - F(\tilde{s})| \leq \epsilon$. The uniform convergence (2.47) can be shown similarly as in the first proof of (2.9), with $F_n(s)$ and $F(s)$ corresponding to $n^{-1} \sum_{i=1}^n \rho(Z_i; \theta)$ and $E\{\rho(Z; \theta)\}$ respectively. The term Δ_{1j} is controlled by stochastic equi-continuity [similarly to the control by *strong* stochastic equi-continuity in the first proof of (2.9)], Δ_{2j} is controlled by pointwise convergence of $F_n(s)$ to $F(s)$, and Δ_{3j} is controlled by uniform continuity of $F(s)$. Details for the proof are left as Exercise 2.5.

(i) $\Rightarrow$ *(ii)*. Suppose that $F(\cdot)$ is continuous on $\mathcal{S}$ and (2.47) holds. The latter directly implies the pointwise convergence of $F_n(s)$ to $F(s)$. From the former and the compactness of $\mathcal{S}$, $F(\cdot)$ is uniformly continuous on $\mathcal{S}$. For any $\epsilon > 0$, there exists $\delta > 0$ such that $|F(s) - F(\tilde{s})| \leq \frac{\epsilon}{3}$ for any $s, \tilde{s} \in \mathcal{S}$ satisfying $\|s - \tilde{s}\| \leq \delta$. Then

$$\begin{aligned} &P\left(\sup_{\|s-\tilde{s}\| \leq \delta} |F_n(s) - F_n(\tilde{s})| > \epsilon\right) \\ &\leq P\left(\sup_{\|s-\tilde{s}\| \leq \delta} |F(s) - F_n(s)| > \frac{\epsilon}{3}\right) + P\left(\sup_{\|s-\tilde{s}\| \leq \delta} |F(\tilde{s}) - F_n(\tilde{s})| > \frac{\epsilon}{3}\right), \end{aligned}$$

which tends to 0 by the uniform convergence (2.47). □

It is interesting that the strong and standard versions of stochastic equi-continuity, although in general distinct, are interchangeable with each other in the context of Theorem 2.5 [i.e., compact set $\mathcal{S}$ and pointwise convergence of a sequence $F_n(s)$ to non-random $F(s)$]. Similarly to Lemma 2.4, Lemma 2.5 can be generalized by relaxing $\mathcal{S}$ to be a bounded, possibly non-closed set and strengthening $F(\cdot)$ to be a uniformly continuous, non-random function on $\mathcal{S}$. Moreover, the equivalence between (i) and (iii), but not that between (i) and (ii), can be generalized by allowing $F(\cdot)$ to be a non-random function and requiring in (i) that $F(\cdot)$ satisfies a stochastic version of uniform continuity. See Exercise 2.6. Therefore, the interchangeability between the two versions of stochastic equi-continuity in Lemma 2.5 is attributable to the assumption that $F(\cdot)$ is a non-random function.

The following result gives Lipschitz conditions sufficient for the two versions of stochastic equi-continuity, which together with pointwise convergence lead to uniform convergence. The proof is left as an exercise. See Newey and McFadden (1994), Lemma 2.9, for a related result.

Lemma 2.6 (Lipschitz conditions for stochastic equi-continuity). *Let $F_n(\cdot)$, $n \geq 1$, be a sequence of random functions on a set $\mathcal{S} \subset \mathbb{R}^d$.*

(i) If there exist random variables D_n, $n \geq 1$, such that $(D_n)_{n\geq 1}$ is strongly asymptotically bounded in probability, and $|F_n(s) - F_n(\tilde{s})| \leq D_n \|s - \tilde{s}\|$ for any $s, \tilde{s} \in \mathcal{S}$. Then $F_n(\cdot)$, $n \geq 1$, is strongly stochastically equi-continuous.

(ii) If there exist random variables D_n, $n \geq 1$, such that $D_n = O_p(1)$ and $|F_n(s) - F_n(\tilde{s})| \leq D_n \|s - \tilde{s}\|$ for any $s, \tilde{s} \in \mathcal{S}$. Then $F_n(\cdot)$, $n \geq 1$, is stochastically equi-continuous.

2.6.2 Proof of convexity lemma for uniform convergence

The uniform convergence of random convex functions as in Lemma 2.1 is fundamental to the asymptotic theory based on convexity in Section 2.5. Proofs of this lemma can be found in Pollard (1991), Niemiro (1992), and references therein. To facilitate understanding, we present two proofs of Lemma 2.1, both relying on continuity properties for (non-random) convex functions. The first proof establishes strong stochastic equi-continuity as an intermediate step, whereas the second proof, adapted from Pollard (1991), is more direct without proving strong or standard stochastic equi-continuity.

First, we present a basic, powerful lemma, which underlies the continuity of convex functions. See Rockafellar (1970, Section 10) and Hiriart-Urruty and Lemaréchal (2001, Section B.3.1). A function $f : \mathcal{T} \to \mathbb{R}$ is said to be convex on a possibly non-convex set $\mathcal{T} \subset \mathbb{R}^d$ if for any $s, t \in \mathcal{T}$ and any $0 \leq \lambda \leq 1$ satisfying $\lambda s + (1 - \lambda)t \in \mathcal{T}$, we have

$$f(\lambda s + (1 - \lambda)t) \leq \lambda f(s) + (1 - \lambda) f(t).$$

Here a generalization is that $\mathcal{T}$ can be a non-convex set, although conventionally a convex function is defined on a convex set (Rockafellar, 1970; Hiriart-Urruty & Lemaréchal, 2001). [Question: Can a convex function $f : \mathcal{T} \to \mathbb{R}$ on a non-convex set $\mathcal{T}$ always be extended to a convex function on the convex hull of $\mathcal{T}$?]

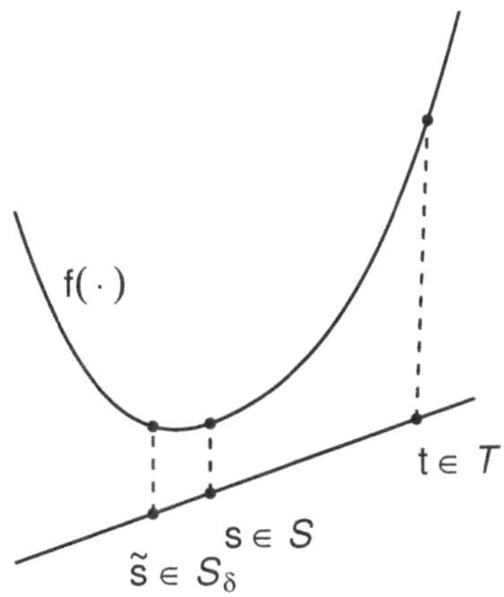

FIGURE 2.9
Illustration of three points $s, t, \tilde{s}$ in the proof of Lemma 2.7.

Lemma 2.7 (Convexity bounds). *Let $f : \mathcal{T} \to \mathbb{R}$ be a convex function on a set $\mathcal{T} \subset \mathbb{R}^d$ (possibly non-open, non-convex). For a subset $\mathcal{S} \subset \mathcal{T}$ and $\delta > 0$, suppose that $\mathcal{S}_\delta \subset \mathcal{T}$, where $\mathcal{S}_\delta = \cup_{s \in \mathcal{S}} B(s, \delta) = \{s + t : s \in \mathcal{S}, \|t\| \le \delta\}$ with $B(s, \delta) = \{t \in \mathbb{R}^d : \|t - s\| \le \delta\}$. Then*

$$|f(s) - f(t)| \le \frac{M - m}{\delta} \|s - t\|, \quad s, t \in \mathcal{S}, \tag{2.48}$$

where $M = \sup_{s \in \mathcal{S}_\delta} f(s)$ and $m = \inf_{s \in \mathcal{S}_\delta} f(s)$. Note that (2.48) indicates that $f(\cdot)$ is Lipschitz on $\mathcal{S}$ provided that $M - m$ is finite. Moreover,

$$f(t) \ge f(s) + \frac{\|t - s\|}{\delta} (f(s) - M), \quad s \in \mathcal{S}, t \in \mathcal{T}. \tag{2.49}$$

Proof. For any distinct points $s \in \mathcal{S}$ and $t \in \mathcal{T}$, let

$$\tilde{s} = s + \frac{\delta}{\|s - t\|}(s - t),$$

satisfying $\|\tilde{s} - s\| = \delta$. See Figure 2.9 for an illustration. Then $\tilde{s} \in \mathcal{S}_\delta \subset \mathcal{T}$ and

$$s = (1 - \lambda)t + \lambda \tilde{s}, \quad \lambda = \frac{\|s - t\|}{\delta + \|s - t\|}.$$

By the convexity of f, we have

$$f(s) \le (1 - \lambda) f(t) + \lambda f(\tilde{s}),$$

or equivalently

$$f(t) \geq \frac{f(s) - \lambda f(\tilde{s})}{1 - \lambda} = f(s) + \frac{\|t - s\|}{\delta}(f(s) - f(\tilde{s})). \tag{2.50}$$

This implies inequality (2.49). Moreover, by restricting $t \in \mathcal{S}$, inequality (2.50) implies that

$$f(s) - f(t) \leq \frac{M - m}{\delta}\|s - t\|, \quad s, t \in \mathcal{S},$$

which also holds with the left-hand side replaced by $f(t) - f(s)$ by exchanging s and t. Therefore, (2.48) is obtained. □

By applying Lemma 2.7 locally, a convex function can be shown to be continuous on an open set.

Lemma 2.8 (Continuity of convex functions). *Let $f : \mathcal{T} \to \mathbb{R}$ be a convex function on an open set $\mathcal{T}$. Then $f(\cdot)$ is continuous on $\mathcal{T}$.*

Proof. We take $\|\cdot\|$ to be the L_∞ norm $\|\cdot\|_\infty$, which is equivalent to L_2 norm with fixed d, in the sense of (1.3) as discussed in Section 1.1. Rewrite $B(s, \delta)$ as $C(s, \delta) = \{t \in \mathbb{R} : \|t - s\|_\infty \leq \delta\}$, which is a cube in Euclidean geometry. Fix any $s_0 \in \mathcal{T}$. Pick $\delta > 0$ sufficiently small such that $C(s_0, 2\delta) \subset \mathcal{T}$. We apply Lemma 2.7 with $\mathcal{S} = C(s_0, \delta)$ and $\mathcal{S}_\delta = C(s_0, 2\delta)$. Then (2.48) indicates that $f(\cdot)$ is Lipschitz on $C(s_0, \delta)$, and hence $f(\cdot)$ is continuous at s_0, provided $M = \sup_{s \in C(s_0, 2\delta)} f(s)$ and $m = \inf_{s \in C(s_0, 2\delta)} f(s)$ are finite, i.e., $f(\cdot)$ is bounded from above and below on $C(s_0, 2\delta)$.

In fact, any $s \in C(s_0, 2\delta)$ can be written as a convex combination $s = \sum_j \lambda_j s_j$, where s_j's are the vertices of $C(s_0, 2\delta)$, and $\lambda_j \geq 0$ with $\sum_j \lambda_j = 1$. [Such vertices would not exist if $\|\cdot\|$ is the L_2 norm.] Then by the convexity of $f(\cdot)$, $f(s) \leq \sum_j \lambda_j f(s_j) \leq \max_j f(s_j)$, which shows that M is finite. Moreover, from (2.49) with $s = s_0$, we have $f(t) \geq f(s_0) + 2(f(s_0) - M)$ for $t \in C(s_0, 2\delta)$, which shows that m is finite. □

Returning to Lemma 2.7, if $\mathcal{S}$ is compact with $\mathcal{S}_\delta \subset \mathcal{T}$, then $\mathcal{S}_\delta$ is also compact and $M = \sup_{s \in \mathcal{S}_\delta} f(s)$ and $m = \inf_{s \in \mathcal{S}_\delta} f(s)$ are finite, because from Lemma 2.8, $f(\cdot)$ is continuous on $\mathcal{T}$ and hence on the compact subset $\mathcal{S}_\delta$. [A continuous function on a compact set attains a (finite) maximum and a (finite) minimum.] With finite M and m, (2.48) then indicates that $f(\cdot)$ is Lipschitz on $\mathcal{S}$ (not necessarily on $\mathcal{T}$).

Corollary 2.3 (Lipschitz continuity of convex functions). *Let $f : \mathcal{T} \to \mathbb{R}$ be a convex function on an open set $\mathcal{T}$. If $\mathcal{S}$ is compact with $\mathcal{S}_\delta \subset \mathcal{T}$ for $\delta > 0$, then $f(\cdot)$ is Lipschitz-L on $\mathcal{S}$ with $L = \frac{M-m}{\delta}$.*

The following lemma is a non-random analogue of Lemma 2.1. By Lemma 2.4 (Uniform convergence and equi-continuity), the proof involves showing that the sequence of functions $f_n(\cdot)$ is equi-continuous on $\mathcal{S}$, which follows from Corollary 2.3 and the fact that the sequence of functions $f_n(\cdot)$ is uniformly bounded from above and below on $\mathcal{S}_\delta\,(\subset \mathcal{T})$ for some $\delta > 0$. See Rockafellar (1970), Section 10, for details.

Lemma 2.9 (Uniform convergence of convex functions). *Let $f_n : \mathcal{T} \to \mathbb{R}$, $n \geq 1$, be a sequence of convex functions on an open set $\mathcal{T}$, and $f : \mathcal{T} \to \mathbb{R}$ be a function on $\mathcal{T}$ such that $f_n(s) \to f(s)$ for each $s \in \mathcal{T}$. Then $f(\cdot)$ is convex on $\mathcal{T}$, and for each compact set $\mathcal{S} \subset \mathcal{T}$,*

$$\sup_{s \in \mathcal{S}} |f_n(s) - f(s)| \to 0.$$

While Lemma 2.9 for non-random convex functions can be called to give a proof of Lemma 2.1 for random convex functions through a sub-sequencing argument as in Niemiro (1992), we present two direct proofs of Lemma 2.1. The first proof obtains uniform convergence by proving strong stochastic equi-continuity and calling Lemma 2.5. From this proof, strong stochastic continuity can be seen to underlie both ULLN based on smoothness and dominance (Theorem 2.5) and ULLN based on convexity (i.e., application of Lemma 2.1 to random convex functions which are sample averages).

First proof of Lemma 2.1. The convexity of $F(\cdot)$ follows from the monotonicity in the limits of sequences converging in probability (Exercise 1.2) and the non-randomness of $F(\cdot)$. To facilitate the proof of uniform convergence, we take $\|\cdot\|$ to be the L_∞ norm $\|\cdot\|_\infty$, and rewrite $B(s,\delta)$ as $C(s,\delta)$, representing a cube in Euclidean geometry. By the pointwise convergence $F_n(s) \underset{\mathcal{P}}{\to} F(s)$ and Lemma 2.5, it suffices to show that the sequence $F_n(\cdot)$, $n \geq 1$, is *strongly* stochastically equi-continuous on $\mathcal{S}$.

Pick $\delta > 0$ sufficiently small such that $\mathcal{S}_{3\delta} \subset \mathcal{T}$. From Corollary 2.3 (Lipschitz continuity), we have

$$|F_n(s) - F_n(t)| \leq \frac{M_n - m_n}{\delta}\|s-t\|_\infty, \quad s,t \in \mathcal{S}, n \geq 1,$$

where $M_n = \sup_{s\in\mathcal{S}_\delta} F_n(s)$ and $m_n = \inf_{s\in\mathcal{S}_\delta} F_n(s)$, which are finite for each n. Then the strong stochastic equi-continuity holds for $F_n(s)$ on $\mathcal{S}$ by Lemma 2.6(i) provided that $(M_n - m_n)_{n\geq 1}$ is *strongly* asymptotically bounded in probability. [Question: Alternatively, can you directly show that $\sup_{n\geq 1} M_n$ is *finite* almost surely? This would then imply that $(M_n)_{n\geq 1}$ is asymptotically bounded from above in probability and, through (2.52), $(m_n)_{n\geq 1}$ is asymptotically bounded from below in probability, so that $(M_n - m_n)_{n\geq 1}$ is asymptotically bounded in probability. A sequence of scalar random variables $(X_n)_{n\geq 1}$ is said to be asymptotically bounded from above or below in probability if for any $\eta > 0$, there exist $R > 0$ and $N \geq 1$ such that $P(X_n > R) \leq \eta$ or, respectively, $P(X_n < R) \leq \eta$ for $n \geq N$.]

First, we show that $(M_n)_{n\geq 1}$ is asymptotically bounded from above in probability by the constant $M_* = \sup_{s\in\mathcal{S}_\delta} F(s) + 1$, i.e., for any $\eta > 0$, there

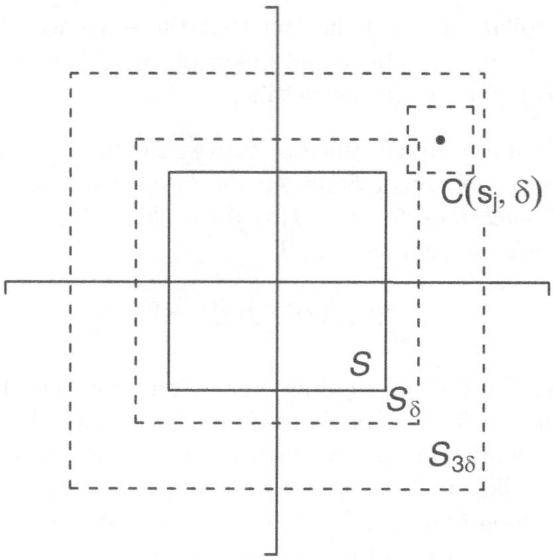

FIGURE 2.10
Two-dimensional illustration of $\mathcal{S}_\delta \subset \cup_{j=1}^{J} C(s_j, \delta) \subset \mathcal{S}_{3\delta}$ in the special case where $\mathcal{S}$ itself is a cube.

exists $N_1 \geq 1$ such that for $n \geq N_1$,

$$P(M_n > M_*) \leq \eta. \tag{2.51}$$

By the compactness (more precisely, the boundedness) of $\mathcal{S}_\delta$, let $\{C(s_j, \delta) : j = 1, \ldots, J\}$ be a finite collection of cubes such that $\mathcal{S}_\delta \subset \cup_{j=1}^{J} C(s_j, \delta)$. Without loss of generality, assume that $\mathcal{S}_\delta \cap C(s_j, \delta) \neq \emptyset$ for each j; otherwise $C(s_j, \delta)$ can be removed from the collection. Then $\cup_{j=1}^{J} C(s_j, \delta) \subset \mathcal{S}_{3\delta} \subset \mathcal{T}$ (Figure 2.10). To see this, fix any j and pick $s \in \mathcal{S}_\delta \cap C(s_j, \delta)$. By definition, there exists $\tilde{s} \in \mathcal{S}$ such that $\|s - \tilde{s}\|_\infty \leq \delta$. Then for any $t \in C(s_j, \delta)$, $\|t - s\|_\infty \leq \|t - s_j\|_\infty + \|s_j - s\|_\infty \leq \delta + \delta = 2\delta$ and $\|t - \tilde{s}\|_\infty \leq 3\delta$, that is, $t \in \mathcal{S}_{3\delta}$. Denote as $V(s_j, \delta)$ the set of 2^d vertices of the cube $C(s_j, \delta)$. Then $\cup_{j=1}^{J} V(s_j, \delta) \subset \mathcal{T}$. By the pointwise convergence of $F_n(\cdot)$ to $F(\cdot)$ on $\mathcal{T}$, there exists $N_1 \geq 1$ such that for $n \geq N_1$,

$$P\left(\max_{j=1,\ldots,J} \max_{t \in V(s_j, \delta)} |F_n(t) - F(t)| > 1 \right) \leq \eta.$$

For any $s \in \mathcal{S}_\delta$, we pick j such that $s \in C(s_j, \delta)$. Then by the convexity of $F_n(\cdot)$ on $\mathcal{T}$ and the fact that s is a convex combination of the vertices in $V(s_j, \delta)$ similarly as in the proof of Lemma 2.8,

$$\begin{aligned} F_n(s) &\leq \max_{t \in V(s_j, \delta)} F_n(t) \\ &\leq \sup_{t \in \mathcal{S}_\delta} F(t) + \max_{t \in V(s_j, \delta)} |F_n(t) - F(t)|. \end{aligned}$$

Taking the supremum over s and j yields

$$M_n \le \sup_{t\in\mathcal{S}_\delta} F(t) + \max_{j=1,\dots,J} \max_{t\in V(s_j,\delta)} |F_n(t) - F(t)|,$$

which leads to inequality (2.51) for $n \ge N_1$:

$$\begin{aligned} &P\left(M_n > \sup_{s\in\mathcal{S}_\delta} F(s) + 1\right) \\ &\le P\left(\max_{j=1,\dots,J} \max_{t\in V(s_j,\delta)} |F_n(t) - F(t)| > 1\right) \le \eta, \end{aligned}$$

because the event on the first line implies the event on the second line.

Second, fix any $s_0 \in \mathcal{S}$. We show that $(m_n)_{n\ge1}$ is asymptotically bounded from below in probability by the constant

$$m_* = F(s_0) - 1 + \left(\sup_{t\in\mathcal{S}_\delta} \frac{\|t - s_0\|_\infty}{\delta}\right)(F(s_0) - 1 - M_*),$$

i.e., for any $\eta > 0$, there exists $N_2 \ge 1$ such that for $n \ge N_2$,

$$P(m_n < m_*) \le \eta. \tag{2.52}$$

In fact, from (2.49) with $s = s_0$, we have for $t \in \mathcal{S}_\delta$,

$$F_n(t) \ge F_n(s_0) + \frac{\|t - s_0\|_\infty}{\delta}(F_n(s_0) - M_n).$$

Taking the infimum over t and noting $F_n(s_0) \le M_n$ yields

$$m_n \ge F_n(s_0) + \left(\sup_{t\in\mathcal{S}_\delta} \frac{\|t - s_0\|_\infty}{\delta}\right)(F_n(s_0) - M_n), \tag{2.53}$$

which in the event $\{M_n \le M_*\} \cap \{|F_n(s_0) - F(s_0)| \le 1\}$ implies

$$m_n \ge F(s_0) - 1 + \left(\sup_{t\in\mathcal{S}_\delta} \frac{\|t - s_0\|_\infty}{\delta}\right)(F(s_0) - 1 - M_*).$$

By (2.51) with η replaced by $\frac{\eta}{2}$ and the pointwise convergence of $F_n(s_0)$ to $F(s_0)$, there exists $N_2 \ge 1$ such that for $n \ge N_2$,

$$P(M_n > M_*) \le \frac{\eta}{2}, \quad P(|F_n(s_0) - F(s_0)| > 1) \le \frac{\eta}{2}.$$

Then the desired inequality (2.52) holds for $n \ge N_2$:

$$\begin{aligned} &P\left(m_n < F(s_0) - 1 + \left(\sup_{t\in\mathcal{S}_\delta} \frac{\|t - s_0\|_\infty}{\delta}\right)(F(s_0) - 1 - M_*)\right) \\ &\le P(M_n > M_*) + P(|F_n(s_0) - F(s_0)| > 1) \le \eta, \end{aligned}$$

because, from the earlier discussion, the event on the first line implies the union $\{M_n > M_*\} \cup \{|F_n(s_0) - F(s_0)| > 1\}$.

Combining (2.51) and (2.52) shows that $(M_n - m_n)_{n\geq 1}$ is asymptotically bounded in probability by the constant

$$M_* - m_* = \left(\sup_{t\in\mathcal{S}_\delta} \frac{\|t - s_0\|_\infty}{\delta} + 1\right)(M_* + 1 - F(s_0)).$$

□

For pedagogical purposes, we present another proof of Lemma 2.1 without directly proving stochastic equi-continuity. Nevertheless, the two proofs can be seen to share some common ideas.

Second proof of Lemma 2.1. As in the first proof, we take $\|\cdot\| = \|\cdot\|_\infty$ and rewrite $B(s,\delta)$ as $C(s,\delta)$. It is desired to show that for any $\epsilon > 0$ and $\eta > 0$, there exists $N \geq 1$ such that for $n \geq N$,

$$P\left(\sup_{s\in\mathcal{S}} |F_n(s) - F(s)| > \epsilon\right) \leq \frac{\eta}{2}.$$

By the convexity of $F(\cdot)$ on open set $\mathcal{T}$, $F(\cdot)$ is continuous and hence uniformly continuous on the compact subset $\mathcal{S}$ (Heine–Cantor theorem). Then we pick $\delta > 0$ such that $|F(s) - F(t)| \leq \frac{\epsilon}{4}$ for any $s, t \in \mathcal{S}$ satisfying $\|s - t\|_\infty \leq \delta$.

We proceed with a construction of cubes similarly as in the proof of (2.51), but with $\mathcal{S}_\delta$ replaced by $\mathcal{S}$. By the compactness (more precisely, the boundedness) of $\mathcal{S}$, let $\{C(s_j,\delta) : j = 1,\ldots,J\}$ be a finite collection of cubes such that $\mathcal{S} \subset \cup_{j=1}^J C(s_j,\delta)$. Without loss of generality, assume that $\mathcal{S} \cap C(s_j,\delta) \neq \emptyset$ for each j. Then $\cup_{j=1}^J C(s_j,\delta) \subset \mathcal{S}_{2\delta} \subset \mathcal{T}$. Denote as $V(s_j,\delta)$ the set of 2^d vertices of the cube $C(s_j,\delta)$. Then $\cup_{j=1}^J V(s_j,\delta) \subset \mathcal{T}$. By the pointwise convergence of $F_n(\cdot)$ to $F(\cdot)$ on $\mathcal{T}$, there exists $N \geq 1$ such that for $n \geq N$,

$$P\left(\max_{j=1,\ldots,J} \max_{t\in\{s_j\}\cup V(s_j,\delta)} |F_n(t) - F(t)| > \frac{\epsilon}{6}\right) \leq \frac{\eta}{2}.$$

First, we show that for $n \geq N$,

$$P\left(\sup_{s\in\mathcal{S}}\{F_n(s) - F(s)\} > \epsilon\right) \leq \frac{\eta}{2}. \tag{2.54}$$

For any $s \in \mathcal{S}$, we pick j such that $s \in C(s_j,\delta)$. Then by the convexity of $F_n(\cdot)$ on $\mathcal{T}$ and the fact that s is a convex combination of the vertices in $V(s_j,\delta)$ similarly as in the proof of Lemma 2.8,

$$\begin{aligned} F_n(s) &\leq \max_{t\in V(s_j,\delta)} F_n(t) \\ &\leq F(s) + \max_{t\in V(s_j,\delta)} |F(t) - F(s)| + \max_{t\in V(s_j,\delta)} |F_n(t) - F(t)| \end{aligned}$$

$$\leq F(s) + \frac{\epsilon}{4} + \max_{t \in V(s_j,\delta)} |F_n(t) - F(t)|,$$

where the last line uses $|F(t) - F(s)| \leq \frac{\epsilon}{4}$ for $t \in V(s_j, \delta)$. Rearranging and taking the supremum over s and j yields

$$\sup_{s \in \mathcal{S}} \{F_n(s) - F(s)\} \leq \frac{\epsilon}{4} + \max_{j=1,\ldots,J} \max_{t \in V(s_j,\delta)} |F_n(t) - F(t)|,$$

which leads to inequality (2.54) for $n \geq N$:

$$\begin{aligned} &P\left(\sup_{s \in \mathcal{S}} \{F_n(s) - F(s)\} > \epsilon\right) \\ &\leq P\left(\max_{j=1,\ldots,J} \max_{t \in V(s_j,\delta)} |F_n(t) - F(t)| > \frac{\epsilon}{4}\right) \leq \frac{\eta}{2}. \end{aligned}$$

Second we show that for $n \geq N$,

$$P\left(\inf_{s \in \mathcal{S}} \{F_n(s) - F(s)\} < -\epsilon\right) \leq \frac{\eta}{2}. \tag{2.55}$$

For any $t \in \mathcal{S}$, we pick j such that $t \in C(s_j, \delta)$. Instead of (2.49), we apply (2.50) with $s = s_j$ and $\tilde{s}_j \in C(s_j, \delta)$ and use $F_n(\tilde{s}_j) \leq \max_{s \in V(s_j,\delta)} F_n(s)$ and $\|t - s_j\|_\infty \leq \delta$ to obtain

$$\begin{aligned} F_n(t) &\geq F_n(s_j) + \frac{\|t - s_j\|_\infty}{\delta}(F_n(s_j) - F_n(\tilde{s}_j)) \\ &\geq F_n(s_j) + (F_n(s_j) - \max_{s \in V(s_j,\delta)} F_n(s)) \\ &\geq F(t) - |F(t) - F(s_j)| - \max_{s \in V(s_j,\delta)} |F(s_j) - F(s)| \\ &\quad - 2|F_n(s_j) - F(s_j)| - \max_{s \in V(s_j,\delta)} |F_n(s) - F(s)| \\ &\geq F(t) - \frac{\epsilon}{2} - 3 \max_{s \in \{s_j\} \cup V(s_j,\delta)} |F_n(s) - F(s)|, \end{aligned}$$

where the last line uses $|F(t) - F(s_j)| \leq \frac{\epsilon}{4}$ and $|F(s_j) - F(s)| \leq \frac{\epsilon}{4}$ for $s \in V(s_j, \delta)$. Rearranging and taking the infimum over t and j yields

$$\inf_{t \in \mathcal{S}} \{F_n(t) - F(t)\} \geq -\frac{\epsilon}{2} - 3 \max_{j=1,\ldots,J} \max_{s \in \{s_j\} \cup V(s_j,\delta)} |F_n(s) - F(s)|,$$

which leads to inequality (2.55) for $n \geq N$:

$$\begin{aligned} &P\left(\inf_{t \in \mathcal{S}} \{F_n(t) - F(t)\} < -\epsilon\right) \\ &\leq P\left(\max_{j=1,\ldots,J} \max_{t \in \{s_j\} \cup V(s_j,\delta)} |F_n(t) - F(t)| > \frac{\epsilon}{6}\right) \leq \frac{\eta}{2}. \end{aligned}$$

Combining (2.54) and (2.55) completes the proof. □

2.6.3 Exchanging differentiation and expectation

Let Z be a random vector and $g(z;\theta)$ be a scalar function of z, parameterized by $\theta \in \Theta$, an open set in $\mathbb{R}^d$. Exchanging differentiation and expectation can be justified under a dominance condition (e.g., Casella & Berger, 2001, Corollary 2.4.4).

Proposition 2.1 (Exchanging under dominance). *Suppose that $g(z;\theta)$ is differentiable in $\theta \in \Theta$ for each z, and there exists a nonnegative function $M(z)$ such that $|\frac{\partial}{\partial\theta} g(z;\theta)| \le M(z)$ for each z and $\theta \in \Theta$ and $E\{M(Z)\} < \infty$. Then $E\{g(Z;\theta)\}$ is differentiable in $\theta \in \Theta$ and*

$$\frac{\partial}{\partial\theta} E\{g(Z;\theta)\} = E\{\frac{\partial}{\partial\theta} g(Z;\theta)\}.$$

In the case of $g(z;\theta)$ convex in θ, the requirement of a dominating function $M(z)$ can be relaxed (Bertsekas (1973), Proposition 2.2). Denote by $\partial_\theta g(z;\theta)$ the set of sub-gradients of $g(z;\theta)$ in θ for each z.

Proposition 2.2 (Exchanging under convexity). *Suppose that $g(z;\theta)$ is convex in $\theta \in \Theta$ for almost all z under the distribution of Z, and $E|g(Z;\theta)| < \infty$ is finite for $\theta \in \Theta$. Then $E\{g(Z;\theta)\}$ is convex in $\theta \in \Theta$ and*

$$\partial_\theta E\{g(Z;\theta)\} = E\{\partial_\theta g(Z;\theta)\},$$

i.e., for any sub-gradient $h(z;\theta)$ of $g(z;\theta)$, $E\{h(Z;\theta)\}$ is finite and a sub-gradient of $E\{g(Z;\theta)\}$; conversely, any sub-gradient of $E\{g(Z;\theta)\}$ can be obtained as $E\{h(Z;\theta)\}$ for a sub-gradient $h(z;\theta)$ of $g(z;\theta)$.

The following result (Bertsekas, 1973, Proposition 2.3) gives a necessary and sufficient condition for $E\{g(Z;\theta)\}$ to be differentiable at a point θ_0.

Proposition 2.3 (Exchanging under convexity and smoothness). *In the setting of Proposition 2.2, $E\{g(Z;\theta)\}$ is differentiable at θ if and only if $g(Z;\theta)$ is differentiable at θ almost surely, i.e., $P(Z \in D_\theta) = 1$, where $D_\theta = \{z : g(z;\theta)$ is differentiable at $\theta\}$. In this case,*

$$\frac{\partial}{\partial\theta} E\{g(Z;\theta)\} = E\{h(Z;\theta)\} \tag{2.56}$$

for any sub-gradient $h(z;\theta)$ of $g(z;\theta)$.

Note that (2.56) can also be interpreted as a necessary and sufficient condition. If $E\{g(Z;\theta)\}$ is differentiable at θ, then (2.56) holds for any sub-gradient $h(z;\theta)$ of $g(z;\theta)$. Conversely, if for any sub-gradients $h(z;\theta)$ and $\tilde{h}(z;\theta)$, $E\{h(Z;\theta)\} = E\{\tilde{h}(Z;\theta)\}$, then $E\{g(Z;\theta)\}$ is differentiable at θ and (2.56) holds.

Proof of sufficiency in (2.56)**.** By the definition of sub-gradients, for any θ and s,

$$g(z;\theta+s) - g(s;\theta) \ge s^{\mathrm{T}} h(z;\theta).$$

For any $t \in \mathbb{R}^d$ and $\delta > 0$, taking $s = \delta t$ in the above inequality gives

$$\frac{g(z;\theta + \delta t) - g(z;\theta)}{\delta} \geq t^{\mathrm{T}} h(z;\theta).$$

Taking $s = -\delta t$ gives

$$\frac{g(z;\theta - \delta t) - g(z;\theta)}{-\delta} \leq t^{\mathrm{T}} h(z;\theta).$$

Taking expectations on both sides of these inequalities gives

$$\begin{aligned}\frac{E\{g(Z;\theta - \delta t)\} - E\{g(Z;\theta)\}}{-\delta} &\leq t^{\mathrm{T}} E\{h(Z;\theta)\} \\ &\leq \frac{E\{g(Z;\theta + \delta t)\} - E\{g(Z;\theta)\}}{\delta}.\end{aligned}$$

Letting $\delta \to 0+$ gives

$$t^{\mathrm{T}} E\{h(Z;\theta)\} = t^{\mathrm{T}} \frac{\partial}{\partial \theta} E\{g(Z;\theta)\},$$

which yields (2.56) because $t \in \mathbb{R}^d$ is arbitrary. □

2.6.4 Asymptotic efficiency of maximum likelihood estimation

We discuss the asymptotic efficiency of the MLE with a correctly specified parametric model. Suppose that $Z_1, \ldots, Z_n$ are IID as Z, from a density function $f(\cdot;\theta^*)$ with some (unknown) true value θ^*, which is assumed to lie in the interior of a parameter set Θ. For $\theta \in \Theta$, we denote by $E_\theta(\cdot)$ and $\mathrm{var}_\theta(\cdot)$ the expectation and variance with $Z \sim f(\cdot;\theta)$ or $Z_1, \ldots, Z_n \sim f(\cdot;\theta)$.

Under regularity conditions based on Theorems 2.6 and 2.10 and under the second Bartlett identity (2.29), the MLE $\hat{\theta}_n$ satisfies

$$\hat{\theta}_n \underset{\mathcal{P}}{\to} \theta^*, \quad \sqrt{n}(\hat{\theta}_n - \theta^*) \underset{\mathcal{D}}{\to} N(0, I_{\theta^*}^{-1}),$$

where $I_\theta = \mathrm{var}_\theta\{s(Z;\theta)\}$ is the Fisher information matrix, and $s(z;\theta) = -\frac{\partial}{\partial \theta} \log f(z;\theta)$ is the negative score function for an individual datum. [The discussion can also be carried out with suitable changes, by taking $s(z;\theta) = \frac{\partial}{\partial \theta} \log f(z;\theta)$ without the minus sign.] By replacing θ^* with any value $\theta \in \Theta$, the preceding result indicates that for each $\theta \in \Theta$,

$$\hat{\theta}_n \underset{\mathcal{P}}{\to} \theta, \quad \sqrt{n}(\hat{\theta}_n - \theta) \underset{\mathcal{D}}{\to} N(0, I_\theta^{-1}) \quad \text{under } f(\cdot;\theta), \tag{2.57}$$

provided that the regularity conditions are satisfied accordingly. Note that (2.57) is obtained (under regularity conditions) for each θ, by applying Theorems 2.6 and 2.10 with the true value θ^* set to θ. This application appears obvious, but internally relies on the two identities $E_\theta\{s(Z;\theta)\} = 0$

and $\text{var}_\theta\{s(Z;\theta)\} = E_\theta\{\frac{\partial}{\partial\theta^T} s(Z;\theta)\}$ for each θ, as discussed in Section 2.4.2. The first identity is responsible for the consistency of $\hat{\theta}_n$ and indicates that $s(z;\theta)$ is an unbiased estimating function as defined later by (2.59), whereas the second identity is responsible for the variance matrix being simplified to I_θ^{-1} in (2.57).

The unscaled variance matrix, $\frac{I_\theta^{-1}}{n}$, for the MLE $\hat{\theta}_n$ in (2.57) is also known as the Cramer–Rao (CR) lower variance bound for unbiased estimation. For any unbiased estimator $\tilde{\theta}_n$, as a function of $(Z_1, \ldots, Z_n)$ only, satisfying

$$E_\theta(\tilde{\theta}_n) = \theta \quad \text{for each } \theta \in \Theta, \tag{2.58}$$

it holds under suitable regularity conditions that

$$\text{var}_\theta(\tilde{\theta}_n) \geq \frac{I_\theta^{-1}}{n},$$

where for two (variance) matrices A_1 and A_2, the inequality $A_1 \geq A_2$ indicates that $A_1 - A_2$ is nonnegative definite. While the CR bound is fundamental, its applicability may be limited because unbiased estimation may often be difficult to achieve. Hence a natural question is whether an asymptotic version of the CR bound holds as follows: If an estimator $\tilde{\theta}_n$ satisfies $\tilde{\theta}_n \xrightarrow[\mathcal{P}]{} \theta$ and $\sqrt{n}(\hat{\theta}_n - \theta) \xrightarrow[\mathcal{D}]{} L_\theta$ under $f(\cdot;\theta)$ for each θ, where L_θ is a probability distribution with variance matrix Σ_θ, then $\Sigma_\theta \geq I_\theta^{-1}$ for each θ?

Unfortunately, the answer to the preceding question is negative. A counter-example is provided by Hodges' estimator as follows, in the context of estimating a normal mean. Suppose that $Z_1, \ldots, Z_n$ are IID from $\text{N}(\mu, 1)$. The MLE is $\hat{\mu}_n = \bar{Z}_n$, the sample mean. The Fisher information is $I_\mu = 1$. Consider an alternative estimator

$$\tilde{\mu}_n = \begin{cases} 0, & \text{if } |\hat{\mu}_n| \leq n^{-1/4}, \\ \bar{Z}_n & \text{if } |\hat{\mu}_n| > n^{-1/4}. \end{cases}$$

It can be verified that for each $\mu \in \mathbb{R}$, $\tilde{\mu}_n \xrightarrow[\mathcal{P}]{} \mu$ and

$$\sqrt{n}(\tilde{\mu}_n - \mu) \xrightarrow[\mathcal{D}]{} \begin{cases} \text{N}(0,0), & \text{if } \mu = 0, \\ \text{N}(0,1) & \text{if } \mu \neq 0, \end{cases}$$

where $\text{N}(0,0)$ is a degenerate normal distribution, placing probability 1 at 0. Hence the rescaled asymptotic variance Σ_μ is $0\,(< I_\mu^{-1})$ if $\mu = 0$ or $1\,(= I_\mu^{-1})$ if $\mu \neq 0$. The estimator $\tilde{\mu}_n$ is said to be super-efficient. Motivated by such examples, formal theory was developed by Le Cam and others to clarify the asymptotic efficiency of MLE.

We provide an asymptotic version of the CR bound, from a seemingly more concrete perspective by restricting possible estimators to Z-estimation. Suppose that $\tilde{\theta}_n$ is a Z-estimator defined as a solution to $T_n(\theta) = 0$, with $T_n(\theta) = \frac{1}{n}\sum_{i=1}^n \psi(Z_i;\theta)$, where $\psi(z;\theta)$ is an unbiased estimating function such that

$$E_\theta\{\psi(Z;\theta)\} = 0 \quad \text{for each } \theta \in \Theta, \tag{2.59}$$

which is similar to the unbiasedness condition (2.58). It is important that (2.59) holds for each θ, not just a single true value. Under the regularity conditions in Theorems 2.7 and 2.9, $\tilde{\theta}_n$ satisfies for each $\theta \in \Theta$,

$$\tilde{\theta}_n \underset{\mathcal{P}}{\to} \theta, \quad \sqrt{n}(\tilde{\theta}_n - \theta) \underset{\mathcal{D}}{\to} \mathrm{N}(0, \Sigma_\theta) \quad \text{under } f(\cdot;\theta), \tag{2.60}$$

where $\Sigma_\theta = H_\theta^{-1} G_\theta H_\theta^{-\mathrm{T}}$ with $G_\theta = \mathrm{var}_\theta\{\psi(Z;\theta)\}$ and $H_\theta = E_\theta\{\frac{\partial}{\partial\theta^{\mathrm{T}}}\psi(Z;\theta)\}$. It is worth noting that, although (2.59) requires $E_\theta\{\psi(Z;\theta)\} = 0$ for all $\theta \in \Theta$ simultaneously, (2.60) is obtained by applying Theorems 2.7 and 2.9 pointwise for each individual θ satisfying $E_\theta\{\psi(Z;\theta)\} = 0$, independently of whether $E_{\theta'}\{\psi(Z;\theta')\} = 0$ for $\theta' \neq \theta$. In other words, there is a subtle distinction: asymptotic properties as in Theorems 2.7 and 2.9 are about an estimator under a single data distribution, whereas the unbiasedness as in (2.59) concerns an estimator under a set of possible data distributions. In addition, the unbiasedness as in (2.59) is only relevant when model $f(\cdot;\theta)$ is correctly specified.

Proposition 2.4 (Asymptotic efficiency of MLE). *In the preceding setting, it holds that for each $\theta \in \Theta$,*

$$H_\theta^{-1} G_\theta H_\theta^{-\mathrm{T}} \geq I_\theta^{-1},$$

provided that $\|\frac{\partial}{\partial\theta^{\mathrm{T}}}\{\psi(z;\theta)f(z;\theta)\}\|$ is uniformly bounded in a neighborhood of θ by a nonnegative, integrable function of z.

Proof. We use the matrix version of the Cauchy–Schwartz inequality: for two random vectors Y_1 and Y_2 with $\mathrm{var}(Y_2)$ invertible,

$$\mathrm{var}(Y_1) \geq \mathrm{cov}(Y_1, Y_2)\mathrm{var}^{-1}(Y_2)\mathrm{cov}(Y_2, Y_1).$$

This can be proved directly by noting

$$\begin{aligned}&\mathrm{var}\{Y_1 - \mathrm{cov}(Y_1, Y_2)\mathrm{var}^{-1}(Y_2)Y_2\}\\ &= \mathrm{var}(Y_1) - \mathrm{cov}(Y_1, Y_2)\mathrm{var}^{-1}(Y_2)\mathrm{cov}(Y_2, Y_1).\end{aligned}$$

Taking $Y_1 = \psi(Z;\theta)$ and $Y_2 = s(Z;\theta)$ under $f(\cdot;\theta)$ yields

$$G_\theta \geq \mathrm{cov}_\theta\{\psi(Z;\theta), s(Z;\theta)\} I_\theta^{-1} \mathrm{cov}_\theta\{s(Z;\theta), \psi(Z;\theta)\},$$

where I_θ is invertible by assumption. It suffices to show that

$$H_\theta = \mathrm{cov}_\theta\{\psi(Z;\theta), s(Z;\theta)\} \quad \text{for each } \theta. \tag{2.61}$$

This identity can be obtained as follows:

$$\begin{aligned}&\mathrm{cov}_\theta\{\psi(Z;\theta), s(Z;\theta)\}\\ &= E_\theta\{\psi(Z;\theta)s(Z;\theta)^{\mathrm{T}}\} - \underbrace{E_\theta\{\psi(Z;\theta)\}E_\theta\{s(Z;\theta)^{\mathrm{T}}\}}_{=0}\end{aligned}$$

$$
\begin{aligned}
&= -\int \psi(z;\theta)\frac{\partial}{\partial\theta^{\mathrm{T}}}f(z;\theta)\,\mathrm{d}z \\
&= -\underbrace{\int \frac{\partial}{\partial\theta^{\mathrm{T}}}\{\psi(z;\theta)f(z;\theta)\}\,\mathrm{d}z}_{=0} + \int \frac{\partial}{\partial\theta^{\mathrm{T}}}\{\psi(z;\theta)\}f(z;\theta)\,\mathrm{d}z \\
&= E_\theta\left\{\frac{\partial}{\partial\theta^{\mathrm{T}}}\psi(Z;\theta)\right\} = H_\theta.
\end{aligned}
$$

The equality $E_\theta\{\psi(Z;\theta)\}E_\theta\{s(Z;\theta)^{\mathrm{T}}\} = 0$ follows from either $E_\theta\{s(Z;\theta)\} = 0$ or $E_\theta\{\psi(Z;\theta)\} = 0$ as in (2.59). The fourth line follows from the third line by the rule of differentiating a product. The equality $\int \frac{\partial}{\partial\theta^{\mathrm{T}}}\{\psi(z;\theta)f(z;\theta)\}\,\mathrm{d}z = 0$ follows from the calculation:

$$
\begin{aligned}
\int \frac{\partial}{\partial\theta^{\mathrm{T}}}\{\psi(z;\theta)f(z;\theta)\}\,\mathrm{d}z &= \frac{\partial}{\partial\theta^{\mathrm{T}}}\int \psi(z;\theta)f(z;\theta)\,\mathrm{d}z \\
&= \frac{\partial}{\partial\theta^{\mathrm{T}}}E_\theta\{\psi(Z;\theta)\} \\
&= 0,
\end{aligned}
$$

where the first line follows by exchanging the differentiation and integration under the condition on $\|\frac{\partial}{\partial\theta^{\mathrm{T}}}\{\psi(z;\theta)f(z;\theta)\}\|$, and the third line follows from the unbiasedness condition (2.59). □

The asymptotic efficiency of MLE $\hat{\theta}_n$, compared with Z-estimator $\tilde{\theta}_n$, can also be explained as follows, owing to the key identity (2.61). In fact, from the asymptotic expansions (2.30) for MLE and (2.19) for Z-estimator, the pair $(\hat{\theta}_n, \tilde{\theta}_n)$ is jointly asymptotically normal under $f(\cdot;\theta)$:

$$
\sqrt{n}\begin{pmatrix}\hat{\theta}_n - \theta \\ \tilde{\theta}_n - \theta\end{pmatrix} \xrightarrow[\mathcal{D}]{} \mathrm{N}\left(0, \begin{pmatrix} I_\theta^{-1} & C_\theta^{\mathrm{T}} \\ C_\theta & H_\theta^{-1}G_\theta H_\theta^{-\mathrm{T}}\end{pmatrix}\right),
$$

where C_θ can be calculated as

$$
\begin{aligned}
C_\theta &= \mathrm{cov}_\theta\left\{-H_\theta^{-1}\psi(Z;\theta), -I_\theta^{-1}s(Z;\theta)\right\} \\
&= H_\theta^{-1}\mathrm{cov}_\theta\left\{\psi(Z;\theta), s(Z;\theta)\right\} I_\theta^{-1} \\
&= I_\theta^{-1}.
\end{aligned}
$$

The last step follows from the identity (2.61). Therefore,

$$
\sqrt{n}\begin{pmatrix}\hat{\theta}_n - \theta \\ \tilde{\theta}_n - \theta\end{pmatrix} \xrightarrow[\mathcal{D}]{} \mathrm{N}\left(0, \begin{pmatrix} I_\theta^{-1} & I_\theta^{-1} \\ I_\theta^{-1} & H_\theta^{-1}G_\theta H_\theta^{-\mathrm{T}}\end{pmatrix}\right), \tag{2.62}
$$

which indicates that $\mathrm{var}_\theta(\hat{\theta}_n) \approx \mathrm{cov}_\theta(\hat{\theta}_n, \tilde{\theta}_n)$ or equivalently

$$
\mathrm{cov}_\theta(\hat{\theta}_n, \tilde{\theta}_n - \hat{\theta}_n) \approx 0.
$$

In other words, $\hat{\theta}_n$ and $\tilde{\theta}_n - \hat{\theta}_n$ are asymptotically uncorrelated so that

$$\mathrm{var}_\theta(\tilde{\theta}_n) \approx \mathrm{var}_\theta(\hat{\theta}_n) + \mathrm{var}_\theta(\tilde{\theta}_n - \hat{\theta}_n),$$

confirming the asymptotic efficiency of MLE $\hat{\theta}_n$.

Example 2.9 (Efficiency comparison for exponential mean)**.** Let $X_1, \ldots, X_n$ be IID from Exponential(θ), with mean θ and variance θ^2. The MLE of θ is $\hat{\theta}_n = \bar{X}_n$, the sample mean. Given the fact that variance of Exponential(θ) is θ^2, two alternative estimators of θ are

$$\tilde{\theta}_{n1} = \left(\frac{1}{n}\sum_{i=1}^n (X_i - \bar{X}_n)^2\right)^{1/2},$$

$$\tilde{\theta}_{n2} = \left(\frac{1}{2n}\sum_{i=1}^n X_i^2\right)^{1/2}.$$

The estimator $\tilde{\theta}_{n1}$ corresponds to $\tilde{s}_n$ studied in Example 2.5. From the calculation there, we have

$$\sqrt{n}\begin{pmatrix}\hat{\theta}_n - \theta \\ \tilde{\theta}_{n1} - \theta\end{pmatrix} \xrightarrow[\mathcal{D}]{} \mathrm{N}\left(0, \begin{pmatrix}\theta^2 & \theta^2 \\ \theta^2 & 2\theta^2\end{pmatrix}\right).$$

By similar calculation [Exercise],

$$\sqrt{n}\begin{pmatrix}\hat{\theta}_n - \theta \\ \tilde{\theta}_{n2} - \theta\end{pmatrix} \xrightarrow[\mathcal{D}]{} \mathrm{N}\left(0, \begin{pmatrix}\theta^2 & \theta^2 \\ \theta^2 & \frac{5}{4}\theta^2\end{pmatrix}\right).$$

The estimator $\hat{\theta}_{n2}$ is a Z-estimator, satisfying Definition 2.2, whereas $\hat{\theta}_{n1}$ does not satisfy Definition 2.2. [The stacked estimator $(\hat{\theta}_n, \tilde{\theta}_{n1})$ can be treated as a Z-estimator as in Example 2.5.] Nevertheless, both asymptotic variance matrices above satisfy the pattern in (2.62), with $I_\theta^{-1} = \theta^2$.

Exercises

Ex 2.1.

(i) For a random variable Z with $E|Z| < \infty$, let $K(\nu) = E(|Z - \nu|)$. Show that ν^* is a minimizer of $K(\nu)$ if and only if $P(Z \geq \nu^*) \geq \frac{1}{2}$ and $P(Z \leq \nu^*) \geq \frac{1}{2}$.

(ii) For any real numbers $z_1, \ldots, z_n$, let $K_n(\nu) = \frac{1}{n}\sum_{i=1}^n |z_i - \nu|$. Find a (simple) necessary and sufficient condition for a minimizer of $K_n(\nu)$.

Ex 2.2.

(i) Find an example of real functions $g(\theta)$ and $g_n(\theta)$, $n \geq 1$ defined on an open interval $\Theta \subset \mathbb{R}$ such that $g(\theta)$ and $g_n(\theta)$, $n \geq 1$, are continuous in $\theta \in \Theta$, $\lim_{n\to\infty} g_n(\theta) = g(\theta)$ for each $\theta \in \Theta$, but $\sup_{\theta\in\Theta} |g_n(\theta) - g(\theta)| \not\to 0$ as $n \to \infty$.

(ii) Can you find an example as in (i), except with Θ a closed interval?

Ex 2.3. Suppose that $\Theta \subset \mathbb{R}^d$ is compact. Show that dominance condition (iii) and local dominance condition (iii†) in Section 2.3.1 (ULLN) are equivalent to each other. [Hint: Use the Heine–Cantor theorem.]

Ex 2.4 (Generalization of Lemma 2.4)**.** Let $f_n : \mathcal{S} \to \mathbb{R}$, $n \geq 1$, be a sequence of functions on a bounded, possibly non-closed set $\mathcal{S} \subset \mathbb{R}^d$, and $f : \mathcal{S} \to \mathbb{R}$ be a function on $\mathcal{S}$. Show that the following two statements are equivalent:

(i) $f(\cdot)$ is uniformly continuous on $\mathcal{S}$ and

$$\sup_{s\in\mathcal{S}} |f_n(s) - f(s)| \to 0,$$

(ii) $f_n(s) \to f(s)$ for each $s \in \mathcal{S}$, and the sequence $f_n(\cdot)$, $n \geq 1$, is equi-continuous on $\mathcal{S}$.

Ex 2.5. Complete the proof of *(iii)* $\Rightarrow$ *(i)* for Lemma 2.5 (Uniform convergence and stochastic equi-continuity).

Ex 2.6 (Generalization of Lemma 2.5)**.** Let $F_n : \mathcal{S} \to \mathbb{R}$, $n \geq 1$, be a sequence of random functions on a bounded, possibly non-closed set $\mathcal{S} \subset \mathbb{R}^d$, and $F : \mathcal{S} \to \mathbb{R}$ be a random function on $\mathcal{S}$. Show that the following two statements are equivalent:

(i) $F(\cdot)$ is stochastically uniformly continuous on $\mathcal{S}$ and

$$\sup_{s\in\mathcal{S}} |F_n(s) - F(s)| \xrightarrow[\mathcal{P}]{} 0.$$

(ii) $F_n(s) \xrightarrow[\mathcal{P}]{} F(s)$ for each $s \in \mathcal{S}$, and the sequence $F_n(\cdot)$ is stochastically equi-continuous on $\mathcal{S}$.

Here $F(\cdot)$ is said to be stochastically uniformly continuous on $\mathcal{S}$ if for any $\epsilon > 0$ and $\eta > 0$, there exists $\delta > 0$ such that $P(\sup_{\|s-\tilde{s}\|\leq\delta} |F(s) - F(\tilde{s})| > \epsilon) \leq \eta$.

Ex 2.7. Provide a proof of Theorem 2.7 (Consistency of Z-estimation under smoothness and dominance).

Ex 2.8. Let $(Z_1, \ldots, Z_n)$ be IID variables as Z. Assume that $\psi(z;\theta)$ is a scalar function, twice differentiable in $\theta = (\theta^{(1)}, \ldots, \theta^{(d)})^{\mathrm{T}}$ for each z, and there exists a nonnegative function $M(z)$ such that $E\{M(Z)\} < \infty$ and

$$\left| \frac{\partial^2}{\partial\theta^{(j)}\partial\theta^{(k)}} \psi(z;\theta) \right| \leq M(z),$$

for $j, k = 1, \ldots, d$, each z, and each θ in a neighborhood of $\bar{\theta}$. Show that if $\hat{\theta}_n \underset{P}{\rightarrow} \bar{\theta}$, then

$$\frac{1}{n} \sum_{i=1}^{n} \frac{\partial^2}{\partial\theta\partial\theta^{\mathrm{T}}} \psi(Z_i; \tilde{\theta}_n) = O_p(1),$$

where $\tilde{\theta}_n$ lies between $\hat{\theta}_n$ and $\bar{\theta}$.

Ex 2.9. Let H be a $d \times d$ nonsingular matrix. Show that

$$(H + o_p(1))^{-1} = H^{-1} + o_p(1).$$

Ex 2.10. For a Z-estimator $\hat{\theta}_n$ based on estimating function $\psi(\cdot;\theta)$, it is shown in Section 2.4.1 that under several conditions,

$$\sqrt{n}(\hat{\theta}_n - \bar{\theta}) \underset{\mathcal{D}}{\rightarrow} N(0, \Sigma),$$

where $\Sigma = H^{-1}GH^{-\mathrm{T}}$ with $G = \mathrm{var}\{\psi(X;\bar{\theta})\}$ and $H = E\{\frac{\partial}{\partial\theta^{\mathrm{T}}}\psi(X;\bar{\theta})\}$. Consider the sandwich variance estimator Σ_n, and determine technical conditions (as weak as you can) to ensure the consistency, $\Sigma_n \underset{P}{\rightarrow} \Sigma$.

Ex 2.11. Consider an IID sample $(X_1, \ldots, X_n)$ from a Gaussian mixture model: $X \sim f(x;\theta)$ with $\theta = (\alpha, \mu_1, \sigma_1, \mu_2, \sigma_2)$ and

$$f(x;\theta) = \alpha\phi(x;\mu_1,\sigma_1) + (1-\alpha)\phi(x;\mu_2,\sigma_2),$$

where $\phi(x;\mu_j,\sigma_j)$ denotes the probability density function of $\mathrm{N}(\mu_j, \sigma_j^2)$.

(i) Let the parameter set be

$$\Theta = \{(\alpha, \mu_1, \sigma_1, \mu_2, \sigma_2) : 0 \leq \alpha \leq 1, \mu_1 \in \mathbb{R}, \sigma_1 > 0, \mu_2 \in \mathbb{R}, \sigma_2 > 0\}.$$

Show that the MLE of θ in Θ is inconsistent.

(ii) Define a parameter set $\Theta_1 \subset \Theta$ such that the MLE of θ in Θ_1 is consistent. Verify the conditions you use to prove the consistency.

Ex 2.12 (Proof of Lemma 2.2). Let $F_n : \mathcal{T} \rightarrow \mathbb{R}$, $n \geq 1$, be a sequence of random convex functions on an open set $\mathcal{T} \subset \mathbb{R}^d$, and $F : \mathcal{T} \rightarrow \mathbb{R}$ be a non-random function on $\mathcal{T}$ such that $F_n(s) \underset{P}{\rightarrow} F(s)$ for each $s \in \mathcal{T}$. In addition, let $\alpha_n = \mathrm{argmin}_{s\in\mathcal{T}} F_n(s)$ and $\alpha = \mathrm{argmin}_{s\in\mathcal{T}} F(s)$. Show that if α is a unique minimizer of $F(\cdot)$ on $\mathcal{T}$, then $\alpha_n \underset{P}{\rightarrow} \alpha$.

Ex 2.13. For Theorem 2.11 (Consistency of M-estimation under convexity), show that under conditions (i)–(iii), $K(\theta) = E\{\rho(Z;\theta)\}$ exists on Θ, being either finite or ∞. [Hint: Use inequality (2.50).]

Ex 2.14. Suppose that $K_n(\theta)$ is a convex function on an open set Θ, and $\bar{\theta}$ is any point in Θ. Show that for any $\epsilon > 0$, if $\inf_{\theta\in\Theta:\|\theta-\bar{\theta}\|=\epsilon} K_n(\theta) > K_n(\bar{\theta})$, then $K_n(\theta) > K_n(\bar{\theta})$ for any $\theta \in \Theta$ satisfying $\|\theta - \bar{\theta}\| > \epsilon$.

Ex 2.15. Suppose that $K(\theta)$ is a convex function on an open set Θ, and $\bar{\theta}$ is a unique minimizer of $K(\theta)$ in Θ. Show that for any $\epsilon > 0$,

$$\inf_{\theta\in\Theta:\|\theta-\bar{\theta}\|\geq\epsilon} K(\theta) > K(\bar{\theta}).$$

Note that Θ is not compact. Otherwise the result would follow directly from the fact that $\{\theta \in \Theta : \|\theta - \bar{\theta}\| \geq \epsilon\}$ is compact and $K(\cdot)$ is continuous due to convexity on Θ.

Ex 2.16. For Lemma 2.3 (Closeness of random minimizers under convexity), complete the proof in the general case where $\mathcal{T}_n \subset \mathbb{R}^d$ is assumed to be asymptotically unrestricted, i.e., for any $M > 0$ there exists $N \geq 1$ such that $\{s \in \mathbb{R}^d : \|s\| \leq M\} \subset \mathcal{T}_n$ for $n \geq N$.

Ex 2.17. Suppose that $K(\theta)$ achieves a minimum at $\bar{\theta}$, and is twice differentiable at $\bar{\theta}$. Show that for any fixed s, as $n \to \infty$.

$$n\left\{K\left(\bar{\theta} + \frac{s}{\sqrt{n}}\right) - K(\bar{\theta})\right\} \to \frac{1}{2}s^{\mathrm{T}}Hs,$$

where $H = \frac{\partial^2}{\partial\theta\partial\theta^{\mathrm{T}}}K(\bar{\theta})$.

Ex 2.18. Let (X_i, Y_i), $i = 1, \ldots, n$, be an IID sample, satisfying logistic regression model

$$P(Y_i = 1|X_i) = \frac{\mathrm{e}^{X_i^{\mathrm{T}}\beta^*}}{1 + \mathrm{e}^{X_i^{\mathrm{T}}\beta^*}},$$

where β^* is a vector of unknown parameters.

(i) Let $\hat{\beta}_{\mathrm{ML}}$ be the maximum likelihood estimator of β^*. Derive a set of simple conditions (as weak as you can) such that $\hat{\theta}_{\mathrm{ML}}$ is consistent and asymptotically normal. Find the asymptotic variance.

(ii) Alternatively, let $\hat{\beta}_{\mathrm{CAL}}$ be an estimator of β^*, defined as a minimizer of

$$K_n(\beta) = \frac{1}{n}\sum_i \left\{Y_i\mathrm{e}^{-X_i^{\mathrm{T}}\beta} + (1 - Y_i)X_i^{\mathrm{T}}\beta\right\}.$$

The estimator $\hat{\beta}_{\text{CAL}}$ is called the calibrated estimator; see Example 5.4 for further information. Derive a set of simple conditions (as weak as you can) such that $\hat{\theta}_{\text{CAL}}$ is consistent and asymptotically normal. Find the asymptotic variance.

(iii) Compare the asymptotic variances obtained from (i) and (ii).

Ex 2.19. Fix $\kappa > 0$ and $\bar{\theta} \in \mathbb{R}$. Let, as in (2.42),

$$g(z;\theta) = \kappa 1\{z - \theta < -\kappa\} - (z - \theta)1\{|z - \theta| \leq \kappa\} - \kappa 1\{z - \theta > \kappa\}.$$

For a random variable Z with CDF $F(\cdot)$, show that if $F(\cdot)$ is continuous at $\bar{\theta} \pm \kappa$, then $E\{g(Z;\theta)\}$ is differentiable at $\bar{\theta}$ with

$$\frac{\partial}{\partial\theta} E\{g(Z;\theta)\}\Big|_{\theta=\bar{\theta}} = E[1\{|Z - \bar{\theta}| \leq \kappa\}].$$

[Hint: Let $g_0(x) = -\kappa 1\{x < -\kappa\} + x1\{|x| \leq \kappa\} + \kappa 1\{x > \kappa\}$ such that $g(z;\theta) = -g_0(z - \theta)$. Write $g_0(x) = g_1(x) - g_2(x)$, where $g_1(x) = -\kappa 1\{x < -\kappa\} + x1\{x \geq -\kappa\}$ and $g_2(x) = (x - \kappa)1\{x \geq \kappa\}$. Apply Proposition 2.3 to $E\{g_1(Z - \theta)\}$ and $E\{g_2(Z - \theta)\}$ separately.]

3

Concentration inequalities

In this chapter, we study concentration inequalities through the following topics. First, we give a review of basic probability inequalities. Second, we discuss sub-Gaussian and sub-Exponential variables and the concentration inequalities for the sample means of such independent random variables including Hoeffding and Bernstein inequalities. Third, we discuss the uniform concentration of a class of sample means around their population means, which involves not only more advanced concentration inequalities, including Azuma and McDiarmid inequalities and a functional Hoeffding inequality, but also additional techniques referred to as symmetrization and contraction inequalities. As an application, we study the uniform concentration of Lipschitz loss functions.

The classical large-sample theory for M- and Z-estimation in Chapter 2 relies on the law of large numbers and the central limit theorem for IID random vectors of a fixed dimension (usually corresponding to the parameter dimension), as the sample size tends to infinity. In this chapter, we study concentration inequalities which give finite-sample (or non-asymptotic) tail probability bounds for the average or sum of independent variables, subject to some condition on the tail behavior of each variable such as sub-Gaussian or sub-Exponential tails. These concentration inequalities will play a fundamental role in high-dimensional statistical theory (where the parameter dimension may increase as the sample size increases), in a similar manner to the LLN and CLT in the classical large-sample theory. We also study concentration inequalities for a class of sample means around their population means in a uniform sense as reflected by the uniform law of large numbers.

We discuss only a small, hopefully representative subset of the large literature on concentration inequalities. A comprehensive reference is Boucheron et al. (2013), and the subject is also discussed in relevant sections of other books (e.g., Bühlmann & van de Geer, 2011; Vershynin, 2018; Wainwright, 2019). The notation used may vary from book to book, and care is needed when comparing results between these books.

DOI: 10.1201/9781003680734-3

3.1 Basic probability inequalities

We present several basic probabilities inequalities. The Markov inequality gives an upper bound on the right-tail probability of a nonnegative random variable through its mean.

Theorem 3.1 (Markov inequality). *For a nonnegative random variable X,*

$$P(X \geq t) \leq \frac{E(X)}{t}, \quad t > 0.$$

Proof. The result follows by taking expectations on both sides of the inequality

$$t1\{X \geq t\} \leq X, \quad t > 0,$$

for nonnegative random variable X. □

The Chebyshev inequality gives an upper bound on the tail probability of a random variable deviating from its mean through its variance.

Theorem 3.2 (Chebyshev inequality). *For a random variable X with finite $\mu = E(X)$,*

$$P(|X - \mu| \geq t) \leq \frac{\text{var}(X)}{t^2}, \quad t > 0.$$

Proof. The result follows by applying the Markov inequality to $Y = (X - \mu)^2$:

$$P(|X - \mu| \geq t) = P(Y \geq t^2) \leq \frac{E(Y^2)}{t^2} = \frac{\text{var}(X)}{t^2}, \quad t > 0.$$

□

The previous two results require only the first or second moment. Chernoff's technique provides probability inequalities by using a moment generating function (MGF).

Theorem 3.3 (Chernoff inequality). *For a random variable X,*

$$P(X \geq t) \leq \inf_{\lambda \geq 0} \mathrm{e}^{-\lambda t} E(\mathrm{e}^{\lambda X}), \quad t \in \mathbb{R},$$

where $E(\mathrm{e}^{\lambda X})$ is the moment generating function of X. Similarly,

$$P(X \leq -t) \leq \inf_{\lambda \geq 0} \mathrm{e}^{-\lambda t} E(\mathrm{e}^{-\lambda X}), \quad t \in \mathbb{R}.$$

The inequalities are valid for any $t \in \mathbb{R}$ but only meaningful for $t \geq 0$.

Proof. For any $\lambda \geq 0$, applying the Markov inequality to $Y = e^{\lambda X}$ gives

$$P(X \geq t) \leq P(e^{\lambda X} \geq e^{\lambda t}) \leq \frac{E(e^{\lambda X})}{e^{\lambda t}}, \quad t \in \mathbb{R}.$$

The first inequality reduces to equality for $\lambda > 0$. Taking the infimum over $\lambda \geq 0$ yields the desired result. □

Example 3.1. Let $X \sim N(\mu, \sigma^2)$, with mean μ and variance σ^2. The moment generating function of X is $E(e^{\lambda X}) = e^{\mu\lambda + \frac{\sigma^2}{2}\lambda^2}$ for $\lambda \in \mathbb{R}$. Then $E(e^{\lambda(X-\mu)}) = e^{\frac{\sigma^2}{2}\lambda^2}$ for $\lambda \in \mathbb{R}$. Applying the Chernoff inequality to $X - \mu$ gives

$$P(X - \mu \geq t) \leq \inf_{\lambda \geq 0} e^{-\lambda t + \frac{\sigma^2}{2}\lambda^2} = \begin{cases} e^{-\frac{t^2}{2\sigma^2}}, & \text{if } t \geq 0, \\ 1, & \text{if } t < 0. \end{cases}$$

Hence the meaningful result is

$$P(X - \mu \geq t) \leq e^{-\frac{t^2}{2\sigma^2}}, \quad t \geq 0. \tag{3.1}$$

[Question: What happens if applying the Chernoff inequality to X?] Similarly using $E(e^{-\lambda(X-\mu)}) = e^{\frac{\sigma^2}{2}\lambda^2}$ for $\lambda \geq 0$, it also holds that

$$P(X - \mu \leq -t) \leq e^{-\frac{t^2}{2\sigma^2}}, \quad t \geq 0, \tag{3.2}$$

and hence

$$P(|X - \mu| \geq t) \leq 2e^{-\frac{t^2}{2\sigma^2}}, \quad t \geq 0. \tag{3.3}$$

3.2 Sub-Gaussian random variables

Example 3.1 motivates the introduction of sub-Gaussian random variables.

Definition 3.1. A random variable X is said to be sub-Gaussian if $\mu = E(X)$ is finite and for a constant $\sigma > 0$,

$$E(e^{\lambda(X-\mu)}) \leq e^{\frac{\sigma^2\lambda^2}{2}}, \quad \lambda \in \mathbb{R}, \tag{3.4}$$

In this case, X is also said to be sub-Gaussian with scale parameter σ. [Note that by this definition, the scale parameter for a sub-Gaussian variable is not unique. A sub-Gaussian variable with scale parameter σ is also sub-Gaussian with scale parameter $\tilde{\sigma} \geq \sigma$.]

The condition (3.4) can be equivalently stated as

$$K(\lambda) \leq \mu\lambda + \frac{\sigma^2\lambda^2}{2}, \quad \lambda \in \mathbb{R}, \tag{3.5}$$

where $K(\lambda) = \log E(e^{\lambda X})$ is the cumulant generating function of X. Similar reasoning as in Example 3.1 leads to the following result.

Lemma 3.1 (Sub-Gaussian tail inequality). *A sub-Gaussian variable with scale parameter σ satisfies the one-sided probability inequalities* (3.1) *and* (3.2) *and hence the two-sided probability inequality* (3.3).

A Gaussian (or normal) variable with standard deviation σ is automatically sub-Gaussian with scale parameter σ. Another simple but important example of sub-Gaussian variables are bounded variables.

Example 3.2. Let X be a random variable with $\mu = E(X)$ and $X \in [a, b]$ (almost surely) with $a, b \in \mathbb{R}$. Then X is sub-Gaussian with $\sigma = \frac{b-a}{2}$. In fact, by a Taylor expansion

$$K(\lambda) = K(0) + K'(0)\lambda + K''(\tilde{\lambda})\frac{\lambda^2}{2},$$

where $K(0) = 0$, and $\tilde{\lambda}$ lies between 0 and λ. By MGF properties (3.49) and (3.50) in Section 3.5.1,

$$K'(0) = E(X) = \mu, \quad K''(\tilde{\lambda}) = \text{var}(X_{\tilde{\lambda}}),$$

where X_t is a random variable with density function $\propto e^{tx} f_0(x)$, and $f_0(\cdot)$ denotes the density function of X. Note that X_t also lies in $[a, b]$ almost surely. Then by the fact (Exercise 3.3) that for any random variable $Y \in [a, b]$ almost surely, $\text{var}(Y) \leq \frac{(b-a)^2}{4}$, we have

$$K(\lambda) = \text{var}(X_{\tilde{\lambda}})\frac{\lambda^2}{2} \leq \frac{(b-a)^2}{4}\frac{\lambda^2}{2},$$

i.e., condition (3.5) holds with $\sigma = \frac{b-a}{2}$.

The following lemma shows that the sum of independent, sub-Gaussian random variables is sub-Gaussian, and the sum of squared scale parameters gives the latter squared scale parameter (i.e., the squared scale parameters add up).

Lemma 3.2 (Additivity of independent Sub-Gaussian variables). *Suppose that $(X_1, \ldots, X_n)$ are independent, and X_i is sub-Gaussian with scale parameters σ_i for $i = 1, \ldots, n$. Then $\sum_{i=1}^n X_i$ is sub-Gaussian with scale parameter $\sqrt{\sum_{i=1}^n \sigma_i^2}$.*

Proof. It suffices to show the result for $n = 2$. Without loss of generality, assume that $E(X_1) = E(X_2) = 0$. Then

$$E(e^{\lambda(X_1+X_2)}) = E(e^{\lambda X_1})E(e^{\lambda X_2}) \le e^{\frac{\sigma_1^2\lambda^2}{2}} e^{\frac{\sigma_2^2\lambda^2}{2}} = e^{\frac{(\sigma_1^2+\sigma_2^2)\lambda^2}{2}},$$

by the independence of X_1 and X_2 and then the sub-Gaussianity of X_1 and X_2. □

Combining Lemmas 3.1 and 3.2 gives the Hoeffding inequality.

Theorem 3.4 (Hoeffding inequality for Sub-Gaussian variables). *Suppose that $(X_1, \dots, X_n)$ are independent, and X_i is sub-Gaussian with scale parameters σ_i for $i = 1, \dots, n$. Then with $\mu_i = E(X_i)$,*

$$P\left\{\left|\sum_{i=1}^n (X_i - \mu_i)\right| \ge t\right\} \le 2\exp\left(-\frac{t^2}{2\sum_{i=1}^n \sigma_i^2}\right), \quad t \ge 0. \tag{3.6}$$

Similar results hold for one-sided probabilities as in (3.1) *and* (3.2).

Proof. By Lemma 3.2, $\sum_{i=1}^n X_i$ is sub-Gaussian with scale parameter $\sqrt{\sum_{i=1}^n \sigma_i^2}$. The result follows by applying (3.3) to $\sum_{i=1}^n X_i$. □

It is helpful to restate the Hoeffding inequality in terms of the sample mean, even though $(X_1, \dots, X_n)$ are independent but may not be identically distributed and hence not a sample of observations from a certain population (or distribution). For convenience, we ignore such a technical distinction. Taking $t = ns$ in (3.6) gives

$$P\left\{\left|\frac{1}{n}\sum_{i=1}^n (X_i - \mu_i)\right| \ge s\right\} \le 2\exp\left(-\frac{n^2s^2}{2\sum_{i=1}^n \sigma_i^2}\right), \quad s \ge 0.$$

In the special case where $\mu_i = \mu$ and $\sigma_i = \sigma$ for $i = 1, \dots, n$, the inequality becomes

$$P\left\{\left|\bar{X}_n - \mu\right| \ge s\right\} \le 2\exp\left(-\frac{ns^2}{2\sigma^2}\right), \quad s \ge 0,$$

or equivalently

$$P\left\{\left|\sqrt{n}(\bar{X}_n - \mu)\right| \ge s\right\} \le 2\exp\left(-\frac{s^2}{2\sigma^2}\right), \quad s \ge 0.$$

The central limit theorem says that for IID random variables $X_1, \dots, X_n$ with finite $\mu = E(X_1)$ and $\sigma^2 = \mathrm{var}(X_1)$,

$$P\left\{\left|\sqrt{n}(\bar{X}_n - \mu)\right| \ge s\right\} \to P(|Z| \ge s) \le 2\exp\left(-\frac{s^2}{2\sigma^2}\right), \quad s \ge 0,$$

where $Z \sim \mathrm{N}(0, \sigma^2)$. Hence the CLT provides the exact limit of the tail probability as $n \to \infty$ while (only) requiring a finite variance for each variable X_i, whereas the Hoeffding inequality gives (only) an upper bound of the tail probability for any fixed n, but requires that the variables $(X_1, \ldots, X_n)$ are sub-Gaussian.

The Hoeffding inequality can be directly applied to bounded, independent variables.

Corollary 3.1 (Hoeffding inequality for bounded variables). *Suppose that $(X_1, \ldots, X_n)$ are independent, and X_i is bounded in $[a_i, b_i]$ almost surely with constants $a_i, b_i \in \mathbb{R}$ for $i = 1, \ldots, n$. Then with $\mu_i = E(X_i)$,*

$$P\left\{\left|\sum_{i=1}^{n}(X_i - \mu_i)\right| \geq t\right\} \leq 2\exp\left(-\frac{2t^2}{\sum_{i=1}^{n}(b_i - a_i)^2}\right), \quad t \geq 0.$$

While the use of MGFs to prove Hoeffding inequality for sub-Gaussian variables seems natural due to the sub-Gaussian definition through MGFs. the use of MGFs to prove Hoeffding inequality for bounded variables would be less expected (hence an important advance). On a related note, MGFs can also be used to prove the central limit theorem.

The use of MGFs is also instrumental to obtaining the following result about the maximum of sub-Gaussian variables (possibly non-independent). In fact, it is standard to apply the union bound and obtain $P(\max_{i=1,\ldots,n}|X_i| \geq t) \leq 2n\mathrm{e}^{-\frac{t^2}{2\sigma^2}}$ for $t \geq 0$. However, using this probability bound and the identity (3.10) and proceeding similarly as in the proof *(ii)* $\Rightarrow$ *(iii)* for Theorem 3.6 does not seem to yield inequality (3.8).

Theorem 3.5 (Sub-Gaussian maximum). *Suppose that $X_1, \ldots, X_n$ are mean-zero, sub-Gaussian variables with scale parameter σ. Then*

$$E(\max_{i=1,\ldots,n} X_i) \leq \sigma\sqrt{2\log n}. \tag{3.7}$$

The inequality also holds with the left-hand side replaced by $E\max_{i=1,\ldots,n}(-X_i)$. Moreover,

$$E(\max_{i=1,\ldots,n} |X_i|) \leq \sigma\sqrt{2\log(2n)}. \tag{3.8}$$

[Note: the bound (3.7) is asymptotically sharp if $X_1, \ldots, X_n \sim \mathrm{N}(0, \sigma^2)$. See Exercise 2.17 in Boucheron et al. (2013).]

Proof. By Jensen's inequality, for $\lambda > 0$,

$$\begin{aligned} \mathrm{e}^{\lambda E(\max_{i=1,\ldots,n} X_i)} &\leq E\mathrm{e}^{\lambda(\max_{i=1,\ldots,n} X_i)} = E(\max_{i=1,\ldots,n} \mathrm{e}^{\lambda X_i}) \\ &\leq \sum_{i=1}^{n} E(\mathrm{e}^{\lambda X_i}) \leq n\mathrm{e}^{\frac{\sigma^2\lambda^2}{2}}, \end{aligned}$$

where the last inequality uses $Ee^{\lambda X_i} \le e^{\frac{\sigma^2\lambda^2}{2}}$ for $\lambda > 0$ by assumption. Taking logarithms on both sides gives

$$E(\max_{i=1,\ldots,n} X_i) \le \frac{\log n}{\lambda} + \frac{\sigma^2\lambda}{2}. \tag{3.9}$$

To minimize the upper bound, we choose $\lambda = \sqrt{\frac{2\log n}{\sigma^2}}$ and then obtain (3.7) provided $n \ge 2$. Note that (3.7) holds trivially by the mean-zero assumption when $n = 1$. [Exercise: Show that (3.7) remains valid even if the mean-zero assumption is dropped. In other words, show that if $X_1, \ldots, X_n$ satisfy $Ee^{\lambda X_i} \le e^{\frac{\sigma^2\lambda^2}{2}}$ for $\lambda > 0$ and $i = 1, \ldots, n$, then (3.7) holds for $n \ge 1$.]

To prove (3.8), we obtain by Jensen's inequality for $\lambda > 0$,

$$\begin{aligned} e^{\lambda E(\max_{i=1,\ldots,n} |X_i|)} &\le Ee^{\lambda(\max_{i=1,\ldots,n} |X_i|)} \le \sum_{i=1}^n E(e^{\lambda |X_i|}) \\ &\le \sum_{i=1}^n E(e^{\lambda X_i} + e^{-\lambda X_i}) \le 2ne^{\frac{\sigma^2\lambda^2}{2}}, \end{aligned}$$

and then proceed similarly as above. □

The following result gives several equivalent characterizations of sub-Gaussian random variables. A random variable X is sub-Gaussian if and only if $\mu = E(X)$ is finite and the centered variable $X - \mu$ satisfies any of properties (i)–(iv) below, where c_1 is the scale parameter σ and c_2, c_3, and c_4 can be chosen within an absolute constant factor of σ. As discussed later, properties (ii)–(iv) can be used to directly characterize sub-Gaussian variables without the need for centering by the mean.

Theorem 3.6 (Equivalent characterizations of Sub-Gaussian). *The following properties (i)–(iv) are equivalent for a random variable X with $E(X) = 0$, i.e., if one of the properties holds, then the other properties also hold. Moreover, properties (ii)–(iv) are also equivalent even if $E(X) \ne 0$.*

(i) There exists a constant $c_1 > 0$ such that $E(e^{\lambda X}) \le e^{\frac{c_1^2\lambda^2}{2}}$ for $\lambda \in \mathbb{R}$;

(ii) There exists a constant $c_2 > 0$ such that $P(|X| \ge t) \le 2e^{-\frac{t^2}{c_2^2}}$ for $t \ge 0$;

(iii) There exists a constant $c_3 > 0$ such that $E(|X|^p)^{1/p} \le c_3 p^{1/2}$ for $p \ge 1$;

(iv) There exists a constant $c_4 > 0$ such that $E(e^{X^2/c_4^2}) \le 2$.

The constants c_1, c_2, c_3, and c_4 differ from each other by at most a constant factor independent of X, even though c_1, c_2, c_3, or c_4 itself depends on X.

Proof. We first show the equivalence of (ii)–(iv), whether or not $E(X) \ne 0$.

(ii) ⇒ (iii). Let $Y = X/c_2$. Then $P(|Y| \geq s) = P(|X| \geq c_2 s) \leq 2e^{-s^2}$ for $s \geq 0$, and $E(|X|^p)^{1/p} = c_2 E(|Y|^p)^{1/p}$. Next, we use the fact (Exercise 3.4) that for any nonnegative random variable Z,

$$E(Z) = \int_0^\infty P(Z \geq z)\,\mathrm{d}z, \tag{3.10}$$

and proceed as follows:

$$\begin{aligned} E(|Y|^p) &= \int_0^\infty P(|Y|^p \geq u)\,\mathrm{d}u \overset{u=t^p}{=} \int_0^\infty P(|Y| \geq t) p t^{p-1}\,\mathrm{d}t \\ &\leq \int_0^\infty 2e^{-t^2} p t^{p-1}\,\mathrm{d}t \overset{s=t^2}{=} p \int_0^\infty e^{-s} s^{\frac{p}{2}-1}\,\mathrm{d}s = p\Gamma\left(\frac{p}{2}\right) = 2\Gamma\left(\frac{p}{2}+1\right), \end{aligned}$$

where $\Gamma(\cdot)$ is the Gamma function, defined as $\Gamma(u) = \int_0^\infty e^{-x} x^{u-1}\,\mathrm{d}x$ for $u > 0$. Then

$$E(|Y|^p)^{1/p} \leq \left\{2\Gamma\left(\frac{p}{2}+1\right)\right\}^{1/p} \leq \{2\sqrt{2}(p/2)^{p/2}\}^{1/p} \leq 2p^{1/2}.$$

The second inequality follows because $\Gamma(x+1) \leq \sqrt{2}x^x$ for $x \geq \frac{1}{2}$ (Exercise 3.5). The third inequality follows because $(2\sqrt{2})^{1/p} \leq 2\sqrt{2}$ for $p \geq 1$. Hence $E(|X|^p)^{1/p} \leq (2c_2)p^{1/2}$.

(iii) ⇒ (iv). We use the power series expansion, $e^x = 1 + \sum_{k=1}^\infty \frac{x^k}{k!}$, and obtain by the Fubini theorem

$$E(e^{\lambda^2 X^2}) = E\left\{1 + \sum_{k=1}^\infty \frac{1}{k!}(\lambda^2 X^2)^k\right\} = 1 + \sum_{k=1}^\infty \frac{1}{k!}\lambda^{2k} E(X^{2k}).$$

Then we apply (iii) and use the inequality $k^k \leq e^k k!$ for $k \geq 1$ (Exercise 3.5):

$$\begin{aligned} E(e^{\lambda^2 X^2}) &\leq 1 + \sum_{k=1}^\infty \frac{1}{k!}\lambda^{2k}(c_3\sqrt{2k})^{2k} \\ &\leq 1 + \sum_{k=1}^\infty \frac{1}{k!}\lambda^{2k} c_3^{2k} 2^k \cdot e^k k! = 1 + \sum_{k=1}^\infty \lambda^{2k} c_3^{2k} 2^k e^k = \frac{1}{1 - 2ec_3^2\lambda^2}, \end{aligned}$$

provided $2ec_3^2\lambda^2 < 1$. Taking $\lambda^2 = 1/(4ec_3^2)$ gives $E(e^{X^2/(4ec_3^2)}) \leq 2$.

(iv) ⇒ (ii). We use the Markov inequality and apply (iv):

$$\begin{aligned} P(|X| \geq t) &= P(e^{X^2/c_4^2} \geq e^{t^2/c_4^2}) \\ &\leq e^{-t^2/c_4^2} E(e^{X^2/c_4^2}) \leq 2e^{-t^2/c_4^2}. \end{aligned}$$

Next we show the equivalence between (i) and (ii)–(iv) provided $E(X) = 0$.

(i) ⇒ (ii). This holds with $c_2 = \sqrt{2}c_1$ by the Chernoff inequality as in Example 3.1. In fact, this direction is valid even if $E(X) \neq 0$.

(iii) ⇒ (i). We use the power series expansion, $e^x = 1 + \sum_{k=1}^{\infty} \frac{x^k}{k!}$, and obtain

$$E(e^{\lambda X}) = E\left\{1 + \lambda X + \sum_{k=2}^{\infty} \frac{1}{k!}(\lambda X)^k\right\} \overset{E(X)=0}{=} 1 + E\left\{\sum_{k=2}^{\infty} \frac{1}{k!}(\lambda X)^k\right\}$$
$$\le 1 + E\left\{\sum_{k=2}^{\infty} \frac{1}{k!}|\lambda|^k|X|^k\right\} = 1 + \sum_{k=2}^{\infty} \frac{1}{k!}|\lambda|^k E(|X|^k),$$

where the last step uses the Fubini theorem. Then we split the summation over k into over even and odds numbers and apply (iii):

$$E(e^{\lambda X})$$
$$\le 1+\sum_{m=1}^{\infty}\left\{\frac{1}{(2m)!}|\lambda|^{2m}E(|X|^{2m})+\frac{1}{(2m+1)!}|\lambda|^{2m+1}E(|X|^{2m+1})\right\} \quad (3.11)$$
$$\le 1+\sum_{m=1}^{\infty}\left\{\frac{1}{(2m)!}|\lambda|^{2m}(c_3\sqrt{2m})^{2m}+\frac{1}{(2m+1)!}|\lambda|^{2m+1}(c_3\sqrt{2m+1})^{2m+1}\right\}.$$

It is left as Exercise 3.6 to show from the above that $E(e^{\lambda X}) \le e^{\frac{c_1^2\lambda^2}{2}}$ for $c_1 = 2c_3 e$. [Hint: Consult the proof of Corollary 3.2 below.]

□

Property (iv) in Theorem 3.6 leads to the following definition: The ψ_2 norm of a random variable X, possibly with $E(X) \ne 0$, is

$$\|X\|_{\psi_2} = \inf\{t > 0 : E(e^{X^2/t^2}) \le 2\},$$

where the infimum of an empty set is ∞. Then property (iv) is equivalent to $\|X\|_{\psi_2}$ being finite. As discussed later, for sub-Gaussian X with $E(X) \ne 0$, the ψ_2 norm depends on both the mean μ and the sale parameter σ of X. For convenience, we provide a variation of the Hoeffding inequality in terms of the ψ_2 norms for mean-zero, sub-Gaussian variables.

Corollary 3.2 (Hoeffding inequality with ψ_2 norms). *Suppose that $(X_1, \ldots, X_n)$ are independent, and $X_i - \mu_i$ is sub-Gaussian with ψ_2 norm upper bounded by τ_i, $E(e^{(X_i-\mu_i)^2/\tau_i^2}) \le 2$, for $i = 1, \ldots, n$, where $\mu_i = E(X_i)$. Then*

$$P\left\{\left|\sum_{i=1}^{n}(X_i - \mu_i)\right| \ge x\right\} \le 2\exp\left(-\frac{x^2}{8\sum_{i=1}^{n}\tau_i^2}\right), \quad x \ge 0. \quad (3.12)$$

Similar results hold for one-sided probabilities.
[Exercise: Suppose that $(X_1, \ldots, X_n)$ are independent, and the ψ_2 norm of non-centered X_i is upper bounded by τ_i, $E(e^{X_i^2/\tau_i^2}) \le 2$, for $i = 1, \ldots, n$.

Show that inequality (3.12) *can be modified as*

$$P\left\{\left|\sum_{i=1}^{n}(X_i-\mu_i)\right|\geq x\right\}\leq 2\exp\left(-\frac{x^2}{c\sum_{i=1}^{n}\tau_i^2}\right),\quad x\geq 0,$$

where $c>0$ is an absolute constant. See Theorem 3.7 later.]

Proof. It suffices to show that for a random variable X, if $E(X)=0$ and $E(\mathrm{e}^{X^2/\tau^2})\leq 2$, then $E(\mathrm{e}^{\lambda X})\leq \mathrm{e}^{2\tau^2\lambda^2}$ for $\lambda\in\mathbb{R}$, i.e., X is sub-Gaussian with scale parameter $\sigma=2\tau$. This amounts to a direct proof of (iv) $\Rightarrow$ (i) with $E(X)=0$ in Theorem 3.6.

First, from $E(\mathrm{e}^{X^2/\tau^2})\leq 2$, we have $P(|X|\geq x)\leq 2\mathrm{e}^{-\frac{x^2}{\tau^2}}$ for $x\geq 0$ and $E(|X|^p)\leq \tau^p p\Gamma(\frac{p}{2})$ for $p\geq 1$, by the proof of *(iv)* $\Rightarrow$ *(ii)* and *(ii)* $\Rightarrow$ *(iii)* in Theorem 3.6. Then we return to (3.11). Applying the moment bound and using $\Gamma(k+1)=k!$ for integer $k\geq 1$ yields

$$\begin{aligned}
E(\mathrm{e}^{\lambda X})&\leq 1+\sum_{m=1}^{\infty}\left\{\frac{1}{(2m)!}|\lambda|^{2m}E(|X|^{2m})+\frac{1}{(2m+1)!}|\lambda|^{2m+1}E(|X|^{2m+1})\right\}\\
&\leq 1+\sum_{m=1}^{\infty}\left\{|\lambda|^{2m}\tau^{2m}\frac{\Gamma(m)}{\Gamma(2m)}+|\lambda|^{2m+1}\tau^{2m+1}\frac{\Gamma(m+\frac{1}{2})}{\Gamma(2m+1)}\right\}.
\end{aligned}$$

Note that for $k\geq 2$,

$$\frac{\Gamma(\frac{k}{2})}{\Gamma(k)}=\frac{1}{\Gamma(\frac{k}{2})}\cdot\frac{\Gamma(\frac{k}{2})\Gamma(\frac{k}{2})}{\Gamma(k)}=\frac{1}{\Gamma(\frac{k}{2})}B(\frac{k}{2},\frac{k}{2})\leq\frac{(\frac{1}{4})^{\frac{k}{2}-1}}{\Gamma(\frac{k}{2})},$$

where $B(\frac{k}{2},\frac{k}{2})=\int_0^1 x^{\frac{k}{2}-1}(1-x)^{\frac{k}{2}-1}\,\mathrm{d}x\leq(\frac{1}{4})^{\frac{k}{2}-1}$. Then we continue from the preceding bound of $E(\mathrm{e}^{\lambda X})$ as follows:

$$\begin{aligned}
E(\mathrm{e}^{\lambda X})&\leq 1+\sum_{m=1}^{\infty}\left\{\frac{|\lambda|^{2m}\tau^{2m}}{\Gamma(m)}\left(\frac{1}{4}\right)^{m-1}+\frac{|\lambda|^{2m+1}\tau^{2m+1}}{\Gamma(m+\frac{1}{2})}\left(\frac{1}{4}\right)^{m-\frac{1}{2}}\right\}\\
&\leq 1+\sum_{m=1}^{\infty}\left\{\frac{|\lambda|^{2m}\tau^{2m}}{m!}\frac{m}{4^{m-1}}+\frac{|\lambda|^{2m+1}\tau^{2m+1}}{m!}\frac{1}{4^{m-\frac{1}{2}}}\right\}\\
&\leq 1+\sum_{m=1}^{\infty}\left\{\frac{|\lambda|^{2m}\tau^{2m}}{m!}(1+|\lambda|\tau)\right\}=1+(\mathrm{e}^{\lambda^2\tau^2}-1)(1+|\lambda|\tau)\\
&\leq 1+(\mathrm{e}^{\lambda^2\tau^2}-1)(1+\mathrm{e}^{\lambda^2\tau^2})=\mathrm{e}^{2\lambda^2\tau^2},
\end{aligned}$$

where the second line uses $\Gamma(m+\frac{1}{2})\leq\Gamma(m+1)=m!$ for integer $m\geq 1$, and the fourth line uses $\mathrm{e}^{x^2}\geq x^2+1\geq x$ for $x\geq 0$. $\square$

We discuss equivalent definitions of sub-Gaussian variables, *without* involving centering by the mean. A random variable X is sub-Gaussian as stated in Definition 3.1 if and only if X satisfies any of properties (ii)–(iv) in Theorem 3.6, which are equivalent to each other even with $E(X) \neq 0$. For example, a random variable X is sub-Gaussian if and only if $E(e^{X^2/c_4^2}) \leq 2$ for a (finite) constant $c_4 > 0$ [i.e., if and only if $\|X\|_{\psi_2} < \infty$]. However, the constants c_2, c_3, and c_4 in properties (ii)–(iv) depend on both the mean $\mu = E(X)$ and the scale parameter σ, and hence may not be chosen within an absolute constant factor of σ alone. In fact, the following relationship holds:

(i) If X is sub-Gaussian with $\mu = E(X)$ and scale parameter σ, then $E(e^{X^2/\tau^2}) \leq 2$ with $\tau = k_1(|\mu| + \sigma)$ for a constant $k_1 > 0$ independent of X.

(ii) If $E(e^{X^2/\tau^2}) \leq 2$ for a constant $\tau > 0$, then X is sub-Gaussian with $|E(X)| \leq k_2\tau$ and scale parameter $k_3\tau$ for constants $k_2, k_3 > 0$ independent of X.

This can be shown by using Theorem 3.7 below in terms of tail probabilities, together with using the relations established in Theorem 3.6.

Theorem 3.7 (Sub-Gaussian tails with or without centering).

(i) If a random variable X with finite $\mu = E(X)$ satisfies (3.3), *i.e., for a constant $\sigma > 0$,*

$$P(|X - \mu| \geq t) \leq 2\exp\left(-\frac{t^2}{2\sigma^2}\right), \quad t \geq 0, \tag{3.13}$$

then with $\tau = |\mu| + \sigma$,

$$P(|X| \geq x) \leq 2\exp\left(-\frac{x^2}{2\tau^2}\right), \quad x \geq 0. \tag{3.14}$$

(ii) If a random variable X satisfies (3.14) *for a constant $\tau > 0$, then X also satisfies* (3.13), *with finite $\mu = E(X)$ and $\sigma = (2\sqrt{2} + 1)\tau$.*

Proof.

(i) For X satisfying (3.13), we have by the triangle inequality,

$$P(|X| \geq |\mu| + t) \leq P(|X - \mu| \geq t) \leq 2\exp\left(-\frac{t^2}{2\sigma^2}\right), \quad t \geq 0.$$

Then X satisfies (3.14) with $\tau = |\mu| + \sigma$, by Lemma 3.3 below.

(ii) For X satisfying (3.14) [i.e., property (ii) with $c_2 = \sqrt{2}\tau$ in Theorem 3.6], we have $|\mu| = |E(X)| \leq E|X| \leq 2c_2 = 2\sqrt{2}\tau$ by the proof of *(ii) ⇒ (iii)* in Theorem 3.6. Then

$$\begin{aligned} P(|X - \mu| \geq 2\sqrt{2}\tau + x) &\leq P(|X - \mu| \geq |\mu| + x) \\ &\leq P(|X| \geq x) \leq 2\exp\left(-\frac{x^2}{2\tau^2}\right), \quad x \geq 0, \end{aligned}$$

where the second step follows from the triangle inequality. Hence X satisfies (3.13) with $\sigma = (2\sqrt{2}+1)\tau$, by Lemma 3.3 below. □

The following result shows that being sub-Gaussian for a random variable depends only on tail probabilities beyond a threshold. The proof is left as Exercise 3.7.

Lemma 3.3 (Sub-Gaussian tails beyond a threshold). *If, for some constants $a, \sigma > 0$, a random variable X satisfies*

$$P(|X| \geq a + t) \leq 2\exp\left(-\frac{t^2}{2\sigma^2}\right), \quad t \geq 0,$$

then with $\tilde{\sigma} = a + \sigma$,

$$P(|X| \geq x) \leq 2\exp\left(-\frac{x^2}{2\tilde{\sigma}^2}\right), \quad x \geq 0.$$

3.3 Sub-Exponential random variables

Sub-Gaussian variables have a squared exponential decay in tail probabilities. A class of random variables with exponential tail decay can be characterized in terms of MGFs as follows.

Definition 3.2. A random variable X is said to be sub-Exponential if $\mu = E(X)$ is finite and for constants $\nu > 0$ and $\rho > 0$,

$$E(e^{\lambda(X-\mu)}) \leq e^{\frac{\nu^2\lambda^2}{2}}, \quad |\lambda| < \frac{1}{\rho}. \tag{3.15}$$

In this case, X is said to be sub-Exponential with parameters (ν, ρ). [Note that by this definition, the parameter pair (ν, ρ) for a sub-Exponential variable is not unique. A sub-Exponential variable with parameter pair (ν, ρ) is also sub-Exponential with parameter pair $(\tilde{\nu}, \tilde{\rho})$ satisfying $\tilde{\nu} \geq \nu$ and $\tilde{\rho} \geq \rho$.]

Example 3.3. Let X be a Gamma (α, β) random variable with density function

$$f(x) = \frac{1}{\Gamma(\alpha)\beta^\alpha} x^{\alpha-1} e^{-x/\beta}, \quad x > 0,$$

where $\alpha, \beta > 0$. Then $\mu = E(X) = \alpha\beta$, $\text{var}(X) = \alpha\beta^2$, and the moment generating function of X is $E(e^{\lambda X}) = \frac{1}{(1-\beta\lambda)^\alpha}$ for $\lambda < \frac{1}{\beta}$. The moment generating function of $X - \mu$ is

$$E(e^{\lambda(X-\mu)}) = e^{-\alpha\beta\lambda - \alpha\log(1-\beta\lambda)}, \quad \lambda < \frac{1}{\beta},$$

and can be upper bounded such that

$$E(e^{\lambda(X-\mu)}) \leq e^{-\frac{\alpha\beta^2\lambda^2}{2(1-\beta|\lambda|)}}, \quad |\lambda| < \frac{1}{\beta}, \tag{3.16}$$

using the inequality that $u + \log(1-u) \geq \frac{-u^2}{2(1-|u|)}$ for $|u| < 1$ and then $u + \log(1-u) \geq -u^2$ for $|u| < \frac{1}{2}$ [Exercise]. Hence

$$E(e^{\lambda(X-\mu)}) \leq e^{-\frac{(2\alpha\beta^2)\lambda^2}{2}}, \quad |\lambda| < \frac{1}{2\beta}, \tag{3.17}$$

that is, X is sub-Exponential with parameters $(\sqrt{2\alpha\beta^2}, 2\beta)$.

By Chernoff's technique, the tail probabilities of a sub-Exponential variable, $P(|X-\mu| \geq t)$, can be upper bounded with the exponent quadratic in t (like sub-Gaussian ones) for small t, whereas with the exponent linear in t for large t.

Lemma 3.4 (Sub-Exponential tail inequality). *A sub-Exponential (ν, ρ) random variable X satisfies*

$$P(X - \mu \geq t) \leq e^{-\min\left(\frac{t^2}{2\nu^2}, \frac{t}{2\rho}\right)}, \quad t \geq 0.$$

The inequality also holds with the left-hand side replaced by $P(X - \mu < -t)$, and hence

$$P(|X - \mu| \geq t) \leq 2e^{-\min\left(\frac{t^2}{2\nu^2}, \frac{t}{2\rho}\right)}, \quad t \geq 0.$$

Proof. Applying the Chernoff inequality to $X - \mu$ gives

$$\begin{aligned} P(X-\mu \geq t) &\leq \inf_{0 \leq \lambda < \frac{1}{\rho}} e^{-\lambda t + \frac{\nu^2\lambda^2}{2}} = \begin{cases} e^{-\frac{t^2}{2\nu^2}}, & \text{if } 0 \leq t \leq \frac{\nu^2}{\rho}, \\ e^{-\frac{t}{2\rho}}, & \text{if } t > \frac{\nu^2}{\rho}, \end{cases} \\ &= e^{-\min\left(\frac{t^2}{2\nu^2}, \frac{t}{2\rho}\right)}, \quad t \geq 0. \end{aligned}$$

□

For a Gamma (α, β) variable X, by the MGF bound (3.17) and Lemma 3.4, we have

$$P(|X - \mu| \geq t) \leq 2e^{-\min\left(\frac{t^2}{2\nu^2}, \frac{t}{4\beta}\right)}, \quad t \geq 0, \tag{3.18}$$

where $\nu^2 = 2\alpha\beta^2$. Alternatively, applying Chernoff's technique directly with the MGF bound (3.16) yields

$$P(X - \mu \geq t) \leq \inf_{0 \leq \lambda < \frac{1}{\beta}} e^{-\lambda t + \frac{(\nu^2/2)\lambda^2}{2(1-\beta\lambda)}}$$

$$\leq e^{-\lambda t + \frac{(\nu^2/2)\lambda^2}{2(1-\beta\lambda)}}\Big|_{\lambda = \frac{t}{(\nu^2/2)+\beta t}} = e^{-\frac{t^2}{2((\nu^2/2)+\beta t)}}, \quad t \geq 0.$$

The two-sided probability bound gives

$$P(|X - \mu| \geq t) \leq 2e^{-\frac{t^2}{2((\nu^2/2)+\beta t)}}, \quad t \geq 0. \tag{3.19}$$

This also leads to (hence is slightly sharper than) (3.19) because $e^{-\frac{t^2}{2((\nu^2/2)+\beta t)}} \leq e^{-\min(\frac{t^2}{2\nu^2}, \frac{t}{4\beta})}$. An MGF bound in the form (3.16) and the associated probability bound (3.19) can also be used to characterize sub-Exponential variables.

Lemma 3.5 (Additivity of independent Sub-Exponential variables). *Suppose that $(X_1, \ldots, X_n)$ are independent, and X_i is sub-Exponential with parameters (ν_i, ρ_i) for $i = 1, \ldots, n$. Then $\sum_{i=1}^n X_i$ is sub-Exponential with parameter $(\sqrt{\sum_{i=1}^n \nu_i^2}, \max_{i=1}^n \rho_i)$.*

Proof. It suffices to show the result for $n = 2$. Without loss of generality, assume that $E(X_1) = E(X_2) = 0$. Then

$$\begin{aligned} E(e^{\lambda(X_1+X_2)}) &= E(e^{\lambda X_1})E(e^{\lambda X_2}) \\ &\leq e^{\frac{\nu_1^2\lambda^2}{2}} e^{\frac{\nu_2^2\lambda^2}{2}} = e^{\frac{(\nu_1^2+\nu_2^2)\lambda^2}{2}}, \quad |\lambda| < \frac{1}{\max(\rho_1, \rho_2)}, \end{aligned}$$

by the independence of X_1 and X_2 and then the sub-Exponentiality of X_1 and X_2. □

Combining Lemmas 3.4 and 3.5 gives the Bernstein inequality.

Theorem 3.8 (Bernstein inequality). *Suppose that $(X_1, \ldots, X_n)$ are independent, and X_i is sub-Exponential with parameters (ν_i, ρ_i) for $i = 1, \ldots, n$. Then with $\mu_i = E(X_i)$,*

$$P\left\{\left|\sum_{i=1}^n (X_i - \mu_i)\right| \geq t\right\} \leq 2\exp\left\{-\min\left(\frac{t^2}{2\sum_{i=1}^n \nu_i^2}, \frac{t}{2\max_{i=1}^n \rho_i}\right)\right\}, \quad t \geq 0. \tag{3.20}$$

In the special case where $\nu_i = \nu$ and $\rho_i = \rho$ for $i = 1, \ldots, n$, taking $t = ns$ in the Bernstein inequality yields

$$P\left\{\left|\bar{X}_n - \mu\right| \geq s\right\} \leq 2\exp\left\{-n\min\left(\frac{s^2}{2\nu^2}, \frac{s}{2\rho}\right)\right\}, \quad s \geq 0. \tag{3.21}$$

or equivalently

$$P\left\{\left|\sqrt{n}(\bar{X}_n - \mu)\right| \geq s\right\} \leq 2\exp\left\{-\min\left(\frac{s^2}{2\nu^2}, \frac{s\sqrt{n}}{2\rho}\right)\right\}$$

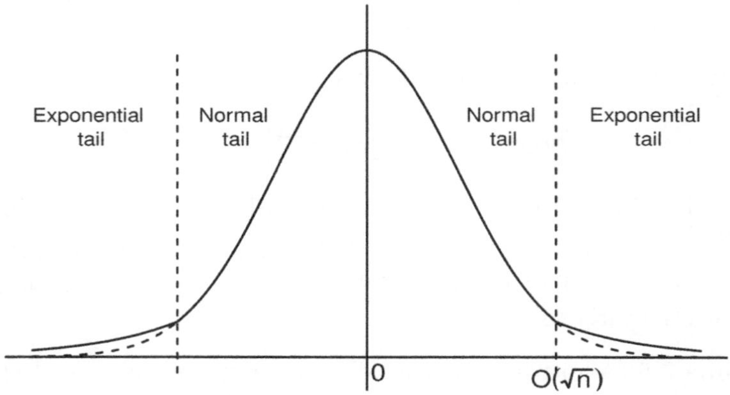

FIGURE 3.1
Illustration of two regimes of tail decays for $\sqrt{n}(\bar{X}_n - \mu)$.

$$= \begin{cases} 2\exp\left(-\frac{s^2}{2\nu^2}\right), & 0 \le s \le \frac{\nu^2}{\rho}\sqrt{n}, \\ 2\exp\left(-\frac{s\sqrt{n}}{2\rho}\right), & s \ge \frac{\nu^2}{\rho}\sqrt{n}, \end{cases}$$

which indicates two regimes: a normal (squared exponential) decay for small s and an exponential decay for large s, depending on the cutoff $\frac{\nu^2}{\rho}\sqrt{n}$. See Figure 3.1 for an illustration. Nevertheless, as $n \to \infty$ for any fixed $s \ge 0$, the small-s regime takes effect, resulting in

$$P\left\{\left|\sqrt{n}(\bar{X}_n - \mu)\right| \ge s\right\} \le 2\mathrm{e}^{-\frac{s^2}{2\nu^2}}, \quad s \ge 0,$$

which is consistent with the CLT for IID variables $X_1, \ldots, X_n$ with mean μ and variance ν^2. Moreover, (3.21) can also be stated as

$$P\left\{\left|\bar{X}_n - \mu\right| \ge \max(\sqrt{2\nu^2 s}, 2\rho s)\right\} \le 2\mathrm{e}^{-ns}, \quad s \ge 0,$$

which implies that

$$P\left\{\left|\bar{X}_n - \mu\right| \ge \sqrt{2\nu^2 s} + 2\rho s\right\} \le 2\mathrm{e}^{-ns}, \quad s \ge 0.$$

The last inequality can be convenient in relating the tail cutoff, $\sqrt{2\nu^2 s} + 2\rho s$, to the probability bound, $2\mathrm{e}^{-ns}$, in a simple manner.

Similarly to Theorem 3.6 about sub-Gaussian variables, the following result gives several equivalent characterizations of sub-Exponential random variables. For simplicity, the two parameters (ν, ρ) for a sub-Exponential variable can be reduced to one parameter: a sub-Exponential variable with parameters (ν, ρ) is also sub-Exponential $(\tilde{\rho}, \tilde{\rho})$, where $\tilde{\rho} = \max(\nu, \rho)$.

Theorem 3.9 (Equivalent characterizations of Sub-Exponential). *The following properties (i)–(iv) are equivalent for a random variable X with $E(X) = 0$, i.e., if one of the properties holds, then the other properties also hold. Moreover, properties (ii)–(iv) are also equivalent even if $E(X) \neq 0$.*

(i) There exists a constant $c_1 > 0$ such that $E(\mathrm{e}^{\lambda X}) \leq \mathrm{e}^{\frac{c_1^2\lambda^2}{2}}$ for $|\lambda| < \frac{1}{c_1}$;

(ii) There exists a constant $c_2 > 0$ such that $P(|X| \geq t) \leq 2\mathrm{e}^{-\frac{t}{c_2}}$ for $t \geq 0$;

(iii) There exists a constant $c_3 > 0$ such that $E(|X|^p)^{1/p} \leq c_3 p$ for $p \geq 1$;

(iv) There exists a constant $c_4 > 0$ such that $E(\mathrm{e}^{|X|/c_4}) \leq 2$.

The constants c_1, c_2, c_3, and c_4 differ from each other by at most a constant factor independent of X, even though c_1, c_2, c_3, or c_4 itself depends on X.

Proof. We first show the equivalence of (ii)–(iv), whether or not $E(X) \neq 0$.

(ii) ⇒ (iii). Let $Y = X/c_2$. Then $P(|Y| \geq s) = P(|X| \geq c_2 s) \leq 2\mathrm{e}^{-s}$ for $s \geq 0$, and $E(|X|^p)^{1/p} = c_2 E(|Y|^p)^{1/p}$. Next, we proceed as follows:

$$
\begin{aligned}
E(|Y|^p) &= \int_0^\infty P(|Y|^p \geq u)\,\mathrm{d}u \overset{u=t^p}{=} \int_0^\infty P(|Y| \geq t) p t^{p-1}\,\mathrm{d}t \\
&\leq \int_0^\infty 2\mathrm{e}^{-t} p t^{p-1}\,\mathrm{d}t = 2p\Gamma(p) = 2\Gamma(p+1).
\end{aligned}
$$

Then

$$
E(|Y|^p)^{1/p} \leq \{2\Gamma(p+1)\}^{1/p} \leq (2p^p)^{1/p} \leq 2p.
$$

The second inequality follows because $\Gamma(x+1) \leq x^x$ for $x \geq 1$ (Exercise 3.5). The third inequality follows because $2^{1/p} \leq 2$ for $p \geq 1$. Hence $E(|X|^p)^{1/p} \leq (2c_2)p$.

(iii) ⇒ (iv). We use the power series expansion, $\mathrm{e}^x = 1 + \sum_{k=1}^\infty \frac{x^k}{k!}$, and obtain by the Fubini theorem

$$
E(\mathrm{e}^{\lambda|X|}) = E\left\{1 + \sum_{k=1}^\infty \frac{1}{k!}(\lambda|X|)^k\right\} = 1 + \sum_{k=1}^\infty \frac{1}{k!}\lambda^k E(|X|^k), \quad \lambda \geq 0.
$$

Then we apply (iii) and use the inequality $k^k \leq \mathrm{e}^k k!$ for $k \geq 1$ (Exercise 3.5):

$$
\begin{aligned}
E(\mathrm{e}^{\lambda|X|}) &\leq 1 + \sum_{k=1}^\infty \frac{1}{k!}\lambda^k (c_3 k)^k \\
&\leq 1 + \sum_{k=1}^\infty \frac{1}{k!}\lambda^k c_3^k \cdot \mathrm{e}^k k! = 1 + \sum_{k=1}^\infty \lambda^k c_3^k \mathrm{e}^k = \frac{1}{1 - \mathrm{e}c_3\lambda},
\end{aligned}
$$

provided $0 \leq \lambda < \frac{1}{\mathrm{e}c_3}$. Taking $\lambda = 1/(2\mathrm{e}c_3)$ gives $E(\mathrm{e}^{|X|/(2\mathrm{e}c_3)}) \leq 2$.

(iv) ⇒ (ii). We use the Markov inequality and apply (iv):

$$P(|X| \geq t) = P(e^{|X|/c_4} \geq e^{t/c_4}) \\ \leq e^{-t/c_4} E(e^{|X|/c_4}) \leq 2e^{-t/c_4}.$$

Next we show the equivalence between (i) and (ii)–(iv) provided $E(X) = 0$.
(i) ⇒ (ii). By Lemma 3.4, property (i) implies that

$$P(|X| \geq t) \leq 2e^{-\min(\frac{t^2}{2c_1^2}, \frac{t}{2c_1})} = \begin{cases} 2e^{-\frac{t^2}{2c_1^2}}, & \text{if } 0 \leq t \leq c_1, \\ 2e^{-\frac{t}{2c_1}}, & \text{if } t > c_1. \end{cases}$$

However, if $0 \leq t \leq c_1$, then it holds trivially that $P(|X| \geq t) \leq 1 \leq 2e^{-\frac{t}{2c_1}}$, because $2e^{-1/2} \approx 1.2 > 1$. Hence property (ii) holds with $c_2 = 2c_1$.

(iii) ⇒ (i). We use the power series expansion, $e^x = 1 + \sum_{k=1}^{\infty} \frac{x^k}{k!}$, and obtain

$$E(e^{\lambda X}) = E\left\{1 + \lambda X + \sum_{k=2}^{\infty} \frac{1}{k!}(\lambda X)^k\right\} \overset{E(X)=0}{=} 1 + E\left\{\sum_{k=2}^{\infty} \frac{1}{k!}(\lambda X)^k\right\} \\ \leq 1 + E\left\{\sum_{k=2}^{\infty} \frac{1}{k!}|\lambda|^k |X|^k\right\} = 1 + \sum_{k=2}^{\infty} \frac{1}{k!}|\lambda|^k E(|X|^k), \tag{3.22}$$

where the last step uses the Fubini theorem. Then we apply (iii) and use the inequality $k^k \leq e^k k!$ for $k \geq 1$:

$$E(e^{\lambda X}) \leq 1 + \sum_{k=2}^{\infty} \frac{1}{k!}|\lambda|^k (c_3 k)^k \\ \leq 1 + \sum_{k=2}^{\infty} \frac{1}{k!}|\lambda|^k c_3^k \cdot e^k k! = 1 + \sum_{k=2}^{\infty} |\lambda|^k c_3^k e^k = 1 + \frac{e^2 c_3^2 \lambda^2}{1 - ec_3|\lambda|},$$

provided $ec_3|\lambda| < 1$. Hence for $|\lambda| \leq \frac{1}{2ec_3}$,

$$E(e^{\lambda X}) \leq 1 + 2e^2 c_3^2 \lambda^2 \leq e^{2e^2 c_3^2 \lambda^2},$$

that is, property (i) holds with $c_1 = 2ec_3$. □

Property (iv) in Theorem 3.9 leads to the following definition: The ψ_1 norm of a random variable X, possibly with $E(X) \neq 0$, is

$$\|X\|_{\psi_1} = \inf\{t > 0 : E(e^{|X|/t}) \leq 2\},$$

where the infimum of an empty set is ∞. For convenience, we provide a variation of the Bernstein inequality in terms of the ψ_1 norms for mean-zero, sub-Exponential variables.

Corollary 3.3 (Bernstein inequality with ψ_1 norms). *Suppose that $(X_1, \dots, X_n)$ are independent, and $X_i - \mu_i$ is sub-Exponential with ψ_1 norm upper bounded by τ_i, $E(e^{|X_i-\mu_i|/\tau_i}) \le 2$, for $i = 1, \dots, n$, where $\mu_i = E(X_i)$. Then*

$$P\left\{\left|\sum_{i=1}^n (X_i - \mu_i)\right| \ge x\right\} \le 2\exp\left\{-\min\left(\frac{x^2}{16\sum_{i=1}^n \tau_i^2}, \frac{x}{4\max_{i=1}^n \tau_i}\right)\right\}, \quad x \ge 0. \tag{3.23}$$

Similar results hold for one-sided probabilities.
[Exercise: Suppose that $(X_1, \dots, X_n)$ are independent, and the ψ_1 norm of non-centered X_i is upper bounded by τ_i, $E(e^{|X_i|/\tau_i}) \le 2$, for $i = 1, \dots, n$. Show that (3.23) *can be modified as*

$$P\left\{\left|\sum_{i=1}^n (X_i - \mu_i)\right| \ge x\right\} \le 2\exp\left\{\frac{-1}{c}\min\left(\frac{x^2}{\sum_{i=1}^n \tau_i^2}, \frac{x}{\max_{i=1}^n \tau_i}\right)\right\}, \quad x \ge 0,$$

where $c > 0$ is an absolute constant.]

Proof. It suffices to show that for a random variable X satisfying $E(X) = 0$ and $E(e^{X^2/\tau^2}) \le 2$, then $E(e^{\lambda X}) \le e^{4\tau^2\lambda^2}$ for $|\lambda| \le \frac{1}{2\tau}$, i.e., X is sub-Exponential with parameters $(\nu, \rho) = (\sqrt{8}\tau, 2\tau)$. This amounts to a direct proof of (iv) $\Rightarrow$ (i) with $E(X) = 0$ in Theorem 3.9.

From $E(e^{|X|/\tau}) \le 2$, we have $P(|X| \ge x) \le 2e^{-\frac{x}{\tau}}$ for $x \ge 0$ and $E(|X|^p) \le 2\tau^p\Gamma(p+1)$ for $p \ge 1$, by the proof of *(iv)* $\Rightarrow$ *(ii)* and *(ii)* $\Rightarrow$ *(iii)* in Theorem 3.9. Then we return to (3.22). Applying the moment bound and using $\Gamma(k+1) = k!$ for integer $k \ge 1$ yields

$$\begin{aligned} E(e^{\lambda X}) &\le 1 + \sum_{k=2}^\infty \frac{1}{k!}|\lambda|^k E(|X|^k) \\ &\le 1 + \sum_{k=2}^\infty \frac{1}{k!}|\lambda|^k \cdot 2\tau^k k! = 1 + \frac{2\tau^2\lambda^2}{1 - \tau|\lambda|} \\ &\le e^{\frac{2\tau^2\lambda^2}{1-\tau|\lambda|}}, \end{aligned}$$

provided $\tau|\lambda| < 1$. Hence $E(e^{\lambda X}) \le e^{4\tau^2\lambda^2}$ for $|\lambda| \le \frac{1}{2\tau}$. □

By the use of ψ_1 and ψ_2 norms, it is convenient to show that the product of two sub-Gaussian variables (not necessarily independent) is sub-Exponential. As a special case, the square of a sub-Gaussian variable is sub-Exponential.

Theorem 3.10 (Product of Sub-Gaussian variables). *For sub-Gaussian variables X and Y, the product XY is sub-Exponential. Moreover, $\|XY\|_{\psi_1} \le \|X\|_{\psi_2}\|Y\|_{\psi_2}$.*

Proof. Without loss of generality, assume that $\|X\|_{\psi_2} = \|Y\|_{\psi_2} = 1$. It suffices to show that if $E(e^{X^2}) \le 2$ and $E(e^{Y^2}) \le 2$, then $E(e^{|XY|}) \le 2$. In fact,

$$\begin{aligned} E(e^{|XY|}) &\le E\left(e^{\frac{X^2}{2}+\frac{Y^2}{2}}\right) \\ &\le \frac{1}{2}E\left(e^{X^2}+e^{Y^2}\right) \\ &\le \frac{1}{2}(2+2) = 2, \end{aligned}$$

where both the first two inequalities follow because $xy \le \frac{1}{2}(x^2+y^2)$ for $x, y \in \mathbb{R}$. □

A caveat of Theorem 3.10 is that the product XY may not be mean-zero (even if X and Y are mean-zero) and hence the Bernstein inequality applied to IID copies of XY involves ψ_1 norms upper bounded for non-centered variables, as indicated in Corollary 3.3.

3.4 Supremum of sample means

The Hoeffding and Bernstein inequalities, as finite-sample counterparts to LLN and CLT, deal with the concentration (or closeness) of a *single* sample mean $n^{-1}\sum_{i=1}^n X_i$ around the population mean $n^{-1}\sum_{i=1} EX_i$, for independent variables $X_1, \ldots, X_n$. [For convenience, we use the phrases "sample mean" and "population mean" even though $(X_1, \ldots, X_n)$ may not be identically distributed and hence not an IID sample from a population (or distribution).] A more challenging problem is to characterize the concentration of a class of sample means around their population means in a uniform sense as demonstrated by the uniform law of large numbers (Theorem 2.5).

Formally, let $X_1, \ldots, X_n$ be independent variables taking values in $\mathcal{X}$, and $\mathcal{H}$ be a class of (measurable) functions $h : \mathcal{X} \to \mathbb{R}$. Define the *supremum statistic*

$$Z_n = \sup_{h\in\mathcal{H}} \sum_{i=1}^n (h(X_i) - Eh(X_i)). \tag{3.24}$$

or the absolute version

$$Z_n = \sup_{h\in\mathcal{H}} \left| \sum_{i=1}^n (h(X_i) - Eh(X_i)) \right|, \tag{3.25}$$

For notational simplicity, the definition involves sample sums instead of sample means. It is of interest to study the probability $P(Z_n \ge t)$ for $t \ge 0$. A major approach involves combining the following two steps:

- Derive bounds on $P(Z_n - E(Z_n) \geq t)$ or $P(|Z_n - E(Z_n)| \geq t)$ for $t \geq 0$, referred to as *concentration inequalities.*
- Derive bounds on $E(Z_n)$, referred to as *complexity inequalities.*

While both types of bounds depend on the definition of the functional class $\mathcal{H}$, the bounds on $E(Z_n)$ are usually more sensitive to the complexity of $\mathcal{H}$ than the probability bounds on the deviation $Z_n - E(Z_n)$.

For illustration, we present a simple example. Let $\mathcal{H} = \{h_1, \ldots, h_p\}$, where all functions h_j are bounded by a constant C_0, i.e., $\max_{j=1,\ldots,p} \sup_{x \in \mathcal{X}} |h_j(x)| \leq C_0$. Then Z_n in (3.25) is

$$Z_n = \max_{j=1,\ldots,p} \left| \sum_{i=1}^{n} (h_j(X_i) - Eh_j(X_i)) \right|.$$

We compare the two-step (concentration-complexity) approach outlined above with a more elementary approach using the union bound.

Union bound. For each j, by the Hoeffding inequality (Corollary 3.1),

$$P\left\{ \left| \sum_{i=1}^{n} (h_j(X_i) - Eh_j(X_i)) \right| \geq t \right\} \leq 2\exp\left(-\frac{t^2}{n(2C_0^2)} \right), \quad t \geq 0.$$

By the union bound, we obtain

$$P(Z_n \geq t) \leq 2p\exp\left(-\frac{t^2}{n(2C_0^2)} \right), \quad t \geq 0.$$

For $0 < \epsilon < 1$, setting the right-hand side to 2ϵ and solving for t yields

$$P\left(\frac{Z_n}{n} \geq C_0 \sqrt{\frac{2\log(p/\epsilon)}{n}} \right) \leq 2\epsilon. \tag{3.26}$$

Two-step approach. First, by the functional Hoeffding inequality (Theorem 3.11) later,

$$P(Z_n - E(Z_n) \geq t) \leq \exp\left(-\frac{t^2}{n(2C_0^2)} \right), \quad t \geq 0.$$

Remarkably, this bound does not depend on the complexity measure p. Second, from Example 3.2, the bounded variables $h_j(X_1), \ldots, h_j(X_n)$ are sub-Gaussian with scale parameter $\frac{C_0-(-C_0)}{2} = C_0$, and hence $\sum_{i=1}^{n}(h_j(X_i) - Eh_j(X_i))$ are mean-zero, sub-Gaussian with scale parameter $\sqrt{nC_0^2}$. By Theorem 3.5 (Sub-Gaussian maximum),

$$E(Z_n) \leq \sqrt{nC_0^2}\sqrt{2\log(2p)},$$

which depends on the complexity measure p through its logarithm. Combining the preceding two inequalities leads to

$$P(Z_n \geq \sqrt{nC_0^2}\sqrt{2\log(2p)} + t) \leq P(Z_n - E(Z_n) \geq t)$$
$$\leq \exp\left(-\frac{t^2}{n(2C_0^2)}\right), \quad t \geq 0.$$

Setting the right-hand side to 2ϵ and solving for t yields

$$P\left(\frac{Z_n}{n} \geq C_0\sqrt{\frac{2\log(2p)}{n}} + C_0\sqrt{\frac{2\log(1/(2\epsilon))}{n}}\right) \leq 2\epsilon,$$

which by the inequality $\sqrt{a} + \sqrt{b} \leq \sqrt{2(a+b)}$ implies

$$P\left(\frac{Z_n}{n} \geq 2C_0\sqrt{\frac{\log(p/\epsilon)}{n}}\right) \leq 2\epsilon. \tag{3.27}$$

The two bounds (3.26) and (3.27) can be matched up to a constant factor in the cutoff, $C_0\sqrt{\frac{2\log(p/\epsilon)}{n}}$ versus $2C_0\sqrt{\frac{\log(p/\epsilon)}{n}}$. The union-bound approach when applicable is often powerful, as used repeatedly in the later chapters (e.g., Lemmas 4.2 and 4.3). The two-step approach is more general and can be applied when the union-bound approach does not directly work, for example, in the case of infinite $\mathcal{H}$. [Nevertheless, the union bound remains an important tool along with other ideas when deriving specific bounds in the two-step approach. See the derivation of Dudley's entropy bound in Talagrand (2005), Section 1.2.] In the following subsections, we discuss various results which can be used in applying the two-step approach.

3.4.1 Controlling the deviation of supremum

Various concentration inequalities can be used to control the deviation of the supremum statistic Z_n or the absolute version from its mean. We present one of such inequalities, known as the Hoeffding functional inequality.

Theorem 3.11 (Hoeffding functional inequality). *Let $X_1, \ldots, X_n$ be independent variables taking values in $\mathcal{X}_1, \ldots, \mathcal{X}_n$ with $\mathcal{X}_i \subset \mathcal{X}$ for $i = 1, \ldots, n$, and $\mathcal{H}$ be a class of (measurable) functions $h : \mathcal{X} \to \mathbb{R}$. Suppose that for $i = 1, \ldots, n$ and $h \in \mathcal{H}$, there exist constants $a_{i,h} < b_{i,h}$ such that $a_{i,h} \leq h(x) \leq b_{i,h}$ for $x \in \mathcal{X}_i$. Then for Z_n defined in* (3.24) *or* (3.25),

$$P(Z_n \geq E(Z_n) + t) \leq \exp(-\frac{2t^2}{V}), \quad t \geq 0, \tag{3.28}$$

where $V = \sum_{i=1}^n \sup_{h\in\mathcal{H}}(b_{i,h} - a_{i,h})^2$. The inequality also holds with the left-hand side replaced by $P(Z_n \leq E(Z_n) - t)$. Hence

$$P(|Z_n - E(Z_n)| \geq t) \leq 2\exp(-\frac{2t^2}{V}), \quad t \geq 0.$$

Proof. We give a proof for the absolute version of Z_n in (3.25). The proof for Z_n in (3.24) is similar. Define $f(x_1, \ldots, x_n) = \sup_{h \in \mathcal{H}} |\sum_{i=1}^n (h(x_i) - Eh(X_i))|$. Then

$$\begin{aligned} & |f(x_1, \ldots, x_n) - f(\tilde{x}_1, \ldots, \tilde{x}_n)| \\ & \leq \sup_{h \in \mathcal{H}} \left| |\sum_{i=1}^n (h(x_i) - Eh(X_i))| - |\sum_{i=1}^n (h(\tilde{x}_i) - Eh(X_i))| \right| \\ & \leq \sup_{h \in \mathcal{H}} \sum_{i=1}^n |h(x_i) - h(\tilde{x}_i)|, \end{aligned}$$

where the first step follows from $|\sup_h u_h - \sup_h \tilde{u}_h| \leq \sup_h |u_h - \tilde{u}_h|$, and second step follows from the triangle inequality $||u| - |\tilde{u}|| \leq |u - \tilde{u}|$. If x and $\tilde{x}$ differ only in the ith component, then $|f(x_1, \ldots, x_n) - f(\tilde{x}_1, \ldots, \tilde{x}_n)| \leq \sup_{h \in \mathcal{H}} (b_{i,h} - a_{i,h})$. Hence the bounded difference property holds for f with $c_i = \sup_{h \in \mathcal{H}} (b_{i,h} - a_{i,h})$, $i = 1, \ldots, n$, as defined in Theorem 3.12 below. The result then follows from the McDiarmid inequality. □

When $\mathcal{H}$ consists of a single function h, Theorem 3.11 recovers the standard Hoeffding inequality for bounded variables (Corollary 3.1):

$$P\left(\left| \sum_{i=1}^n (h(X_i) - Eh(X_i)) \right| \geq t \right) \leq 2 \exp\left(-\frac{2t^2}{V} \right), \quad t \geq 0.$$

where $V = \sum_{i=1}^n (b_{i,h} - a_{i,h})^2$. The significance of Theorem 3.11 is that probability bounds depend on the functional class $\mathcal{H}$ only through V, regardless of the complexity of $\mathcal{H}$ (e.g., smoothness of functions in $\mathcal{H}$).

There is an alternative version of the Hoeffding functional inequality as follows. See Massart (2000), Theorem 9, and Boucheron et al. (2013), Theorem 12.1. In the setting of Theorem 3.11, if $\mathcal{H}$ is countable, then for Z_n defined in (3.24),

$$P(Z_n \geq E(Z_n) + t) \leq \exp\left(-\frac{t^2}{2v} \right), \quad t \geq 0, \tag{3.29}$$

where $v = \sup_{h \in \mathcal{H}} \sum_{i=1}^n (b_{i,h} - a_{i,h})^2$, which is no greater than V and hence preferable over V as a "variance factor". On the other hand, inequality (3.28) involves a worse variance factor V ($\geq v$) but does not require that $\mathcal{H}$ is a countable class.

In the remainder of this subsection, we present the McDiarmid inequality (or bounded difference inequality), which we prove using the Azuma inequality (or martingale difference inequality). The McDiarmid inequality and inequality (3.28) [with a countable $\mathcal{H}$] can also be proved using another technique called the entropy method. In addition, inequality (3.29) [with a countable $\mathcal{H}$] can be proved using the entropy method (although not using the McDiarmid inequality). See Boucheron et al. (2013) for details.

Theorem 3.12 (McDiarmid inequality). *Suppose that $X_1, \ldots, X_n$ are independent variables taking values in $\mathcal{X}_1, \ldots, \mathcal{X}_n$, and $f : \mathcal{X}_1 \times \cdots \times \mathcal{X}_n \to \mathbb{R}$ is a function satisfying the bounded difference property with constants $(c_1, \ldots, c_n)$: for $i = 1, \ldots, n$,*

$$\sup_{x_1 \in \mathcal{X}_1, \ldots, x_n \in \mathcal{X}_n, \tilde{x}_i \in \mathcal{X}_i} |f(x_1, \ldots, x_n) - f(x_1, \ldots, x_{i-1}, \tilde{x}_i, x_{i+1}, \ldots, x_n)| \leq c_i.$$

Denote $Z_n = f(X_1, \ldots, X_n)$. Then

$$P(Z_n \geq E(Z_n) + t) \leq \exp\left(-\frac{2t^2}{\sum_{i=1}^n c_i^2}\right), \quad t \geq 0.$$

The inequality also holds with the left-hand side replaced by $P(Z_n \leq E(Z_n) - t)$. Hence

$$P(|Z_n - E(Z_n)| \geq t) \leq 2\exp\left(-\frac{2t^2}{\sum_{i=1}^n c_i^2}\right), \quad t \geq 0.$$

Proof. Denote $f(X) = f(X_1, \ldots, X_n)$ and $X_{i:j} = (X_i, \ldots, X_j)$. For $i = 1, \ldots, n$, let

$$D_i = E(f(X)|X_{1:i}) - E(f(X)|X_{1:(i-1)}),$$

which reads $D_1 = E(f(X)|X_1) - E(f(X))$ in case of $i = 1$. Then $Z_n - E(Z_n) = \sum_{i=1}^n D_i$. The desired result follows from Azuma inequality (Theorem 3.13) provided we verify the two conditions (i) and (ii) involved. First, condition (i) in Theorem 3.13 is satisfied:

$$\begin{aligned}
&E(D_i|X_{1:(i-1)}) \\
&= E\{E(f(X)|X_{1:i}) - E(f(X)|X_{1:(i-1)})|X_{1:(i-1)}\} \\
&= E\{E(f(X)|X_{1:i})|X_{1:(i-1)}\} - E(f(X)|X_{1:(i-1)}) \\
&= 0,
\end{aligned}$$

by the law of iterated expectations. Second, from the preceding calculation, $D_i = g(X_{1:i}) - E(g(X_{1:i})|X_{1:(i-1)})$, where $g(X_{1:i}) = E(f(X)|X_{1:i})$. Let

$$\begin{aligned}
a_i(X_{1:(i-1)}) &= \inf_{x_i \in \mathcal{X}_i} \left\{g(X_{1:(i-1)}, x_i) - E(g(X_{1:i})|X_{1:(i-1)})\right\}, \\
b_i(X_{1:(i-1)}) &= \sup_{x_i \in \mathcal{X}_i} \left\{g(X_{1:(i-1)}, x_i) - E(g(X_{1:i})|X_{1:(i-1)})\right\}.
\end{aligned}$$

By the independence of $X_1, \ldots, X_n$ or more precisely, independent of X_i with $(X_{1:(i-1)}, X_{(i+1):n})$,

$$\begin{aligned}
g(X_{1:(i-1)}, x_i) &= E\{f(X_{1:(i-1)}, X_i, X_{(i+1):n})|X_{1:(i-1)}, X_i = x_i\} \\
&= E\{f(X_{1:(i-1)}, x_i, X_{(i+1):n})|X_{1:(i-1)}, X_i = x_i\} \\
&= E\{f(X_{1:(i-1)}, x_i, X_{(i+1):n})|X_{1:(i-1)}\}.
\end{aligned}$$

Then

$$\begin{aligned}
& b_i(X_{1:(i-1)}) - a_i(X_{1:(i-1)}) \\
\leq & \sup_{x_i, \tilde{x}_i \in \mathcal{X}_i} \{g(X_{1:(i-1)}, x_i) - g(X_{1:(i-1)}, \tilde{x}_i)\} \\
= & \sup_{x_i, \tilde{x}_i \in \mathcal{X}_i} E\{f(X_{1:(i-1)}, x_i, X_{(i+1):n}) - f(X_{1:(i-1)}, \tilde{x}_i, X_{(i+1):n}) | X_{1:(i-1)}\} \\
\leq & c_i,
\end{aligned}$$

by the bounded difference property of $f(\cdot)$. Hence condition (ii) in Theorem 3.13 is also satisfied. □

The McDiarmid inequality demonstrates the concentration of a *nonlinear* function of independent variables subject to the bounded difference property. For bounded variables $X_i \in [a_i, b_i]$, $i = 1, \ldots, n$, the linear function $f(x_1, \ldots, x_n) = \sum_{i=1}^n (x_i - E(X_i))$ can be easily shown to satisfy the bounded difference property with $c_i = b_i - a_i$ [Exercise]. When applied in this special case, the McDiarmid inequality recovers the standard Hoeffding inequality for bounded variables (Corollary 3.1).

The McDiarmid inequality can be proved using the Azuma inequality, which can be proved through the use of MGFs similarly as in the proof of the Hoeffding inequality, albeit with martingale differences.

Theorem 3.13 (Azuma inequality). *Let $X_1, \ldots, X_n$ be a sequence of random variables and, for $i = 1, \ldots, n$, let D_i be a (measurable) function of $(X_1, \ldots, X_i)$. Suppose that for $i = 1, \ldots, n$,*

(i) $E(D_i | X_1, \ldots, X_{i-1}) = 0$, which reads $E(D_1) = 0$ in case of $i = 1$,

(ii) $D_i \in [a_i(X_1, \ldots, X_{i-1}), b_i(X_1, \ldots, X_{i-1})]$ almost surely given $(X_1, \ldots, X_{i-1})$, and $0 \leq b_i(X_1, \ldots, X_{i-1}) - a_i(X_1, \ldots, X_{i-1}) \leq c_i$ almost surely for a constant $c_i \geq 0$.

Then

$$P\left(\sum_{i=1}^n D_n \geq t\right) \leq \exp\left(-\frac{2t^2}{\sum_{i=1}^n c_i^2}\right), \quad t \geq 0.$$

The inequality also holds with the left-hand side replaced by $P(Z_n \leq E(Z_n) - t)$. Hence

$$P\left(\left|\sum_{i=1}^n D_n\right| \geq t\right) \leq 2\exp\left(-\frac{2t^2}{\sum_{i=1}^n c_i^2}\right), \quad t \geq 0.$$

Proof. Denote $S_i = \sum_{j=1}^i D_j$ for $i = 1, \ldots, n$ and $S_0 = 0$. Then condition (i) indicates that $E(S_i | X_1, \ldots, X_{i-1}) = S_{i-1}$ for $i = 1, \ldots, n$. Such a sequence $(S_1, \ldots, S_n)$ is said to be a martingale with respect to the sequence $(X_1, \ldots, X_n)$, and $(D_1, \ldots, D_n)$ is a sequence of martingale differences.

By the law of iterated expectations, $E(S_n) = 0$ [Exercise]. It suffices to show that S_n is sub-Gaussian with scale parameter $\sqrt{\frac{\sum_{i=1}^n c_i^2}{4}}$, i.e.,

$$E(\mathrm{e}^{\lambda S_n}) \le \mathrm{e}^{\frac{\sum_{i=1}^n c_i^2}{4}\frac{\lambda^2}{2}}, \quad \lambda \in \mathbb{R}.$$

By the law of iterated expectations, we have for $\lambda \in \mathbb{R}$,

$$\begin{aligned} E(\mathrm{e}^{\lambda S_n}) &= E\left\{\mathrm{e}^{\lambda S_{n-1}} E(\mathrm{e}^{\lambda D_n}|X_1,\ldots,X_{n-1})\right\} \\ &\le E\left(\mathrm{e}^{\lambda S_{n-1}}\mathrm{e}^{\frac{c_n^2}{4}\frac{\lambda}{2}}\right) \\ &= \mathrm{e}^{\frac{c_n^2}{4}\frac{\lambda}{2}} E\left(\mathrm{e}^{\lambda S_{n-1}}\right), \end{aligned}$$

where the second line follows from Example 3.2, with $E(D_n|X_1,\ldots,X_{n-1}) = 0$, $D_n \in [a_n(X_1,\ldots,X_{n-1}), b_n(X_1,\ldots,X_{n-1})]$, and $0 \le b_i(X_1,\ldots,X_{n-1}) - a_i(X_1,\ldots,X_{n-1}) \le c_n$. Repeating the procedure gives, for $\lambda \in \mathbb{R}$,

$$\begin{aligned} E(\mathrm{e}^{\lambda S_n}) &\le \mathrm{e}^{\frac{c_{n-1}^2+c_n^2}{4}\frac{\lambda}{2}} E\left(\mathrm{e}^{\lambda S_{n-2}}\right) \\ &\le \cdots \\ &\le \mathrm{e}^{\frac{\sum_{i=1}^n c_i^2}{4}\frac{\lambda^2}{2}}. \end{aligned}$$

□

In the special case where $(X_1,\ldots,X_n)$ are independent, and D_i is a function of X_i only, independent of $(X_1,\ldots,X_{i-1})$ and satisfying $D_i \in [a_i, b_i]$ for $i = 1,\ldots,n$, the Azuma inequality recovers the standard Hoeffding inequality for bounded variables (Corollary 3.1).

3.4.2 Controlling the mean of supremum

To control $E(Z_n)$ for the supremum statistic Z_n in (3.24) or the absolute version (3.25), an important technique is symmetrization inequalities, as shown in the following result. A Rademacher random variable ζ takes values 1 or -1 each with probability 1/2, i.e., $P(\zeta = 1) = P(\zeta = -1) = 1/2$.

Theorem 3.14 (Symmetrization inequality). *Suppose that $X_1,\ldots,X_n$ are independent variables taking values in $\mathcal{X}$, and $\mathcal{H}$ is a class of (measurable) functions on $\mathcal{X}$ such that $E|h(X_i)| < \infty$ for $h \in \mathcal{H}$ and $i = 1,\ldots,n$. Then*

$$E \sup_{h\in\mathcal{H}} \left\{\sum_{i=1}^n (h(X_i) - Eh(X_i))\right\} \le 2E \sup_{h\in\mathcal{H}} \left\{\sum_{i=1}^n \zeta_i h(X_i)\right\}, \tag{3.30}$$

where $\zeta_1,\ldots,\zeta_n$ are IID Rademacher random variables, independently of $X_1,\ldots,X_n$. Moreover,

$$\frac{1}{2}E \sup_{h\in\mathcal{H}} \left|\sum_{i=1}^n \zeta_i(h(X_i) - Eh(X_i))\right| \le E \sup_{h\in\mathcal{H}} \left|\sum_{i=1}^n (h(X_i) - Eh(X_i))\right| \tag{3.31a}$$

$$\le 2E \sup_{h\in\mathcal{H}} \left|\sum_{i=1}^n \zeta_i h(X_i)\right|. \tag{3.31b}$$

Note that the lower bound in (3.31a) involves the product of mean-centered function $h(X_i) - Eh(X_i)$ and Rademacher variable ζ_i. Moreover, the upper bounds in (3.30) and (3.31b) also hold with $h(X_i)$ replaced by $h(X_i) - Eh(X_i)$, as presented in Boucheron et al. (2013), Lemma 11.4. [Exercise: State the result and provide a proof.]

Proof. Let $(\tilde{X}_1, \ldots, \tilde{X}_n)$ be an IID copy of $(X_1, \ldots, X_n)$, independently of $(\zeta_1, \ldots, \zeta_n)$. Then

$$
\begin{aligned}
& E \sup_{h \in \mathcal{H}} \left\{ \sum_{i=1}^n (h(X_i) - Eh(X_i)) \right\} \\
&= E \sup_{h \in \mathcal{H}} E \left\{ \sum_{i=1}^n (h(X_i) - h(\tilde{X}_i)) \middle| X_1, \ldots, X_n \right\} \\
&\le EE \left\{ \sup_{h \in \mathcal{H}} \sum_{i=1}^n (h(X_i) - h(\tilde{X}_i)) \middle| X_1, \ldots, X_n \right\} && (3.32) \\
&= E \sup_{h \in \mathcal{H}} \left\{ \sum_{i=1}^n (h(X_i) - h(\tilde{X}_i)) \right\} \\
&= E \sup_{h \in \mathcal{H}} \left\{ \sum_{i=1}^n \zeta_i (h(X_i) - h(\tilde{X}_i)) \right\} && (3.33) \\
&\le E \sup_{h \in \mathcal{H}} \left\{ \sum_{i=1}^n \zeta_i h(X_i) \right\} + E \sup_{h \in \mathcal{H}} \left\{ \sum_{i=1}^n (-\zeta_i) h(\tilde{X}_i) \right\} && (3.34) \\
&= 2E \sup_{h \in \mathcal{H}} \left\{ \sum_{i=1}^n \zeta_i h(X_i) \right\}.
\end{aligned}
$$

Line (3.32) follows by exchanging the supremum and expectation. Line (3.33) follows because the vector with components $h(X_i) - h(\tilde{X}_i)$ has the same joint distribution as the vector with components $\zeta_i(h(X_i) - h(\tilde{X}_i))$ (Exercise 3.9). Line (3.34) follows by exchanging the supremum and summation. Similarly, the absolute version can be proved, with additional inequalities used:

$$
\begin{aligned}
& E \sup_{h \in \mathcal{H}} \left| \sum_{i=1}^n (h(X_i) - Eh(X_i)) \right| \\
&\le E \sup_{h \in \mathcal{H}} E \left\{ \left| \sum_{i=1}^n (h(X_i) - h(\tilde{X}_i)) \right| \middle| X_1, \ldots, X_n \right\} && (3.35) \\
&\le EE \left\{ \left| \sup_{h \in \mathcal{H}} \sum_{i=1}^n (h(X_i) - h(\tilde{X}_i)) \right| \middle| X_1, \ldots, X_n \right\} \\
&= E \sup_{h \in \mathcal{H}} \left| \sum_{i=1}^n (h(X_i) - h(\tilde{X}_i)) \right|
\end{aligned}
$$

$$
\begin{aligned}
&= E \sup_{h\in\mathcal{H}} \left| \sum_{i=1}^n \zeta_i (h(X_i) - h(\tilde{X}_i)) \right| \\
&\le E \sup_{h\in\mathcal{H}} \left| \sum_{i=1}^n \zeta_i h(X_i) \right| + E \sup_{h\in\mathcal{H}} \left| \sum_{i=1}^n \zeta_i h(\tilde{X}_i) \right| \qquad (3.36) \\
&= 2E \sup_{h\in\mathcal{H}} \left| \sum_{i=1}^n \zeta_i h(X_i) \right| .
\end{aligned}
$$

Line (3.35) follows from Jensen's inequality, $|E(Y)| \le E(|Y|)$. Line (3.36) follows from the triangle inequality and exchanging the supremum and summation. The proof of the lower bound (3.31a) is left as Exercise 3.9. □

As motivated by symmetrization inequalities, the conditional (or empirical) Rademacher complexity of functional class $\mathcal{H}$ given elements $x_1, \ldots, x_n \in \mathcal{X}$ is defined as

$$
\begin{aligned}
&E\left[\sup_{h\in\mathcal{H}} \left\{ \sum_{i=1}^n \zeta_i h(X_i) \right\} \Big| X_1 = x_1, \ldots, X_n = x_n \right] \\
&= E \sup_{h\in\mathcal{H}} \left\{ \sum_{i=1}^n \zeta_i h(x_i) \right\},
\end{aligned}
$$

and the absolute version is

$$
\begin{aligned}
&E\left[\sup_{h\in\mathcal{H}} \left| \sum_{i=1}^n \zeta_i h(X_i) \right| \Big| X_1 = x_1, \ldots, X_n = x_n \right] \\
&= E \sup_{h\in\mathcal{H}} \left| \sum_{i=1}^n \zeta_i h(x_i) \right|,
\end{aligned}
$$

where $(\zeta_1, \ldots, \zeta_n)$ are IID Rademacher random variables, independently of $X_1, \ldots, X_n$. For Z_n in (3.24) or (3.25), $E(Z_n)$ can be upper or lower bounded through upper or lower bounds on the empirical Rademacher complexity. For example, if $E \sup_{h\in\mathcal{H}} |\sum_{i=1}^n \zeta_i h(x_i)| \le R(\mathcal{H})$ for an upper bound $R(\mathcal{H})$ independently of $x_1, \ldots, x_n \in \mathcal{X}$, then immediately

$$
E \sup_{h\in\mathcal{H}} \left| \sum_{i=1}^n (h(X_i) - Eh(X_i)) \right| \le 2R(\mathcal{H}),
$$

by inequality (3.31b) and the law of iterated expectations.

The contraction inequality below provides a useful technique for controlling Rademacher complexities. A function $\phi : \mathbb{R} \to \mathbb{R}$ is said to be Lipschitz-L for a constant $L > 0$ if $|\phi(u) - \phi(\tilde{u})| \le L|u - \tilde{u}|$ for any $u, \tilde{u} \in \mathbb{R}$.

Theorem 3.15 (Contraction inequality). *Let $x_1, \ldots, x_n$ be elements of $\mathcal{X}$ and $\mathcal{H}$ be a class of functions on $\mathcal{X}$, and $\phi_i : \mathbb{R} \to \mathbb{R}$ be a Lipschitz-L function for $i = 1, \ldots, n$. Then for IID Rademacher random variables $\zeta_1, \ldots, \zeta_n$, we have*

$$E \sup_{h \in \mathcal{H}} \left\{ \sum_{i=1}^{n} \zeta_i \phi_i(h(x_i)) \right\} \leq LE \sup_{h \in \mathcal{H}} \left\{ \sum_{i=1}^{n} \zeta_i h(x_i) \right\}.$$

Moreover, if $\phi_i(0) = 0$ for $i = 1, \ldots, n$, then

$$E \sup_{h \in \mathcal{H}} \left| \sum_{i=1}^{n} \zeta_i \phi_i(h(x_i)) \right| \leq 2LE \sup_{h \in \mathcal{H}} \left| \sum_{i=1}^{n} \zeta_i h(x_i) \right|. \tag{3.37}$$

Equivalently, the preceding result can be stated as follows. Let $\mathcal{T}_n$ be a subset of $\mathbb{R}^n$ and denote $t \in \mathcal{T}_n$ as $t = (t_1, \ldots, t_n)^{\mathrm{T}}$. Let $\phi_i : \mathbb{R} \to \mathbb{R}$ be a Lipschitz-L function for $i = 1, \ldots, n$. Then for IID Rademacher random variables $\zeta_1, \ldots, \zeta_n$, we have

$$E \sup_{t \in \mathcal{T}_n} \left(\sum_{i=1}^{n} \zeta_i \phi_i(t_i) \right) \leq LE \sup_{t \in \mathcal{T}_n} \left(\sum_{i=1}^{n} \zeta_i t_i \right). \tag{3.38}$$

Moreover, if $\phi_i(0) = 0$ for $i = 1, \ldots, n$, then

$$E \sup_{t \in \mathcal{T}_n} \left| \sum_{i=1}^{n} \zeta_i \phi_i(t_i) \right| \leq 2LE \sup_{t \in \mathcal{T}_n} \left| \sum_{i=1}^{n} \zeta_i t_i \right|. \tag{3.39}$$

Note that both inequalities (3.38) and (3.39) are presented in Boucheron et al. (2013), Theorem 11.6, for Lipschitz functions ϕ_i satisfying $\phi_i(0) = 0$. However, the raw version (3.38) holds for Lipschitz functions ϕ_i without requiring $\phi_i(0) = 0$, as presented in Meir and Zhang (2003), Theorem 7.

Proof. We present a proof of inequality (3.39) in Section 3.5.2, which is more complicated than that of inequality (3.38) below. Without loss of generality, assume that $L = 1$, i.e., ϕ_i is a Lipschitz-1 function.

To prove inequality (3.38), we show that for any function $a : \mathcal{T}_n \to \mathbb{R}$,

$$E \sup_{t \in \mathcal{T}_n} \left\{ a(t) + \sum_{i=1}^{n} \zeta_i \phi_i(t_i) \right\} \leq E \sup_{t \in \mathcal{T}_n} \left\{ a(t) + \sum_{i=1}^{n} \zeta_i t_i \right\}, \tag{3.40}$$

which is stronger than (3.38). The proof is by induction. For $n = 1$, it is desired to show that for $\mathcal{T}_1 \subset \mathbb{R}$, a Lipschitz-1 function ϕ, and a Rademacher random variable ζ,

$$E \sup_{t \in \mathcal{T}_1} \{a(t) + \zeta\phi(t)\} \leq E \sup_{t \in \mathcal{T}_1} \{a(t) + \zeta t\},$$

or equivalently, after evaluating the expectation,

$$\sup_{t\in\mathcal{T}_1}\{a(t)+\phi(t)\}+\sup_{t\in\mathcal{T}_1}\{a(t)-\phi(t)\}\le\sup_{t\in\mathcal{T}_1}\{a(t)+t\}+\sup_{t\in\mathcal{T}_1}\{a(t)-t\}. \tag{3.41}$$

This can be proved directly:

$$\begin{aligned}
&\sup_{t\in\mathcal{T}_1}\{a(t)+\phi(t)\}+\sup_{t\in\mathcal{T}_1}\{a(t)-\phi(t)\}\\
&=\sup_{t,\tilde{t}\in\mathcal{T}_1}\left\{a(t)+\phi(t)+a(\tilde{t})-\phi(\tilde{t})\right\}\\
&=\sup_{t,\tilde{t}\in\mathcal{T}_1}\left\{a(t)+a(\tilde{t})+|\phi(t)-\phi(\tilde{t})|\right\}\\
&\le\sup_{t,\tilde{t}\in\mathcal{T}_1}\left\{a(t)+a(\tilde{t})+|t-\tilde{t}|\right\}\\
&=\sup_{t\in\mathcal{T}_1}\{a(t)+t\}+\sup_{t\in\mathcal{T}_1}\{a(t)-t\}.
\end{aligned}$$

For $n\ge 2$, we proceed as follows:

$$\begin{aligned}
&E\sup_{t\in\mathcal{T}_n}\left\{a(t)+\sum_{i=1}^{n}\zeta_i\phi_i(t_i)\right\}\\
&=EE\left[\sup_{t\in\mathcal{T}_n}\left\{a(t)+\sum_{i=1}^{n-1}\zeta_i\phi_i(t_i)+\zeta_n\phi_n(t_n)\right\}\middle|\zeta_1,\ldots,\zeta_{n-1}\right]\\
&\le EE\left[\sup_{t\in\mathcal{T}_n}\left\{a(t)+\sum_{i=1}^{n-1}\zeta_i\phi_i(t_i)+\zeta_n t_n\right\}\middle|\zeta_1,\ldots,\zeta_{n-1}\right] && (3.42)\\
&=EE\left[\sup_{t\in\mathcal{T}_n}\left\{a(t)+\zeta_n t_n+\sum_{i=1}^{n-1}\zeta_i\phi_i(t_i)\right\}\middle|\zeta_n\right]\\
&\le EE\left[\sup_{t\in\mathcal{T}_n}\left\{a(t)+\zeta_n t_n+\sum_{i=1}^{n-1}\zeta_i t_i\right\}\middle|\zeta_n\right] && (3.43)\\
&=E\sup_{t\in\mathcal{T}_n}\left\{a(t)+\sum_{i=1}^{n}\zeta_i t_i\right\}.
\end{aligned}$$

The first inequality (3.42) follows from the case of $n=1$:

$$\begin{aligned}
&E\left[\sup_{t\in\mathcal{T}_n}\left\{a(t)+\sum_{i=1}^{n-1}\zeta_i\phi_i(t_i)+\zeta_n\phi_n(t_n)\right\}\middle|\zeta_1,\ldots,\zeta_{n-1}\right]\\
&=E\left[\sup_{t_n\in\mathcal{T}_1}\{\tilde{a}(t_n;\zeta_1,\ldots,\zeta_{n-1})+\zeta_n\phi_n(t_n)\}\middle|\zeta_1,\ldots,\zeta_{n-1}\right]\\
&\le E\left[\sup_{t_n\in\mathcal{T}_1}\{\tilde{a}(t_n;\zeta_1,\ldots,\zeta_{n-1})+\zeta_n t_n\}\middle|\zeta_1,\ldots,\zeta_{n-1}\right]
\end{aligned}$$

$$= E\left[\sup_{t\in\mathcal{T}_n}\left\{a(t) + \sum_{i=1}^{n-1}\zeta_i\phi_i(t_i) + \zeta_n t_n\right\}\middle|\zeta_1,\ldots,\zeta_{n-1}\right],$$

where $\mathcal{T}_1 \subset \mathbb{R}$ is the collection of t_n such that $\{(t_1,\ldots,t_{n-1}) : t \in \mathcal{T}_n\}$ is non-empty, and $\tilde{a}(t_n;\zeta_1,\ldots,\zeta_{n-1}) = \sup_{(t_1,\ldots,t_{n-1}):t\in\mathcal{T}_n}\{a(t)+\sum_{i=1}^{n-1}\zeta_i\phi_i(t_i)\}$. The second inequality (3.43) follows from the induction hypothesis that (3.40) holds for $n-1$ Rademacher random variables. □

To facilitate applications, it is helpful to restate the absolute contraction inequalities (3.37) and (3.39) as follows, with the requirement $\phi_i(0) = 0$ seemingly removed. Let $\bar{h} \in \mathcal{H}$ be fixed. Then for Lipschitz-L functions $\phi_1,\ldots,\phi_n$ and IID Rademacher variables $\zeta,\ldots,\zeta_n$, we have

$$E\sup_{h\in\mathcal{H}}\left|\sum_{i=1}^{n}\zeta_i\{\phi_i(h(x_i)) - \phi_i(\bar{h}(x_i))\}\right| \le 2LE\sup_{h\in\mathcal{H}}\left|\sum_{i=1}^{n}\zeta_i\{h(x_i) - \bar{h}(x_i)\}\right|. \tag{3.44}$$

Similarly, let $\bar{t} = (\bar{t}_1,\ldots,\bar{t}_n)^{\mathrm{T}} \in \mathcal{T}_n$ be fixed. Then for Lipschitz-L functions $\phi_1,\ldots,\phi_n$ and IID Rademacher variables $\zeta,\ldots,\zeta_n$, we have

$$E\sup_{t\in\mathcal{T}_n}\left|\sum_{i=1}^{n}\zeta_i\{\phi_i(t_i) - \phi_i(\bar{t}_i)\}\right| \le 2LE\sup_{t\in\mathcal{T}_n}\left|\sum_{i=1}^{n}\zeta_i\{t_i - \bar{t}_i\}\right|. \tag{3.45}$$

[Exercise: Show that (3.44) and (3.45) can be deduced from (3.37) and (3.39) respectively.]

3.4.3 Applications to Lipschitz loss functions

We discuss an application of the techniques in Sections 3.4.1 and 3.4.2 to study the concentration of Lipshitz loss functions. The main result (Proposition 3.1) is used to study Lasso-penalized non-smooth M-estimation in Section 5.5.2.

Let $(X_1,Y_1),\ldots,(X_n,Y_n)$ be a collection of n observations, where $X_i = (X_{i1},\ldots,X_{ip})^{\mathrm{T}} \in \mathbb{R}^p$ is a covariate vector and $Y_i \in \mathbb{R}$ is a response variable. Consider a sample objective (or loss) function in the form

$$K_n(\beta) = \frac{1}{n}\sum_{i=1}^{n}\rho(Y_i, X_i^{\mathrm{T}}\beta),$$

where $\rho(y,\eta)$ is Lipschitz-L in η, i.e., $|\rho(y,\eta) - \rho(y,\tilde{\eta})| \le L|\eta - \tilde{\eta}|$ for any $\eta,\tilde{\eta} \in \mathbb{R}$. For logistic regression with maximum likelihood estimation, $\rho(y,\eta) = -y\eta + \log(1+\mathrm{e}^{\eta})$, which is Lipschitz-1 in η. For τ-quantile regression (Section 5.5.2), $\rho(y,\eta) = \tau(y-\eta)_+ + (1-\tau)(\eta-y)_+$, which is Lipschitz-$L$ in η with $L = \max(\tau, 1-\tau)$.

For fixed $\bar{\beta}$, for example (but not necessarily) a minimizer of $EK_n(\beta)$, define

$$D_n(\beta,\bar{\beta}) = K_n(\beta) - K_n(\bar{\beta}) - E(K_n(\beta) - K_n(\bar{\beta}))$$

$$= \frac{1}{n}\sum_{i=1}^{n}\{\rho(Y_i, X_i^{\mathrm{T}}\beta) - \rho(Y_i, X_i^{\mathrm{T}}\bar{\beta}) - E(\rho(Y_i, X_i^{\mathrm{T}}\beta) - \rho(Y_i, X_i^{\mathrm{T}}\bar{\beta}))\}. \quad (3.46)$$

Equivalently, $D_n(\beta, \bar{\beta})$ can be written as

$$D_n(\beta, \bar{\beta}) = \frac{1}{n}\sum_{i=1}^{n}(h(X_i, Y_i; \beta) - Eh(X_i, Y_i; \beta)),$$

where $h(X_i, Y_i; \beta) = \rho(Y_i, X_i^{\mathrm{T}}\beta) - \rho(Y_i, X_i^{\mathrm{T}}\bar{\beta})$. For fixed $r > 0$, let

$$\begin{aligned} Z_n &= \sup_{\|\beta-\bar{\beta}\|_1 \le r} |nD_n(\beta, \bar{\beta})| \\ &= \sup_{\|\beta-\bar{\beta}\|_1 \le r} \left|\sum_{i=1}^{n}(h(X_i, Y_i; \beta) - Eh(X_i, Y_i; \beta))\right|, \end{aligned}$$

where the dependency of Z_n on $\bar{\beta}$ and r is suppressed in the notation. In other words, Z_n is a supremum absolute statistic as in (3.25):

$$Z_n = \sup_{h\in\mathcal{H}} \left|\sum_{i=1}^{n}(h(X_i, Y_i) - Eh(X_i, Y_i))\right|$$

where $\mathcal{H} = \{h(x, y; \beta) : \|\beta - \bar{\beta}\|_1 \le r\}$.

Proposition 3.1 (Concentration of Lipschitz loss functions). *Suppose that $(X_1, Y_1), \ldots, (X_n, Y_n)$ are independent observations. Assume that $\rho(y, \eta)$ is Lipschitz-L in η, and the covariate components are bounded: $\max_{j=1,\ldots,p}|X_{ij}| \le C_0$ for $i = 1, \ldots, n$ with a constant $C_0 > 0$. For fixed $r > 0$ and $\epsilon \in (0, 1/2)$, we have with probability at least $1 - 2\epsilon$,*

$$\frac{Z_n}{n} = \sup_{\|\beta-\bar{\beta}\|_1 \le r} |D_n(\beta, \bar{\beta})| \le 8LC_0 r \cdot \sqrt{\frac{\log(p/\epsilon)}{n}},$$

where D_n is defined in (3.46). The result is also valid when $(X_1, \ldots, X_n)$ are fixed and $(Y_1, \ldots, Y_n)$ are independent.

Proof. We proceed in two steps as outlined at the beginning of Section 3.4. First, for β satisfying $\|\beta - \bar{\beta}\|_1 \le r$,

$$\begin{aligned} |h(X_i, Y_i; \beta)| &= |\rho(Y_i, X_i^{\mathrm{T}}\beta) - \rho(Y_i, X_i^{\mathrm{T}}\bar{\beta})| \\ &\le L|X_i^{\mathrm{T}}\beta - X_i^{\mathrm{T}}\bar{\beta}| \\ &\le LC_0\|\beta - \bar{\beta}\|_1 \le LC_0 r, \end{aligned}$$

where the first inequality follows from the Lipschitz-L property of $\rho(y, \eta)$ in η, and the second inequality follows from the Hölder inequality. By the functional Hoeffding inequality (Theorem 3.11),

$$P(Z_n - E(Z_n) \ge t) \le \exp\left(-\frac{t^2}{n(2L^2C_0^2r^2)}\right), \quad t \ge 0.$$

Setting the right-hand side to 2ϵ and solving for t shows that with probability at least $1-2\epsilon$,

$$\frac{Z_n}{n} \leq E(\frac{Z_n}{n}) + LC_0 r\sqrt{\frac{2\log(1/(2\epsilon))}{n}}. \tag{3.47}$$

Second, by the symmetrization inequality (3.31b),

$$E(Z_n) \leq 2E \sup_{\|\beta-\bar\beta\|_1\leq r} \left|\sum_{i=1}^n \zeta_i(\rho(Y_i, X_i^{\mathrm{T}}\beta) - \rho(Y_i, X_i^{\mathrm{T}}\bar\beta))\right|,$$

where $\zeta_1,\ldots,\zeta_n$ are IID Rademacher variables, independent of $(X_1,Y_1),\ldots,(X_n,Y_n)$. Moreover,

$$\begin{aligned}
&E \sup_{\|\beta-\bar\beta\|_1\leq r} \left|\sum_{i=1}^n \zeta_i(\rho(Y_i, X_i^{\mathrm{T}}\beta) - \rho(Y_i, X_i^{\mathrm{T}}\bar\beta))\right| \\
&\leq 2L\cdot E \sup_{\|\beta-\bar\beta\|_1\leq r} \left|\sum_{i=1}^n \zeta_i(X_i^{\mathrm{T}}\beta - X_i^{\mathrm{T}}\bar\beta)\right| \\
&\leq 2Lr\cdot E \max_{j=1,\ldots,p} \left|\sum_{i=1}^n \zeta_i X_{ij}\right|,
\end{aligned}$$

where the first step follows from the law of iterated expectations and the contraction inequality (3.44) with $\phi_i(\eta) = \rho(Y_i,\eta)$ which is Lipschitz-L, and the second step follows from the Hölder inequality, $|\sum_{i=1}^n \zeta_i(X_i^{\mathrm{T}}\beta - X_i^{\mathrm{T}}\bar\beta)| \leq \{\max_{j=1,\ldots,p}|\sum_{i=1}^n \zeta_i X_{ij}|\}\|\beta-\bar\beta\|_1$. For each j, the variables $\zeta_1X_{1j},\ldots,\zeta_nX_{nj}$ are mean-zero by independence of ζ_i and X_{ij} and $E(\zeta_i)=0$, and are sub-Gaussian with scale parameter C_0 by the boundedness of X_{ij}. By Theorem 3.5 (Sub-Gaussian maximum),

$$E \sup_{j=1,\ldots,p} \left|\sum_{i=1}^n \zeta_i X_{ij}\right| \leq \sqrt{nC_0^2}\sqrt{2\log(2p)}.$$

[Exercise: Explain that the preceding two displays hold with all expectations replaced by conditional expectations given $(X_1,Y_1),\ldots,(X_n,Y_n)$.] From the preceding three displays, we obtain

$$E(\frac{Z_n}{n}) \leq 4LC_0 r\sqrt{\frac{2\log(2p)}{n}}. \tag{3.48}$$

Finally, combining (3.47) and (3.48) shows that with probability at least $1-2\epsilon$,

$$Z_n \leq LC_0 r\left\{4\sqrt{\frac{2\log(2p)}{n}} + \sqrt{\frac{2\log(1/(2\epsilon))}{n}}\right\}$$

$$\le LC_0 r \left\{ 4\sqrt{\frac{2\log(2p)}{n}} + 4\sqrt{\frac{2\log(1/(2\epsilon))}{n}} \right\}$$

$$\le LC_0 r \cdot 8\sqrt{\frac{\log(p/\epsilon)}{n}},$$

where the last step uses $\sqrt{a} + \sqrt{b} \le \sqrt{2(a+b)}$. □

To connect with the uniform law of large numbers (Theorem 2.5), the following result can be deduced from Proposition 3.1 for Lipschitz loss functions over any bounded parameter set (measured by L_1 norm). The parameter dimension p may increase to ∞ (with $\bar\beta$ varying along a sequence) as $n \to \infty$. The proof is left as Exercise 3.11.

Corollary 3.4 (ULLN for Lipschitz loss functions). *In the setting of Proposition 3.1, assume further that $(X_1, Y_1), \ldots, (X_n, Y_n)$ are IID and $E|\rho(Y_1, X_1^{\mathrm{T}}\bar\beta)| < \infty$. Show that if $\log(p)/n \to 0$ as $n \to \infty$, then for any $r > 0$,*

$$\sup_{\|\beta - \bar\beta\|_1 \le r} \left| \frac{1}{n} \sum_{i=1}^{n} (\rho(Y_i, X_i^{\mathrm{T}}\beta) - E\rho(Y_i, X_i^{\mathrm{T}}\beta)) \right| \xrightarrow{P} 0, \quad \text{as } n \to \infty.$$

3.5 Appendices

3.5.1 Moment and cumulant generating functions

For a random variable X, its moment generating function (MGF) is defined as $M(t) = E(\mathrm{e}^{tX})$, where the dependency on X is suppressed in the notation (e.g., Casella & Berger, 2001, Theorem 2.3.7).

Proposition 3.2 (Expansion of MGF). *For a random variable X, suppose that $M(t)$ is finite in a (non-empty) neighborhood $\mathcal{N}$ of 0. Then in the neighborhood $\mathcal{N}$,*

$$M(t) = 1 + \sum_{m=1}^{\infty} \frac{E(X^m)}{m!} t^m.$$

Moreover, $\frac{\mathrm{d}^m}{\mathrm{d}t^m} M(t)\Big|_{t=0} = E(X^m)$ for any $m \ge 1$. In particular,

$$M'(0) = \frac{\mathrm{d}}{\mathrm{d}t} M(t)\Big|_{t=0} = E(X), \quad M''(0) = \frac{\mathrm{d}^2}{\mathrm{d}t^2} M(t)\Big|_{t=0} = E(X^2).$$

The cumulant generating function (CGF) is defined as $K(t) = \log M(t)$. In the setting of Proposition 3.2, $K(t)$ also satisfies a power series expansion:

$$K(t) = \sum_{m=1}^{\infty} \frac{\kappa_m(X)}{m!} t^m,$$

where $\kappa_m(X) = \frac{\mathrm{d}^m}{\mathrm{d}t^m} K(t)|_{t=0}$ is called the mth cumulant of X. In particular, the first two cumulants are

$$\kappa_1(X) = K'(0) = E(X), \quad \kappa_2(X) = K''(0) = \mathrm{var}(X). \tag{3.49}$$

[Exercise: Prove these.] In contrast with $M''(0) = E(X^2)$, the second cumulant $\kappa_2(X) = K''(0)$ gives the variance (or second centered moment) of X.

As an extension of (3.49), the derivatives (in various orders) of $K(t)$ at $t \in \mathcal{N}$ (possibly $t \neq 0$) can be related to the cumulants of an exponential tilt distribution with density function

$$f_t(x) = \frac{\mathrm{e}^{tx}}{M(t)} f_0(x) \propto \mathrm{e}^{tx} f_0(x),$$

where $f_0(\cdot)$ denotes the density function of X. The function $\mathrm{e}^{tx} f_0(x)$ represents an un-normalized density and $M(t)$ serves as the normalizing constant such that $\int f_t(x)\,\mathrm{d}x = 1$. In general, without requiring a density function f_0, the exponential tilt distribution, with CDF $F_t(\cdot)$, is defined such that

$$\frac{\mathrm{d}F_t}{\mathrm{d}F_0}(x) = \frac{\mathrm{e}^{tx}}{M(t)} \propto \mathrm{e}^{tx},$$

where $F_0(\cdot)$ denotes the CDF of X.

Proposition 3.3 (Derivatives of CGF). *In the setting of Proposition 3.2, for any $t \in \mathcal{N}$ and $m \geq 1$,*

$$\frac{\mathrm{d}^m}{\mathrm{d}t^m} K(t) = \kappa_m(X_t),$$

where X_t is a random variable with density function equal to $f_t(\cdot)$ or CDF equal to $F_t(\cdot)$. In particular,

$$K'(t) = E(X_t), \quad K''(t) = \mathrm{var}(X_t). \tag{3.50}$$

Proof. Fix $t \in \mathcal{N}$. For any s such that $t + s \in \mathcal{N}$,

$$M(t+s) = \int \mathrm{e}^{(t+s)x} f_0(x)\,\mathrm{d}x = M(t) \int \mathrm{e}^{sx} f_t(x)\,\mathrm{d}x = M(t) E(\mathrm{e}^{sX_t}),$$

or equivalently

$$K(t+s) = K(t) + \log E(\mathrm{e}^{sX_t}).$$

A similar calculation holds using the CDFs $F_0(\cdot)$ and $F_t(\cdot)$. For $m \geq 1$, taking mth derivatives in s and evaluate at $s = 0$ gives

$$\frac{\mathrm{d}^m}{\mathrm{d}t^m} K(t) = \frac{\mathrm{d}^m}{\mathrm{d}s^m} K(t+s)\Big|_{s=0} = \frac{\mathrm{d}^m}{\mathrm{d}t^m} \log E(\mathrm{e}^{sX_t})\Big|_{s=0} = \kappa_m(X_t),$$

where the first equality holds by the chain rule, and the third equality holds by the definition of $\kappa_m(\cdot)$, with $\log E(e^{sX_t})$ being the cumulant generating function of X_t. □

For pedagogical purposes, we also give a direct proof of $K'(t) = E(X_t)$ in (3.50). The corresponding proof of $K''(t) = \text{var}(X_t)$ is left as Exercise 3.2. In fact,

$$\begin{aligned}\frac{\mathrm{d}}{\mathrm{d}t}K(t) &= \frac{\mathrm{d}}{\mathrm{d}t}\log E(e^{tX}) = \frac{\frac{\mathrm{d}}{\mathrm{d}t}E(e^{tX})}{E(e^{tX})} \\ &= \frac{E(Xe^{tX})}{E(e^{tX})} = \frac{\int xe^{tx}f_0(x)\,\mathrm{d}x}{M(t)} = E(X_t).\end{aligned}$$

The third equality follows because $\frac{\mathrm{d}}{\mathrm{d}t}E(e^{tX}) = E(Xe^{tX})$ by exchanging differentiation and expectation. The exchange can be justified using Proposition 2.1 [Exercise].

3.5.2 Proof of contraction inequality

The following result gives generalized contraction inequalities, for Lipschitz functions ϕ_i satisfying $\phi_i(0) = 0$, $i = 1, \ldots, n$, and a convex, non-decreasing function Ψ. Taking $\Psi(u) = u$ in (3.51) and (3.52) lead to (3.38) and (3.39) respectively. [As noted earlier, (3.38) also holds without requiring $\phi_i(0) = 0$.] The absolute version (3.39) does not directly follow from the raw version (3.38). However, by allowing general convex, non-decreasing Ψ, the absolute version (3.52) is proved below using the raw version (3.51).

Theorem 3.16 (Generalized contraction inequality). *Let $\mathcal{T}_n$ be a subset of $\mathbb{R}^n$. Let $\phi_i : \mathbb{R} \to \mathbb{R}$ be a Lipschitz-1 function such that $\phi_i(0) = 0$ for $i = 1, \ldots, n$, and $\Psi : \mathbb{R} \to \mathbb{R}$ be a convex, non-decreasing function. Then for IID Rademacher random variables $\zeta_1, \ldots, \zeta_n$, we have*

$$E\Psi\left[\sup_{t\in\mathcal{T}_n}\left(\sum_{i=1}^n \zeta_i\phi_i(t_i)\right)\right] \le E\Psi\left[\sup_{t\in\mathcal{T}_n}\left(\sum_{i=1}^n \zeta_i t_i\right)\right], \tag{3.51}$$

and

$$E\Psi\left[\frac{1}{2}\sup_{t\in\mathcal{T}_n}\left|\sum_{i=1}^n \zeta_i\phi_i(t_i)\right|\right] \le E\Psi\left[\sup_{t\in\mathcal{T}_n}\left|\sum_{i=1}^n \zeta_i t_i\right|\right]. \tag{3.52}$$

Proof. To prove inequality (3.51), we show that for any function $a : \mathcal{T}_n \to \mathbb{R}$,

$$E\Psi\left[\sup_{t\in\mathcal{T}_n}\left\{a(t) + \sum_{i=1}^n \zeta_i\phi_i(t_i)\right\}\right] \le E\Psi\left[\sup_{t\in\mathcal{T}_n}\left\{a(t) + \sum_{i=1}^n \zeta_i t_i\right\}\right], \tag{3.53}$$

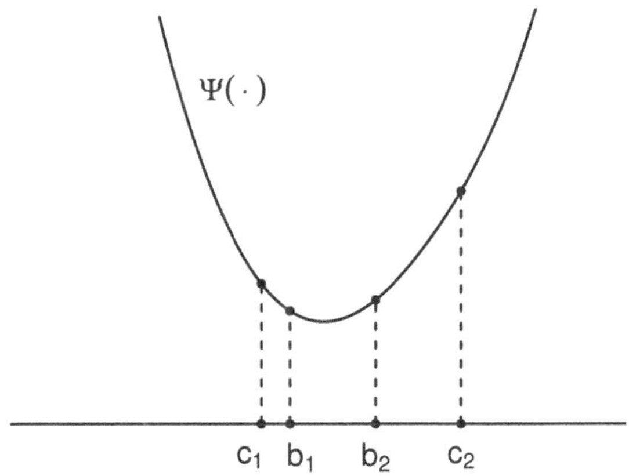

FIGURE 3.2
Illustration of a convex function at four points as in Lemma 3.6.

which is stronger than (3.51). The proof is by induction. The key is to show the case of $n = 1$, i.e., for $\mathcal{T}_1 \subset \mathbb{R}$, a Lipschitz function ϕ satisfying $\phi(0) = 0$, and a Rademacher random variable ζ,

$$E\Psi \left[\sup_{t \in \mathcal{T}_1} \{a(t) + \zeta\phi(t)\} \right] \leq E\Psi \left[\sup_{t \in \mathcal{T}_1} \{a(t) + \zeta t\} \right],$$

or equivalently, after evaluating the expectation,

$$\begin{aligned} &\Psi \left[\sup_{t \in \mathcal{T}_1} \{a(t) + \phi(t)\} \right] + \Psi \left[\sup_{t \in \mathcal{T}_1} \{a(t) - \phi(t)\} \right] \\ &\leq \Psi \left[\sup_{t \in \mathcal{T}_1} \{a(t) + t\} \right] + \Psi \left[\sup_{t \in \mathcal{T}_1} \{a(t) - t\} \right]. \end{aligned} \tag{3.54}$$

We use the following lemma, whose proof is left as Exercise 3.10. A similar result is stated in the proof of Theorem 11.6 in Boucheron et al. (2013), but under a more restrictive assumption that Ψ is convex and non-decreasing.

Lemma 3.6 (Convex function at four points). *Let $\Psi : \mathbb{R} \to \mathbb{R}$ be a convex function. Suppose that b_1, b_2, c_1, c_2 satisfy $c_1 \leq b_1 \leq b_2 \leq c_2$ [the two numbers c_1 and c_2 sandwich b_1 and b_2] and $b_1 + b_2 \leq c_1 + c_2$. Then $\Psi(b_1) + \Psi(b_2) \leq \Psi(c_1) + \Psi(c_2)$. See Figure 3.2.*

To prove (3.54), denote by $\hat{s}$ and $\hat{t}$ the elements of $\mathcal{T}_1$ which achieve the two suprema respectively on the left-hand side. It suffices to show that

$$\Psi\{a(\hat{s}) + \phi(\hat{s})\} + \Psi\{a(\hat{t}) - \phi(\hat{t})\} \leq \Psi\{a(\hat{s}) + \hat{s}\} + \Psi\{a(\hat{t}) - \hat{t}\}. \tag{3.55}$$

By the definition of $\hat{s}$ and $\hat{t}$,

$$a(\hat{s}) + \phi(\hat{s}) \geq a(\hat{t}) + \phi(\hat{t}),$$
$$a(\hat{t}) - \phi(\hat{t}) \geq a(\hat{s}) - \phi(\hat{s}).$$

which imply

$$\phi(\hat{s}) - \phi(\hat{t}) \geq a(\hat{s}) - a(\hat{t}) \geq \phi(\hat{t}) - \phi(\hat{s}).$$

and hence, using the Lipschitz-1 property of ϕ,

$$|a(\hat{s}) - a(\hat{t})| \leq \phi(\hat{s}) - \phi(\hat{t}) \leq |\hat{s} - \hat{t}|.$$

Without loss of generality, assume that $\hat{s} \geq \hat{t}$. Otherwise, we proceed with ϕ replaced by $-\phi$. Then the above display gives $\phi(\hat{s}) - \phi(\hat{t}) \leq \hat{s} - \hat{t}$ and

$$\{a(\hat{s}) + \phi(\hat{s})\} + \{a(\hat{t}) - \phi(\hat{t})\} \leq \{a(\hat{s}) + \hat{s}\} + \{a(\hat{t}) - \hat{t}\}. \tag{3.56}$$

[From this, we obtain an alternative proof of inequality (3.41).] Consider three cases, depending on how $(\hat{s}, \hat{t})$ are compared with 0. Note that the Lipschitz-1 property and $\phi(0) = 0$ imply that $|\phi(t)| \leq |t|$ for $t \in \mathbb{R}$.

(i) $\hat{s} \geq \hat{t} \geq 0$. Then by the property $|\phi(t)| \leq |t|$,

$$\underbrace{a(\hat{s}) + \hat{s}}_{c_2} \geq a(\hat{s}) + \phi(\hat{s}) \geq a(\hat{t}) + \phi(\hat{t}) \geq \underbrace{a(\hat{t}) - \hat{t}}_{c_1},$$
$$a(\hat{t}) - \phi(\hat{t}) \geq a(\hat{t}) - \hat{t}.$$

If $a(\hat{t}) - \phi(\hat{t}) > a(\hat{s}) + \hat{s}$, then this and $a(\hat{s}) + \phi(\hat{s}) \geq a(\hat{t}) - \hat{t}$ from above yield $\{a(\hat{s}) + \phi(\hat{s})\} + \{a(\hat{t}) - \phi(\hat{t})\} > \{a(\hat{s}) + \hat{s}\} + \{a(\hat{t}) - \hat{t}\}$, contradicting (3.56). Therefore $a(\hat{s}) + \hat{s} \geq a(\hat{t}) - \phi(\hat{t})$, which together with the above display implies that the two numbers $a(\hat{s}) + \hat{s}$ and $a(\hat{t}) - \hat{t}$ (taken to be c_2 and c_1 respectively) sandwish $a(\hat{s}) + \phi(\hat{s})$ and $a(\hat{t}) - \phi(\hat{t})$ (taken to be b_1 and b_2 in either order). Applying Lemma 3.6 gives (3.55).

(ii) $\hat{s} \geq 0 \geq \hat{t}$. Then by the property $|\phi(t)| \leq |t|$,

$$a(\hat{s}) + \hat{s} \geq a(\hat{s}) + \phi(\hat{s}), \quad a(\hat{t}) - \hat{t} \geq a(\hat{t}) - \phi(\hat{t}).$$

Then (3.55) follows directly from the non-decreasingness of Ψ.

(iii) $0 \geq \hat{s} \geq \hat{t}$. We proceed with $\phi(t)$ redefined as $\phi(-t)$. Then (3.55) follows from the proof in case (i).

This concludes the proof of (3.54). The induction from $n-1$ to n for proving (3.53) can be completed similarly as in the proof of (3.40) (Exercise 3.10).

To prove inequality (3.52), we use Jensen's inequality by convexity of Ψ:

$$E\Psi\left[\frac{1}{2}\sup_{t \in \mathcal{T}_n}\left|\sum_{i=1}^{n} \zeta_i \phi_i(t_i)\right|\right]$$

$$
\begin{aligned}
&= E\Psi\left[\frac{1}{2}\sup_{t\in\mathcal{T}_n}\left(\sum_{i=1}^n \zeta_i\phi_i(t_i)\right)_+ + \frac{1}{2}\sup_{t\in\mathcal{T}_n}\left(\sum_{i=1}^n (-\zeta_i)\phi_i(t_i)\right)_+\right] \\
&\le \frac{1}{2}E\Psi\left[\sup_{t\in\mathcal{T}_n}\left(\sum_{i=1}^n \zeta_i\phi_i(t_i)\right)_+\right] + \frac{1}{2}E\Psi\left[\sup_{t\in\mathcal{T}_n}\left(\sum_{i=1}^n (-\zeta_i)\phi_i(t_i)\right)_+\right] \\
&= E\Psi\left[\sup_{t\in\mathcal{T}_n}\left(\sum_{i=1}^n \zeta_i\phi_i(t_i)\right)_+\right],
\end{aligned}
$$

where $(c)_+ = \max(c, 0)$ and the last step uses the symmetry between $(\zeta_1, \dots, \zeta_n)$ and $(-\zeta_1, \dots, -\zeta_n)$. Then (3.52) can be obtained as follows, continuing from the preceding display:

$$
\begin{aligned}
& E\Psi\left[\sup_{t\in\mathcal{T}_n}\left(\sum_{i=1}^n \zeta_i\phi_i(t_i)\right)_+\right] = E\Psi\left[\left(\sup_{t\in\mathcal{T}_n}\sum_{i=1}^n \zeta_i\phi_i(t_i)\right)_+\right] \\
&\le E\Psi\left[\left(\sup_{t\in\mathcal{T}_n}\sum_{i=1}^n \zeta_i t_i\right)_+\right] = E\Psi\left[\sup_{t\in\mathcal{T}_n}\left(\sum_{i=1}^n \zeta_i t_i\right)_+\right] \\
&\le E\Psi\left[\sup_{t\in\mathcal{T}_n}\left|\sum_{i=1}^n \zeta_i t_i\right|\right],
\end{aligned}
$$

by exchanging sup and $(\cdot)_+$ and noting $\Psi((\cdot)_+)$ is convex and non-decreasing and applying (3.51). $\square$

Exercises

Ex 3.1.

(i) For a nonnegative random variable X, show that if $E(X)$ is finite, then

$$t\,P(X > t) \to 0 \text{ as } t \to \infty.$$

[Hint: Use $tP(X > t) \le E(X1\{X > t\})$.]

(ii) For a random variable Z, show that if $\mathrm{var}(Z)$ as well as $E(Z)$ is finite, then

$$t^2\,P(|Z - E(Z)| > t) \to 0 \text{ as } t \to \infty.$$

(iii) What is the implication of the results (i) and (ii) on the Markov inequality and Chebyshev inequality?

Ex 3.2. For a random variable X with density function $f_0(X)$, let $K(t) = \log E(e^{tX})$. Show *directly* (instead of calling Proposition 3.3) that under suitable conditions,

$$K''(t) = \mathrm{var}(X_t),$$

where X_t is a random variable with density function $f_t(x) \propto e^{tx} f_0(x)$. Please indicate the conditions you would require.

Ex 3.3. Let X be a random variable satisfying $a \le X \le b$ for some constants $a < b$.

(i) Show that $\mathrm{var}(X) \le (b-a)^2/4$.

(ii) Is the inequality in (i) tight? Justify your answer.

Ex 3.4. For a nonnegative random variable X (continuous or discrete), show that

$$E(X) = \int_0^\infty P(X > x)\mathrm{d}x = \int_0^\infty P(X \ge x)\mathrm{d}x,$$

where the integrals are Lebesgue integrals.

Ex 3.5. The following results are used in the proof of Theorem 3.6 (Equivalent characterizations of Sub-Gaussian).

(i) Show that $\Gamma(x+1) \le \sqrt{2}x^x$ for any real number $x \ge \frac{1}{2}$.

(ii) Show that $m! \ge (m/e)^m$ for any integer $m \ge 1$.

Ex 3.6. Complete the proof of *(iii)* $\Rightarrow$ *(i)* in Theorem 3.6: For a random variable X with $E(X) = 0$, if there exists a constant $c_3 > 0$ such that $E(|X|^p)^{1/p} \le c_3 p^{1/2}$ for $p \ge 1$, then $E(e^{\lambda X}) \le e^{\frac{c_1^2\lambda^2}{2}}$ for $\lambda \in \mathbb{R}$, where $c_1 = 2ec_3$. You may also pick c_1 as another constant within an absolute constant factor of c_3.

Ex 3.7. Provide a proof of Lemma 3.3 (Sub-Gaussian tails beyond a threshold).

Ex 3.8 (Sub-Exponential maximum). Suppose that $X_1, \ldots, X_n$ (possibly non-independent) satisfy

$$E(e^{\lambda X_i}) \le e^{\frac{\nu^2\lambda^2}{2}}, \quad 0 < \lambda < \frac{1}{\rho},$$

for constants $\nu > 0$ and $\rho > 0$ and $i = 1, \ldots, n$. Show that

$$E(\max_{i=1,\ldots,n} X_i) \le \nu\sqrt{2\log n} + 2\rho \log n.$$

[Hint: Derive (3.9) with σ replaced by ν, and choose λ depending on $\sqrt{\frac{2\log n}{\nu^2}} \leq \frac{1}{2\rho}$ or $\sqrt{\frac{2\log n}{\nu^2}} > \frac{1}{2\rho}$. Be careful with the case of $n = 1$.]

Ex 3.9. The following questions are related to the proof of Theorem 3.14 (Symmetrization inequality).

(i) Show that the vector with components $h(X_i) - h(\tilde{X}_i)$ has the same joint distribution as the vector with components $\epsilon_i(h(X_i) - h(\tilde{X}_i))$. [Hint: How is the independence of $X_1, \ldots, X_n$ used?]

(ii) Provide a proof of the lower bound (3.31a) in Theorem 3.14.

Ex 3.10. The following questions are related to the proof of Theorem 3.16 (Generalized contraction inequality).

(i) Provide a proof of Lemma 3.6 (Convex function at four points).

(ii) Complete the induction from $n - 1$ to n for proving inequality (3.53) in Theorem 3.16.

Ex 3.11. Provide a proof of Corollary 3.4 (ULLN for Lipschitz loss functions).

4

High-dimensional linear regression

In this chapter, we study regularized least-squares estimation in high-dimensional, sparse linear regression, where the number of covariates is close to or greater than the sample size, and a small but unknown subset of covariates are associated with nonzero coefficients. We focus on Lasso-penalized estimation and discuss both slow and fast rates for error bounds, in fixed or random design, with correct model specification or with possible misspecification. In addition, we also discuss error bounds for the Dantzig selector in the high-dimensional, sparse setting. The finite-sample error bounds, which are established with specific probabilities, can be converted to O_p statements indicating the asymptotic order.

We study statistical properties of regularized least-squares estimation in high-dimensional linear regression, where the number of covariates (or associated coefficients) is close to or greater than the sample size. We focus on Lasso estimation (Tibshirani, 1996; Hastie et al., 2015), and discuss finite-sample (or non-asymptotic) error bounds (including slow and fast rates), in fixed or random design, with correct model specification or with possible misspecification. We also discuss high-dimensional analysis of the Dantzig selector (Candes & Tao, 2007), which is closely related to that of Lasso estimation (Bickel et al., 2009). The broader subject of high-dimensional statistics, including but not restricted to regression models, is covered by many books (e.g., Bühlmann & van de Geer, 2011; Wainwright, 2019).

4.1 Introduction

Let (X_i, Y_i), $i = 1, \ldots, n$, be a collection of n observations, where $X_i = (X_{i1}, \ldots, X_{ip})^{\mathrm{T}}$ is a $p \times 1$ covariate vector and Y_i is a response variable.

DOI: 10.1201/9781003680734-4

Consider a linear regression model:

$$Y_i = X_i^{\mathrm{T}}\beta^* + \varepsilon_i, \quad i = 1, \dots, n, \tag{4.1}$$

where $\beta^* = (\beta_1^*, \dots, \beta_p^*)^{\mathrm{T}}$ is an unknown coefficient vector and ε_i is a noise variable with mean zero given X_i. In the matrix notation, we write

$$\mathbf{Y} = \begin{pmatrix} Y_1 \\ \vdots \\ Y_n \end{pmatrix}, \quad \mathbf{X} = \begin{pmatrix} X_1^{\mathrm{T}} \\ \vdots \\ X_n^{\mathrm{T}} \end{pmatrix}, \quad \varepsilon = \begin{pmatrix} \varepsilon_1 \\ \vdots \\ \varepsilon_n \end{pmatrix}.$$

The least-squares (LS) estimator is

$$\hat{\beta}_{\mathrm{LS}} = \min_{\beta \in \mathbb{R}^p} \|\mathbf{Y} - \mathbf{X}\beta\|_2^2 = (\mathbf{X}^{\mathrm{T}}\mathbf{X})^{-1}(\mathbf{X}^{\mathrm{T}}\mathbf{Y}),$$

where $\|\cdot\|_2$ denotes the L_2 norm, and the second equality holds provided $\mathbf{X}^{\mathrm{T}}\mathbf{X}$ is non-singular.

As a motivation for Lasso estimation, it is helpful to understand how the prediction performance using the LS estimator depends on the relative magnitudes of n and p. For simplicity, assume that linear regression (4.1) is correctly specified with homoscedastic noises in a fixed design, i.e., $X_1, \dots, X_n$ are fixed and $Y_1, \dots, Y_n$ are random and independent (equivalently $\varepsilon_1, \dots, \varepsilon_n$ are random and independent) with $E(\varepsilon_i) = 0$ and $\mathrm{var}(\varepsilon_i) = \sigma^{*2}$ for $i = 1, \dots, n$, where σ^* is an unknown constant. The in-sample prediction mean squared error (MSE) is defined as

$$\mathrm{MSE}_{\mathrm{in}} = n^{-1}E\|\tilde{\mathbf{Y}} - \mathbf{X}\hat{\beta}_{\mathrm{LS}}\|_2^2,$$

where $\tilde{\mathbf{Y}}$ is a new response vector, independent and identically distributed as $\mathbf{Y}$, and the expectation is taken over both $\tilde{\mathbf{Y}}$ and $\mathbf{Y}$ (through $\hat{\beta}_{\mathrm{LS}}$).

Proposition 4.1 (In-sample prediction MSE). *For linear regression with homoscedastic noises in a fixed design, the in-sample prediction MSE using the LS estimator is*

$$\mathrm{MSE}_{in} = \sigma^{*2} + \frac{p}{n}\sigma^{*2}, \tag{4.2}$$

provided that $p \le n$ and $\mathbf{X}^{\mathrm{T}}\mathbf{X}$ is non-singular.
[For linear regression with homoscedastic noises in a random design, MSE_{in} *is defined as above but with the expectation taken over* $(\mathbf{X}, \tilde{\mathbf{Y}})$ *jointly. The identity* (4.2) *is also valid provided that $p \le n$ and $\mathbf{X}^{\mathrm{T}}\mathbf{X}$ is non-singular almost surely.]*

Proposition 4.1 is a finite-sample result, being valid for any $p \le n$ (provided that $\mathbf{X}^{\mathrm{T}}\mathbf{X}$ is non-singular). There is a simple interpretation for the expression (4.2), in conjunction with (4.3)–(4.5) in the proof. The first term, σ^{*2}, is the oracle MSE that would be obtained if using the linear predictor $\mathbf{X}\beta^*$ with the true coefficient vector β^*, whereas the second term, $\frac{p}{n}\sigma^{*2}$, represents an increase of the prediction error due to the difference $\mathbf{X}\hat{\beta}_{\text{LS}} - \mathbf{X}\beta^*$. In fact, the second term

$$n^{-1}E\|\mathbf{X}\hat{\beta}_{\text{LS}} - \mathbf{X}\beta^*\|_2^2 = \frac{p}{n}\sigma^{*2}$$

is called the excess in-sample prediction MSE for the LS estimator.

As a special case, if $p = n$ and $\mathbf{X}^{\mathrm{T}}\mathbf{X}$ is non-singular, then $\mathbf{X}\hat{\beta}_{\text{LS}} = \mathbf{Y}$ and $\text{MSE}_{\text{in}} = n^{-1}E\|\tilde{\mathbf{Y}} - \mathbf{Y}\|_2^2 = n^{-1}(E\|\tilde{\mathbf{Y}} - \mathbf{X}\beta^*\|_2^2 + E\|\mathbf{Y} - \mathbf{X}\beta^*\|_2^2) = 2\sigma^{*2}$, by the independence of $\tilde{\mathbf{Y}}$ and $\mathbf{Y}$. This is in agreement with the expression (4.2) with $p = n$.

If $p = n$ and $\mathbf{X}^{\mathrm{T}}\mathbf{X}$ is non-singular or if $p > n$, then $\hat{\beta}_{\text{LS}}$ is not unique but satisfies $\mathbf{X}\hat{\beta}_{\text{LS}} = \mathbf{Y}$. In this case, it also follows from the preceding discussion that $\text{MSE}_{\text{in}} = 2\sigma^{*2}$.

Proof. By standard manipulation,

$$\begin{aligned}
&E\|\tilde{\mathbf{Y}} - \mathbf{X}\hat{\beta}_{\text{LS}}\|_2^2 = E\|\tilde{\mathbf{Y}} - \mathbf{X}\beta^* + \mathbf{X}\beta^* - \mathbf{X}\hat{\beta}_{\text{LS}}\|_2^2 \\
&= E\|\tilde{\mathbf{Y}} - \mathbf{X}\beta^*\|_2^2 + E\|\mathbf{X}\beta^* - \mathbf{X}\hat{\beta}_{\text{LS}}\|_2^2 - 2E\{(\tilde{\mathbf{Y}} - \mathbf{X}\beta^*)^{\mathrm{T}}(\mathbf{X}\hat{\beta}_{\text{LS}} - \mathbf{X}\beta^*)\} \\
&= E\|\tilde{\mathbf{Y}} - \mathbf{X}\beta^*\|_2^2 + E\|\mathbf{X}\hat{\beta}_{\text{LS}} - \mathbf{X}\beta^*\|_2^2. \qquad (4.3)
\end{aligned}$$

The last step follows from the fact that

$$E\{(\tilde{\mathbf{Y}} - \mathbf{X}\beta^*)^{\mathrm{T}}(\mathbf{X}\hat{\beta} - \mathbf{X}\beta^*)\} = E(\tilde{\mathbf{Y}} - \mathbf{X}\beta^*)^{\mathrm{T}}E(\mathbf{X}\hat{\beta} - \mathbf{X}\beta^*) = 0$$

because by the independence of $\tilde{\mathbf{Y}}$ and $\mathbf{Y}$, $\tilde{\mathbf{Y}} - \mathbf{X}\beta^*$, a function of $\tilde{\mathbf{Y}}$, and $\mathbf{X}\hat{\beta} - \mathbf{X}\beta^*$, a function of $\mathbf{Y}$, are independent, and $E(\tilde{\mathbf{Y}} - \mathbf{X}\beta^*) = 0$. The first term in (4.3) reduces to

$$E\|\tilde{\mathbf{Y}} - \mathbf{X}\beta^*\|_2^2 = n\sigma^{*2} \qquad (4.4)$$

due to homoscedastic noises. Note that $\mathbf{X}\hat{\beta}_{\text{LS}} - \mathbf{X}\beta^* = \mathbf{X}(\mathbf{X}^{\mathrm{T}}\mathbf{X})^{-1}\mathbf{X}^{\mathrm{T}}\varepsilon$ and $\|\mathbf{X}\hat{\beta}_{\text{LS}} - \mathbf{X}\beta^*\|_2^2 = \varepsilon^{\mathrm{T}}\mathbf{X}(\mathbf{X}^{\mathrm{T}}\mathbf{X})^{-1}\mathbf{X}^{\mathrm{T}}\varepsilon = \text{tr}(\mathbf{X}(\mathbf{X}^{\mathrm{T}}\mathbf{X})^{-1}\mathbf{X}^{\mathrm{T}}\varepsilon\varepsilon^{\mathrm{T}})$, using the fact that $\text{tr}(AB) = \text{tr}(BA)$ for two matrices A and B with appropriate dimensions. Then the second term $E\|\mathbf{X}\hat{\beta}_{\text{LS}} - \mathbf{X}\beta^*\|_2^2$ can be further calculated as

$$\begin{aligned}
&E\|\mathbf{X}\hat{\beta}_{\text{LS}} - \mathbf{X}\beta^*\|_2^2 = \text{tr}\{\mathbf{X}(\mathbf{X}^{\mathrm{T}}\mathbf{X})^{-1}\mathbf{X}^{\mathrm{T}}E(\varepsilon\varepsilon^{\mathrm{T}})\} \\
&= \sigma^{*2}\text{tr}\{\mathbf{X}(\mathbf{X}^{\mathrm{T}}\mathbf{X})^{-1}\mathbf{X}^{\mathrm{T}}\} = \sigma^{*2}\text{tr}\{(\mathbf{X}^{\mathrm{T}}\mathbf{X})^{-1}\mathbf{X}^{\mathrm{T}}\mathbf{X}\} = p\sigma^{*2}, \qquad (4.5)
\end{aligned}$$

where the second equality follows because $E(\varepsilon\varepsilon^{\mathrm{T}}) = \sigma^{*2}I$, and the third equality follows from $\text{tr}(AB) = \text{tr}(BA)$. Combining (4.3)–(4.5) shows that $E\|\tilde{\mathbf{Y}} - \mathbf{X}\hat{\beta}_{\text{LS}}\|_2^2 = n\sigma^{*2} + p\sigma^{*2}$. □

Incidentally, the in-sample prediction MSE is related to the training MSE defined as

$$\text{MSE}_{\text{tr}} = n^{-1}E\|\mathbf{Y} - \mathbf{X}\hat{\beta}_{\text{LS}}\|_2^2,$$

where the predictor error is computed using the "training" response vector $\mathbf{Y}$. By similar calculation as in the proceeding proof, it can be shown that $\text{MSE}_{\text{tr}} = \sigma^{*2} - \frac{p}{n}\sigma^{*2}$, provided that $p \le n$ and $\mathbf{X}^{\text{T}}\mathbf{X}$ is non-singular [Exercise]. Hence

$$\text{MSE}_{\text{in}} = \text{MSE}_{\text{tr}} + 2\frac{p}{n}\sigma^{*2}.$$

Replacing σ^{*2} by an estimator leads to the C_p statistic (Mallows, 1973).

From Proposition 4.1, the prediction performance using the LS estimator quickly deteriorates as p becomes close to or exceeds n. To tackle this issue, it seems a natural approach to incorporate variable selection (i.e., perform linear regression using only a subset of covariates which are deemed important) if p is close to or even greater than n. Underlying this approach is a sparsity assumption that only a small but unknown subset (relative to the sample size) of covariates are associated with nonzero coefficients, i.e.,

$$S_{\beta^*} = \{1 \le j \le p : \beta_j^* \ne 0\} \text{ is of a small size relative to } p.$$

Using the LS estimator based on only the covariates $\{X_{ij} : j \in S_{\beta^*}\}$ would lead to an excess in-sample prediction MSE equal to $\frac{|S_{\beta^*}|}{n}\sigma^{*2}$, which can be considered an oracle excess MSE (that would be obtained if S_{β^*} were known), where $|S|$ denotes the size of a set S. Lasso estimation can be shown to be computationally tractable (implementable by convex optimization) and statistically provable to achieve near the oracle excess MSE under certain conditions.

Lasso estimation is a method of regularized estimation using the L_1 norm of the parameter vector. There are two formally different versions of Lasso estimation. The L_1 restricted LS estimator is defined as

$$\begin{aligned} &\min_{\beta \in \mathbb{R}^p} \ \|\mathbf{Y} - \mathbf{X}\beta\|_2^2 \\ &\text{subject to} \quad \|\beta\|_1 \le \rho, \end{aligned}$$

where $\rho > 0$ is a tuning parameter and $\|\cdot\|_1$ denotes the L_1 norm. The L_1 penalized LS estimator is defined as

$$\min_{\beta \in \mathbb{R}^p} \left\{ \|\mathbf{Y} - \mathbf{X}\beta\|_2^2 + \lambda \|\beta\|_1 \right\}, \tag{4.6}$$

where $\lambda > 0$ is a tuning parameter. The two versions of Lasso estimation achieve similar statistical properties. We mainly study the version of L_1 penalized estimation.

For convenience, we define the Lasso estimator as

$$\hat{\beta}_\lambda = \text{argmin}_{\beta \in \mathbb{R}^p} \left\{ \frac{1}{2}\|\mathbf{Y} - \mathbf{X}\beta\|_n^2 + \lambda \|\beta\|_1 \right\}, \tag{4.7}$$

where $\lambda > 0$ is a tuning parameter and $\|u\|_n = (\sum_{i=1}^n u_i^2)^{1/2}$, called the empirical L_2 norm, for a vector $u = (u_1, \ldots, u_n)^{\text{T}}$, i.e., $\|u\|_n^2 = \|u\|_2^2/n$. The estimator $\hat{\beta}_\lambda$ corresponds to that form (4.6) with a properly re-scaled λ. The stationary condition for $\hat{\beta}_\lambda$ gives

$$\langle \mathbf{Y} - \mathbf{X}\hat{\beta}_\lambda, \mathbf{X}_j \rangle_n = \begin{cases} \lambda, & \text{if } \hat{\beta}_{\lambda,j} > 0, \\ -\lambda, & \text{if } \hat{\beta}_{\lambda,j} < 0, \\ c_j, & \text{if } \hat{\beta}_{\lambda,j} = 0, \end{cases} \tag{4.8}$$

where c_j is some value in $[-\lambda, \lambda]$, and $\langle u, v \rangle_n = u^{\mathrm{T}} v/n$, called the empirical inner product, for two (column) vectors $u, v \in \mathbb{R}^n$.

4.2 Lasso estimation with correct model specification

4.2.1 Fixed-design linear regression

We study Lasso estimation for linear regression, assumed to be correctly specified in a fixed design. For theoretical analysis, consider the following conditions:

(i) [Linear regression in fixed design] Linear regression (4.1) is correctly specified in a fixed design, i.e., $X_1, \ldots, X_n$ are fixed and $\varepsilon_1, \ldots, \varepsilon_n$ are independent with mean zero;

(ii) [Sub-Gaussian noises] $\varepsilon_1, \ldots, \varepsilon_n$ are sub-Gaussian with scale parameter σ;

(iii) [Normalized covariates] $\mathbf{X}_j^{\mathrm{T}} \mathbf{X}_j / n = 1$ for $j = 1, \ldots, p$, where $\mathbf{X}_j = (X_{1j}, \ldots, X_{nj})^{\mathrm{T}}$ is the jth column of $\mathbf{X}$.

Condition (iii) can be relaxed, for example, by assuming that $\max_{j=1,\ldots,p} \mathbf{X}_j^{\mathrm{T}} \mathbf{X}_j / n \le C_0$ for a constant $C_0 > 0$.

For any $0 < \epsilon < 1$, suppose that the tuning parameter λ in (4.7) is specified as $\lambda = A_0 \lambda_0$ with a constant $A_0 > 1$ and

$$\lambda_0 = \sigma \sqrt{\frac{2 \log(p/\epsilon)}{n}},$$

where the dependency of λ_0 and λ on ϵ, as well as (n, p, σ), is suppressed in the notation. The following result gives a slow-rate error bound for Lasso estimation.

Theorem 4.1 (Slow rate in fixed design). *Assume that conditions (i)–(iii) are satisfied. For any $0 < \epsilon < 1$, it holds with probability at least $1 - 2\epsilon$ uniformly over $A_0 > 1$ that the Lasso estimator $\hat{\beta}_\lambda$ with $\lambda = A_0 \lambda_0$ in (4.7) satisfies*

$$\|\mathbf{X}\hat{\beta}_\lambda - \mathbf{X}\beta^*\|_n^2 + (A_0 - 1)\lambda_0 \|\hat{\beta}_\lambda - \beta^*\|_1 \le 2 A_0 \lambda_0 \|\beta^*\|_1, \tag{4.9}$$

i.e., for any $0 < \epsilon < 1$,

$$P\left(\bigcap_{A_0>1} \{ \text{ (4.9) holds with } \lambda = A_0\lambda_0 \ \}\right) \geq 1 - 2\epsilon.$$

The result holds for any choice of β^* *in* (4.1), *which is not unique in the case of* $p > n$.

The error bound (4.9) implies the following bounds on the prediction error (or, more precisely, the excess in-sample prediction MSE) $\|\mathbf{X}\hat{\beta}_\lambda - \mathbf{X}\beta^*\|_n^2$ and the L_1 estimation error $\|\hat{\beta}_\lambda - \beta^*\|_1$: with probability at least $1 - 2\epsilon$,

$$\|\mathbf{X}\hat{\beta}_\lambda - \mathbf{X}\beta^*\|_n^2 \leq 2A_0\lambda_0\|\beta^*\|_1, \tag{4.10a}$$

$$\|\hat{\beta}_\lambda - \beta^*\|_1 \leq \frac{2A_0}{A_0 - 1}\|\beta^*\|_1. \tag{4.10b}$$

Note that a similar identity as (4.3) holds for $\hat{\beta}_\lambda$, $E\|\tilde{\mathbf{Y}} - \mathbf{X}\hat{\beta}_\lambda\|_2^2 = E\|\tilde{\mathbf{Y}} - \mathbf{X}\beta^*\|_2^2 + E\|\mathbf{X}\hat{\beta}_\lambda - \mathbf{X}\beta^*\|_2^2$, where $\tilde{\mathbf{Y}}$ is independent and identically distributed as $\mathbf{Y}$ given $\mathbf{X}$. The error bound (4.9) is referred to as a slow rate mainly because the prediction error in (4.10a) depends on the sample size n in the order $\frac{1}{\sqrt{n}}$, which is slower than $\frac{1}{n}$ as would be expected. See for example the prediction error for the LS estimator in (4.2). In addition, the L_1 estimation error in (4.10b) is loose, even independent of n.

We provide a proof of Theorem 4.1 through two lemmas. The first lemma is a direct consequence of the definition of $\hat{\beta}_\lambda$ in (4.7).

Lemma 4.1 (Basic inequality). *For any* $\beta \in \mathbb{R}^p$,

$$\|\mathbf{X}(\hat{\beta}_\lambda - \beta)\|_n^2 - \langle \mathbf{Y} - \mathbf{X}\beta, \mathbf{X}(\hat{\beta}_\lambda - \beta)\rangle_n + \lambda\|\hat{\beta}_\lambda\|_1 \leq \lambda\|\beta\|_1, \tag{4.11}$$

where $\langle u, v\rangle_n = u^{\mathrm{T}}v/n$, *called the empirical inner product, for two (column) vectors* $u, v \in \mathbb{R}^n$.

Proof. For pedagogical purposes, we first show that for any $\beta \in \mathbb{R}^p$,

$$\frac{1}{2}\|\mathbf{X}(\hat{\beta}_\lambda - \beta)\|_n^2 - \langle \mathbf{Y} - \mathbf{X}\beta, \mathbf{X}(\hat{\beta}_\lambda - \beta)\rangle_n + \lambda\|\hat{\beta}_\lambda\|_1 \leq \lambda\|\beta\|_1, \tag{4.12}$$

which is slightly weaker than (4.11). By the definition of $\hat{\beta}_\lambda$ in (4.7), for any $\beta \in \mathbb{R}^p$,

$$\frac{1}{2}\|\mathbf{Y} - \mathbf{X}\hat{\beta}_\lambda\|_n^2 + \lambda\|\hat{\beta}_\lambda\|_1 \leq \frac{1}{2}\|\mathbf{Y} - \mathbf{X}\beta\|_n^2 + \lambda\|\beta\|_1.$$

By standard manipulation, $\|\mathbf{Y} - \mathbf{X}\hat{\beta}_\lambda\|_n^2 = \|\mathbf{Y} - \mathbf{X}\beta + \mathbf{X}\beta - \mathbf{X}\hat{\beta}_\lambda\|_n^2 = \|\mathbf{Y} - \mathbf{X}\beta\|_n^2 + \|\mathbf{X}\beta - \mathbf{X}\hat{\beta}_\lambda\|_n^2 - 2\langle \mathbf{Y} - \mathbf{X}\beta,\ \mathbf{X}\hat{\beta}_\lambda - \mathbf{X}\beta\rangle_n$, and hence

$$\frac{1}{2}\|\mathbf{Y} - \mathbf{X}\hat{\beta}_\lambda\|_n^2 - \frac{1}{2}\|\mathbf{Y} - \mathbf{X}\beta\|_n^2 = \frac{1}{2}\|\mathbf{X}\hat{\beta}_\lambda - \mathbf{X}\beta\|_n^2 - \langle \mathbf{Y} - \mathbf{X}\beta,\ \mathbf{X}(\hat{\beta}_\lambda - \beta)\rangle_n.$$

Combining the preceding two displays yields (4.12). Using (4.12) together with Lemma 4.2 leads to an error bound which is similar to (4.9) but with $\|\mathbf{X}\hat{\beta}_\lambda - \mathbf{X}\beta^*\|_n^2$ replaced by $\frac{1}{2}\|\mathbf{X}\hat{\beta}_\lambda - \mathbf{X}\beta^*\|_n^2$ on the left-hand side.

Next, we show the refined version (4.11), which is equivalent to

$$-\langle \mathbf{Y} - \mathbf{X}\hat{\beta}_\lambda, \mathbf{X}(\hat{\beta}_\lambda - \beta)\rangle_n \leq \lambda\|\beta\|_1 - \lambda\|\hat{\beta}_\lambda\|_1,$$

that is,

$$-\sum_{j=1}^{p}\langle \mathbf{Y} - \mathbf{X}\hat{\beta}_\lambda, \mathbf{X}_j\rangle_n (\hat{\beta}_{\lambda,j} - \beta_j) \leq \lambda\sum_{j=1}^{p}|\beta_j| - \lambda\sum_{j=1}^{p}|\hat{\beta}_{\lambda,j}|.$$

It suffices to show that for $j = 1, \ldots, p$,

$$-\langle \mathbf{Y} - \mathbf{X}\hat{\beta}_\lambda, \mathbf{X}_j\rangle_n (\hat{\beta}_{\lambda,j} - \beta_j) \leq \lambda|\beta_j| - \lambda|(\hat{\beta}_\lambda)_j|.$$

By the stationary condition (4.8) for $\hat{\beta}_\lambda$, if $\hat{\beta}_{\lambda,j} > 0$, then

$$\begin{aligned}-\langle \mathbf{Y} - \mathbf{X}\hat{\beta}_\lambda, \mathbf{X}_j\rangle_n (\hat{\beta}_{\lambda,j} - \beta_j) &= -\lambda(\hat{\beta}_{\lambda,j} - \beta_j) = \lambda(\beta_j - |\hat{\beta}_{\lambda,j}|)\\ &\leq \lambda(|\beta_j| - |\hat{\beta}_{\lambda,j}|),\end{aligned}$$

If $\hat{\beta}_{\lambda,j} < 0$, then

$$\begin{aligned}-\langle \mathbf{Y} - \mathbf{X}\hat{\beta}_\lambda, \mathbf{X}_j\rangle_n (\hat{\beta}_{\lambda,j} - \beta_j) &= \lambda(\hat{\beta}_{\lambda,j} - \beta_j) = \lambda(-\beta_j - |\hat{\beta}_{\lambda,j}|)\\ &\leq \lambda(|\beta_j| - |\hat{\beta}_{\lambda,j}|),\end{aligned}$$

If $\hat{\beta}_{\lambda,j} = 0$, then

$$\begin{aligned}-\langle \mathbf{Y} - \mathbf{X}\hat{\beta}_\lambda, \mathbf{X}_j\rangle_n (\hat{\beta}_{\lambda,j} - \beta_j) &= -c_j(\hat{\beta}_{\lambda,j} - \beta_j) = c_j\beta_j\\ &\leq \lambda|\beta_j| = \lambda(|\beta_j| - |\hat{\beta}_{\lambda,j}|).\end{aligned}$$

Combining the three cases yields the desired result. See Exercise 4.1 for an alternative proof of inequality (4.11) without directly using the stationary condition for $\hat{\beta}_\lambda$. □

The second lemma provides a probability inequality for controlling the empirical inner products $\langle \mathbf{Y} - \mathbf{X}\beta^*, \mathbf{X}_j\rangle_n$ simultaneously over $j = 1, \ldots, p$.

Lemma 4.2 (Concentration of inner products in fixed design). *For $0 < \epsilon < 1$, let $\Omega_1(\lambda_1)$ be the event that*

$$\max_{j=1,\ldots,p} |\langle \mathbf{Y} - \mathbf{X}\beta^*, \mathbf{X}_j\rangle_n| \leq \lambda_1,$$

where $\mathbf{X}_j$ is the jth column of $\mathbf{X}$. Under conditions (i)–(iii), if $\lambda_1 \geq \sigma\sqrt{\frac{2\log(p/\epsilon)}{n}}$, then $P(\Omega_1(\lambda_1)) \geq 1 - 2\epsilon$.

Proof. For $j = 1, \dots, p$, denote $Z_j = \langle \mathbf{Y} - \mathbf{X}\beta^*, \mathbf{X}_j \rangle_n = n^{-1} \sum_{i=1}^n \varepsilon_i X_{ij}$. By conditions (i) and (ii), the noises $\varepsilon_1, \dots, \varepsilon_n$ are independent, mean-zero, and sub-Gaussian with scale parameter σ. Then $\varepsilon_1 X_{1j}, \dots, \varepsilon_n X_{nj}$ are independent and mean-zero, and $\varepsilon_i X_{ij}$ is sub-Gaussian with scale parameter $\sigma|X_{ij}|$ for $i = 1, \dots, n$. By the Hoeffding inequality (Theorem 3.4),

$$P\left(\left| \sum_{i=1}^n \varepsilon_i X_{ij} \right| \geq t \right) \leq 2e^{-\frac{t^2}{2\tilde{\sigma}^2}} = 2e^{-\frac{t^2}{2n\sigma^2}}, \quad t \geq 0.$$

where $\tilde{\sigma}^2 = \sum_{i=1}^n (\sigma X_{ij})^2 = n\sigma^2$ by condition (iii), $\sum_{i=1}^n X_{ij}^2/n = 1$. Taking $t = ns$ in the preceding inequality gives

$$P(|Z_j| \geq s) \leq 2e^{-\frac{ns^2}{2\sigma^2}}, \quad s \geq 0.$$

By the union bound,

$$\begin{aligned} P(\max_{j=1,\dots,p} |Z_j| \geq s) &= P(\cup_{j=1}^p \{|Z_j| \geq s\}) \\ &\leq \sum_{j=1}^p P(|Z_j| \geq s) \leq 2pe^{-\frac{ns^2}{2\sigma^2}}, \quad s \geq 0. \end{aligned}$$

Take $s = \sigma\sqrt{\frac{2\log(p/\epsilon)}{n}}$ $(= \lambda_0)$ such that $2pe^{-\frac{ns^2}{2\sigma^2}} = 2pe^{-\log(p/\epsilon)} = 2\epsilon$. If $\lambda_1 \geq s$, then the preceding display gives $P(\Omega_1(\lambda_1)) \geq P(\max_{j=1,\dots,p} |Z_j| \leq s) \geq 1 - 2\epsilon$. □

Proof of Theorem 4.1. Taking $\beta = \beta^*$ in the basic inequality (4.11) gives

$$\|\mathbf{X}(\hat{\beta}_\lambda - \beta^*)\|_n^2 - \langle \mathbf{Y} - \mathbf{X}\beta^*, \mathbf{X}(\hat{\beta}_\lambda - \beta^*) \rangle_n + \lambda \|\hat{\beta}_\lambda\|_1 \leq \lambda \|\beta^*\|_1.$$

By the Hölder inequality, $|\sum_j a_j b_j| \leq (\max_j |a_j|)(\sum_j |b_j|)$,

$$\begin{aligned} |\langle \mathbf{Y} - \mathbf{X}\beta^*, \mathbf{X}(\hat{\beta}_\lambda - \beta^*) \rangle_n| &= \left| \sum_{j=1}^p \langle \mathbf{Y} - \mathbf{X}\beta^*, \mathbf{X}_j \rangle_n (\hat{\beta}_{\lambda,j} - \beta^*) \right| \\ &\leq \left\{ \max_{j=1,\dots,p} |\langle \mathbf{Y} - \mathbf{X}\beta^*, \mathbf{X}_j \rangle_n| \right\} \|\hat{\beta}_\lambda - \beta^*\|_1. \end{aligned}$$

Then in the event $\Omega_1 = \Omega_1(\lambda_0)$ with $\lambda_1 = \lambda_0$ in Lemma 4.2,

$$\begin{aligned} &\|\mathbf{X}(\hat{\beta}_\lambda - \beta^*)\|_n^2 + \lambda \|\hat{\beta}_\lambda\|_1 \leq |\langle \mathbf{Y} - \mathbf{X}\beta^*, \mathbf{X}(\hat{\beta}_\lambda - \beta^*) \rangle_n| + \lambda \|\beta^*\|_1 \\ &\leq \left\{ \max_{j=1,\dots,p} |\langle \mathbf{Y} - \mathbf{X}\beta^*, \mathbf{X}_j \rangle_n| \right\} \|\hat{\beta}_\lambda - \beta^*\|_1 + \lambda \|\beta^*\|_1 \\ &\leq \lambda_0 \|\hat{\beta}_\lambda - \beta^*\|_1 + \lambda \|\beta^*\|_1. \end{aligned}$$

Setting $\lambda = A_0 \lambda_0$ gives

$$\|\mathbf{X}(\hat{\beta}_\lambda - \beta^*)\|_n^2 + A_0\lambda_0 \|\hat{\beta}_\lambda\|_1 \leq \lambda_0 \|\hat{\beta}_\lambda - \beta^*\|_1 + A_0\lambda_0 \|\beta^*\|_1. \tag{4.13}$$

Applying the triangle inequality, $\|\hat{\beta}_\lambda\|_1 \geq \|\hat{\beta}_\lambda - \beta^*\|_1 - \|\beta^*\|_1$, to (4.13) yields

$$\|\mathbf{X}(\hat{\beta}_\lambda - \beta^*)\|_n^2 + A_0\lambda_0 \left\{\|\hat{\beta}_\lambda - \beta^*\|_1 - \|\beta^*\|_1\right\}$$
$$\leq \lambda_0\|\hat{\beta}_\lambda - \beta^*\|_1 + A_0\lambda_0\|\beta^*\|_1,$$

which leads to (4.9) after rearrangement. In summary, (4.9) holds in the event Ω_1, which has a probability at least $1 - 2\epsilon$ by Lemma 4.2. □

The slow-rate bound (4.9) can be improved under an additional condition on the matrix $\hat{H} = \mathbf{X}^{\mathrm{T}}\mathbf{X}/n$, which is the (constant) Hessian of the loss function $\frac{1}{2}\|\mathbf{Y} - \mathbf{X}\beta\|_n^2$. The matrix $\hat{H}$ is also called the (scaled) Gram matrix, whose entries are the scaled inner products from the p columns of $\mathbf{X}$.

Definition 4.1. For a $p \times p$ matrix Σ, a compatibility condition (S, ξ_0, ν_0) is said to hold with a subset $S \subset \{1, \ldots, p\}$ and constants $\xi_0 > 0$ and $\nu_0 > 0$ if

$$\nu_0^2\Big(\sum_{j\in S} |b_j|\Big)^2 \leq |S|(b^{\mathrm{T}}\Sigma b) \tag{4.14}$$

for any vector $b = (b_1, \ldots, b_p)^{\mathrm{T}} \in \mathbb{R}^p$ satisfying the cone condition

$$\sum_{j\notin S} |b_j| \leq \xi_0 \sum_{j\in S} |b_j|. \tag{4.15}$$

A restricted eigenvalue condition (S, ξ_0, ν_0) is said to hold with a subset $S \subset \{1, \ldots, p\}$ and constants $\xi_0 > 0$ and $\nu_0 > 0$ if

$$\nu_0^2\Big(\sum_{j\in S} |b_j|^2\Big) \leq b^{\mathrm{T}}\Sigma b \tag{4.16}$$

for any vector $b = (b_1, \ldots, b_p)^{\mathrm{T}} \in \mathbb{R}^p$ satisfying the cone condition (4.15).

Equivalently, the compatibility condition can be stated as

$$\text{(C1)} \qquad \min_{\substack{b\neq 0,\\ \text{satisfying (4.15)}}} \frac{|S|(b^{\mathrm{T}}\Sigma b)}{(\sum_{j\in S} |b_j|)^2} \geq \nu_0^2.$$

The restricted eigenvalue condition can be stated as

$$\text{(C2)} \qquad \min_{\substack{b\neq 0,\\ \text{satisfying (4.15)}}} \frac{b^{\mathrm{T}}\Sigma b}{\sum_{j\in S} |b_j|^2} \geq \nu_0^2.$$

By the Cauchy–Schwartz inequality, $(\sum_{j\in S} |b_j|)^2 \leq |S| \sum_{j\in S} |b_j|^2$, the restricted eigenvalue condition (S, ξ_0, ν_0) implies (hence is more restrictive than) the compatibility condition (S, ξ_0, ν_0), i.e., if the restricted eigenvalue condition (S, ξ_0, ν_0) holds, then the compatibility condition (S, ξ_0, ν_0) also holds.

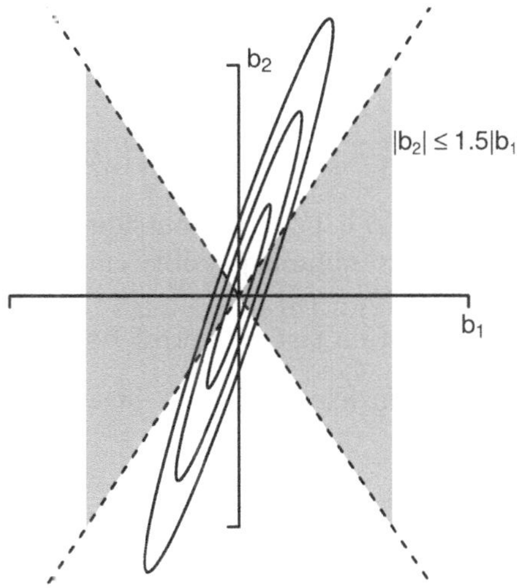

FIGURE 4.1
A quadratic function (contour lines) and a cone (shaded) in two dimensions.

On the other hand, the restricted eigenvalue condition is implied by (hence is less restrictive than)

$$\text{(C3)} \qquad \min_{\substack{b\neq 0,\\ \text{satisfying (4.15)}}} \frac{b^{\mathrm{T}}\Sigma b}{\|b\|_2^2} \geq \nu_0^2,$$

which is in turn implied by (hence is less restrictive than)

$$\text{(C4)} \qquad \min_{b\neq 0} \frac{b^{\mathrm{T}}\Sigma b}{\|b\|_2^2} \geq \nu_0^2.$$

The four conditions, (C1)–(C4), are increasingly more restrictive. For a symmetric matrix Σ, $\min_{b\neq 0} \frac{b^{\mathrm{T}}\Sigma b}{\|b\|_2^2}$ gives the smallest eigenvalue of Σ, and hence (C4) is equivalent to saying that the smallest eigenvalue of Σ is $\geq \nu_0^2$. Condition (C3) is a restricted version of (C4), in saying that the smallest value of $\frac{b^{\mathrm{T}}\Sigma b}{\|b\|_2^2}$ over any b in the cone (4.15) is $\geq \nu_0^2$.

Figure 4.1 provides an illustration in two dimensions, where

$$b^{\mathrm{T}}\Sigma b = \frac{(\frac{1}{3}b_1 + b_2)^2}{16} + (-3b_1 + b_2)^2, \tag{4.17}$$

and the cone is $\{b \in \mathbb{R}^2 : |b_2| \leq 1.5|b_1|\}$ with $S = \{1\}$. Then by some algebra

(Exercise 4.2), it can be shown that

$$\begin{aligned}\min_{b\neq 0,|b_2|\le 1.5|b_1|}\frac{b^{\mathrm{T}}\Sigma b}{b_1^2} &= 2.46 > \min_{b\neq 0}\frac{b^{\mathrm{T}}\Sigma b}{b_1^2} = 0.65,\\ \min_{b\neq 0,|b_2|\le 1.5|b_1|}\frac{b^{\mathrm{T}}\Sigma b}{b_1^2+b_2^2} &= 0.76 > \min_{b\neq 0}\frac{b^{\mathrm{T}}\Sigma b}{b_1^2+b_2^2} = 0.0694.\end{aligned} \tag{4.18}$$

The quadratic function $b^{\mathrm{T}}\Sigma b$ is most flat outside the cone, so that restriction to the cone increases the minimum possible curvature, as measured by the ratio $(b^{\mathrm{T}}\Sigma b)/b_1^2$ or $(b^{\mathrm{T}}\Sigma b)/(b_1^2+b_2^2)$.

The following result gives a fast-rate error bound for Lasso estimation.

Theorem 4.2 (Fast rate in fixed design). *Assume that the following conditions are satisfied:*

(i)–(iii) Same as in Theorem 4.1;

(iv) [Empirical compatibility condition] The compatibility condition $(S_{\beta^*},\xi_0,\nu_0)$ holds for $\hat{H}=\mathbf{X}^{\mathrm{T}}\mathbf{X}/n$, where $S_{\beta^*}=\{1\le j\le p:\beta_j^*\neq 0\}$ and $\xi_0>1$ and $\nu_0>0$ are constants.

For any $0<\epsilon<1$, it holds with probability at least $1-2\epsilon$ uniformly over $A_0\ge\frac{\xi_0+1}{\xi_0-1}$ that the Lasso estimator $\hat\beta_\lambda$ with $\lambda=A_0\lambda_0$ in (4.7) satisfies

$$\|\mathbf{X}\hat\beta_\lambda-\mathbf{X}\beta^*\|_n^2+(A_0-1)\lambda_0\|\hat\beta_\lambda-\beta^*\|_1\le M_0|S_{\beta^*}|\lambda_0^2, \tag{4.19}$$

where $M_0=4A_0^2\nu_0^{-2}$, depending only on (A_0,ν_0).

The error bound (4.19) implies the following bounds: with probability at least $1-2\epsilon$,

$$\|\mathbf{X}\hat\beta_\lambda-\mathbf{X}\beta^*\|_n^2\le M_0|S_{\beta^*}|\lambda_0^2, \tag{4.20a}$$

$$\|\hat\beta_\lambda-\beta^*\|_1\le\frac{M_0}{A_0-1}|S_{\beta^*}|\lambda_0. \tag{4.20b}$$

The prediction error in (4.20a) depends on the sample size n in the order $\frac{1}{n}$ (hence called a fast rate), instead of $\frac{1}{\sqrt{n}}$ in the slow rate (4.10a). In addition, the L_1 estimation error in (4.20b) is improved, depending on n in the order $\frac{1}{\sqrt{n}}$, compared with that in (4.10b).

In principle, Theorem 4.2 holds (i.e., its assumptions imply its conclusion) for any choice of β^* satisfying regression equation (4.1), which is not unique in the case of $p>n$ or in the case of $p\le n$ and linear dependency among p columns of $\mathbf{X}$. On the other hand, $\hat\beta_\lambda$ is expected to converge to β^* in the L_1 norm in probability as $n\to\infty$, and hence β^* must be unique. See Corollary 4.1. This seemingly contradiction can be resolved by noting that if there exists a choice of β^* such that the compatibility condition $(S_{\beta^*},\xi_0,\nu_0)$ is satisfied with $\xi_0\ge 1$ for $\hat{H}=\mathbf{X}^{\mathrm{T}}\mathbf{X}/n$, then such a choice of β^* is unique. See Section 5.5.1 for a proof.

Proof. We return to (4.13) in the event $\Omega_1 = \Omega_1(\lambda_0)$ with $\lambda_1 = \lambda_0$ in Lemma 4.2:

$$\|\mathbf{X}(\hat{\beta}_\lambda - \beta^*)\|_n^2 + A_0\lambda_0\|\hat{\beta}_\lambda\|_1 \le \lambda_0\|\hat{\beta}_\lambda - \beta^*\|_1 + A_0\lambda_0\|\beta^*\|_1.$$

For any subset $S \subset \{1, \dots, p\}$, we apply the triangle inequalities to the above,

$$|\hat{\beta}_{\lambda,j}| \ge |\hat{\beta}_{\lambda,j} - \beta_j^*| - |\beta_j^*|, \quad j \notin S, \tag{4.21a}$$

$$|\hat{\beta}_{\lambda,j}| \ge |\beta_j^*| - |\hat{\beta}_{\lambda,j} - \beta_j^*|, \quad j \in S, \tag{4.21b}$$

and obtain

$$\|\mathbf{X}(\hat{\beta}_\lambda - \beta^*)\|_n^2 + A_0\lambda_0\left\{\sum_{j\notin S}(|\hat{\beta}_{\lambda,j} - \beta_j^*| - |\beta_j^*|) + \sum_{j\in S}(|\beta_j^*| - |\hat{\beta}_{\lambda,j} - \beta_j^*|)\right\}$$
$$\le \lambda_0\|\hat{\beta}_\lambda - \beta^*\|_1 + A_0\lambda_0\|\beta^*\|_1.$$

Rearranging the result gives

$$\|\mathbf{X}(\hat{\beta}_\lambda - \beta^*)\|_n^2 + (A_0 - 1)\lambda_0\|\hat{\beta}_\lambda - \beta^*\|_1$$
$$\le 2A_0\lambda_0\left\{\sum_{j\in S}|\hat{\beta}_{\lambda,j} - \beta_j^*| + \sum_{j\notin S}|\beta_j^*|\right\}. \tag{4.22}$$

Note that taking $S = \emptyset$ and using only the triangle inequality (4.21a) leads back to the slow-rate error bound (4.9). Alternatively, taking $S = S_{\beta^*}$ in (4.22) shows that in the event Ω_1,

$$\|\mathbf{X}(\hat{\beta}_\lambda - \beta^*)\|_n^2 + (A_0 - 1)\lambda_0\|\hat{\beta}_\lambda - \beta^*\|_1 \le 2A_0\lambda_0 \sum_{j\in S_{\beta^*}} |\hat{\beta}_{\lambda,j} - \beta_j^*|. \tag{4.23}$$

In this case, (4.21a) reduces to equality ($\beta_j^* = 0$ for $j \notin S_{\beta^*}$) and only (4.21b) takes effect. Denote $\Delta_n = \|\mathbf{X}(\hat{\beta}_\lambda - \beta^*)\|_n^2 + (A_0 - 1)\lambda_0\|\hat{\beta}_\lambda - \beta^*\|_1$, the left-hand side of (4.23).

It suffices to deduce (4.19) from (4.23) using the empirical compatibility condition. In fact, (4.23) implies that

$$(A_0 - 1)\lambda_0\|\hat{\beta}_\lambda - \beta^*\|_1 \le 2A_0\lambda_0 \sum_{j\in S_{\beta^*}} |\hat{\beta}_{\lambda,j} - \beta_j^*|,$$

or equivalently

$$\sum_{j\notin S_{\beta^*}} |\hat{\beta}_{\lambda,j} - \beta_j^*| \le \frac{A_0 + 1}{A_0 - 1} \sum_{j\in S_{\beta^*}} |\hat{\beta}_{\lambda,j} - \beta_j^*|.$$

Note that $\frac{A_0+1}{A_0-1} \leq \xi_0$ for $A_0 \geq \frac{\xi_0+1}{\xi_0-1}$. Then $\hat{\beta}_\lambda - \beta^*$ satisfies the cone condition, and by the compatibility condition,

$$\begin{aligned}
&\nu_0^2\Big(\sum_{j\in S_{\beta^*}} |\hat{\beta}_{\lambda,j} - \beta_j^*|\Big)^2 \\
&\leq |S_{\beta^*}|(\hat{\beta}_\lambda - \beta^*)^{\mathrm{T}}\hat{H}(\hat{\beta}_\lambda - \beta^*) = |S_{\beta^*}| \cdot \|\mathbf{X}(\hat{\beta}_\lambda - \beta^*)\|_n^2. \qquad (4.24)
\end{aligned}$$

Combining (4.23) and (4.24) yields a quadratic inequality in $\Delta_n^{1/2}$:

$$\begin{aligned}
\Delta_n &\leq 2A_0\lambda_0 \sum_{j\in S_{\beta^*}} |\hat{\beta}_{\lambda,j} - \beta_j^*| \\
&\leq 2A_0\lambda_0 \cdot \nu_0^{-1}|S_{\beta^*}|^{1/2}\|\mathbf{X}(\hat{\beta}_\lambda - \beta^*)\|_n \leq 2A_0\lambda_0\nu_0^{-1}|S_{\beta^*}|^{1/2}\Delta_n^{1/2},
\end{aligned}$$

and hence ("inverting a quadratic inequality")

$$\Delta_n^{1/2} \leq 2A_0\lambda_0\nu_0^{-1}|S_{\beta^*}|^{1/2},$$

that is, $\Delta_n \leq (2A_0\nu_0^{-1})^2|S_{\beta^*}|\lambda_0^2$ as stated in (4.19). In summary, (4.19) holds in the event Ω_1, with probability at least $1 - 2\epsilon$ by Lemma 4.2. □

The error bounds (4.9) and (4.19) with the related probabilities are finite-sample results, which are valid for any (n, p, ϵ) under the stated conditions. A subtle point is that λ_0 in (4.9) and (4.19) is, through ϵ, tied to the probabilities of the error bounds being valid. These results can be converted into O_p statements which make explicit the asymptotic order. For example, the following result can be deduced from Theorem 4.2.

Corollary 4.1 (O_p error rate in fixed design). *Assume that conditions (i)–(iv) in Theorem 4.2 hold. As $n \to \infty$ or $p \to \infty$, the Lasso estimator $\hat{\beta}_\lambda$ in (4.7) with a suitable choice $\lambda = O(1)\sqrt{\frac{\log(n\vee p)}{n}}$ satisfies*

$$\|\mathbf{X}\hat{\beta}_\lambda - \mathbf{X}\beta^*\|_n^2 = O_p(1)|S_{\beta^*}|\frac{\log(n\vee p)}{n},$$

$$\|\hat{\beta}_\lambda - \beta^*\|_1 = O_p(1)|S_{\beta^*}|\sqrt{\frac{\log(n\vee p)}{n}},$$

where the dependency of $|S_{\beta^*}|$ on (n, p) is suppressed in the notation.

Proof. Take $\tilde{A}_0 = A_0\sqrt{\frac{\log(p/\epsilon)}{\log(n\vee p)}}$ for a constant $A_0 \geq \frac{\xi_0+1}{\xi_0-1}$ and $\epsilon = \frac{1}{n\vee p}$, which tends to 0 as $n \to \infty$ or $p \to \infty$. Note that $\tilde{A}_0 \leq \sqrt{2}A_0$. Applying Theorem 4.2 with $\lambda = \tilde{A}_0\sigma\sqrt{\frac{2\log(n\vee p)}{n}} = A_0\sigma\sqrt{\frac{2\log(p/\epsilon)}{n}}$ gives the desired result. Details are left as Exercise 4.3. □

Formally for the O_p statements, there is a triangular array indexed by the row number k. The observations in the kth row are $(X_i, Y_i) = (X_i^{(k)}, Y_i^{(k)})$

with $X_i^{(k)} \in \mathbb{R}^{p_k}$, $Y_i^{(k)} \in \mathbb{R}$ for $i = 1, \ldots, n_k$, where (n_k, p_k) depend on k. In this setup, conditions (i)–(iv) in Theorems 4.1 and 4.2 are imposed row by row. The "constants" (ξ_0, ν_0) may vary with k and hence with (n_k, p_k). From this perspective, the simple O_p result in Corollary 4.1 implicitly assumes that (ξ_0, ν_0) are independent of (n_k, p_k). Nevertheless, more complex O_p statements can still be obtained if (ξ_0, ν_0) are allowed to depend on (n, p). [Question: What modification would be needed in this case?]

4.2.2 Random-design linear regression

For the fixed-design analysis of Lasso estimation to obtain the fast rate (Theorem 4.2), the empirical compatibility condition is imposed on the covariate matrix (or design matrix) $\mathbf{X}$. It is of interest to study Lasso estimation in a random design, where a central issue is to show that the empirical compatibility condition can be satisfied with a high probability under suitable conditions.

For theoretical analysis, consider the following conditions:

(i) [Linear regression in random design] Linear regression (4.1) is correctly specified in a random design, i.e., $(X_1, Y_1), \ldots, (X_n, Y_n)$ are independent with $E(\varepsilon_i | X_i) = 0$ for $i = 1, \ldots, n$;

(ii) [Sub-Gaussian noises] ε_i given X_i is sub-Gaussian with scale parameter σ for $i = 1, \ldots, n$;

(iii) [Bounded covariates] $\max_{j=1,\ldots,p} |X_{ij}| \le C_0$ almost surely for $i = 1, \ldots, n$ with a constant $C_0 > 0$;

(iv) [Theoretical compatibility condition] The compatibility condition $(S_{\beta^*}, \xi_0^*, \nu_0^*)$ holds for $H = E(\mathbf{X}^{\mathrm{T}}\mathbf{X}/n)$, where $S_{\beta^*} = \{1 \le j \le p : \beta_j^* \ne 0\}$ and $\xi_0^* > 1$ and $\nu_0^* > 0$ are constants.

Condition (i) allows $(X_1, Y_1), \ldots, (X_n, Y_n)$ to be independent but non-identically distributed, which presents little difference from the case of IID $(X_1, Y_1), \ldots, (X_n, Y_n)$ when linear regression is correctly specified. Condition (ii) can be weakened to only require that ε_i is marginally sub-Gaussian with scale parameter σ for $i = 1, \ldots, n$. See the proof of Lemma 4.3. Condition (iii) is mainly used to simplify the derivation of concentration results as in Lemmas 4.3 and 4.4. This condition can be relaxed, for example, by allowing individual covariates X_{ij} to be sub-Gaussian with bounded ψ_2 norms. Similar analysis can be conducted, but additional arguments will be needed.

For any $0 < \epsilon < 1$, suppose that the tuning parameter λ in (4.7) is specified as $\lambda = A_0 \lambda_0$ with a constant $A_0 > 1$ and

$$\lambda_0 = \max \left\{ \sigma C_0 \sqrt{\frac{2\log(p/\epsilon)}{n}},\ 2C_0^2 \sqrt{\frac{\log(p/\epsilon)}{n}} \right\}, \tag{4.25}$$

where C_0 is a constant from condition (iii), and the dependency of λ_0 and λ on ϵ, as well as (n, p, σ, C_0), is suppressed in the notation. The following result gives a fast-rate error bound for Lasso estimation in the random design.

Theorem 4.3 (Fast rate in random design). *Assume that conditions (i)–(iv) above are satisfied. For any $0 < \epsilon < 1$, suppose further that*

(v) [Rate condition] $(1+\xi_0^*)^2 \nu_0^{*-2} |S_{\beta^*}| \lambda_0 \leq \eta_0$ for a constant $\eta_0 \in (0,1)$.

Then it holds with probability at least $1 - 2\epsilon - 2\epsilon^2$ uniformly over $A_0 \geq \frac{\xi_0^*+1}{\xi_0^*-1}$ that the Lasso estimator $\hat{\beta}_\lambda$ with $\lambda = A_0 \lambda_0$ in (4.7) satisfies

$$\|\mathbf{X}\hat{\beta}_\lambda - \mathbf{X}\beta^*\|_n^2 + (A_0 - 1)\lambda_0 \|\hat{\beta}_\lambda - \beta^*\|_1 \leq M_0 |S_{\beta^*}| \lambda_0^2, \tag{4.26}$$

where $M_0 = 4A_0^2 \nu_0^{-2}$ with $\nu_0 = \nu_0^*(1-\eta_0)^{1/2}$.

Condition (v), which requires that $|S_{\beta^*}|\lambda_0$ is sufficiently small as $n \to \infty$, is imposed to convert the theoretical compatibility condition to an empirical version (Lemma 4.5). This condition can be relaxed to only require that $|S_{\beta^*}|\lambda_0^2$ is sufficiently small as $n \to \infty$, albeit under other stronger conditions, for example, the covariates $(X_{i1}, \ldots, X_{ip})$ are jointly (not just marginally) sub-Gaussian (Huang & Zhang, 2012; Negahban et al., 2012). On the other hand, condition (v), translated into $|S_{\beta^*}|\sqrt{\frac{\log(n \vee p)}{n}} = o(1)$ for the O_p error bounds (Corollary 4.2), is already weaker than the sparsity condition, $|S_{\beta^*}| \frac{\log(n \vee p)}{\sqrt{n}} = o(1)$, which is needed for obtaining valid confidence intervals for components of β^* by debiased Lasso (Theorem 6.1).

Similarly to the discussion after Theorem 4.2, although H is allowed to be singular (hence linear dependency among the p covariates almost surely) and there may be multiple choices of β^* satisfying regression equation (4.1), the choice of β^* is unique when the compatibility condition $(S_{\beta^*}, \xi_0^*, \nu_0^*)$ is assumed to hold with $\xi_0^* \geq 1$ for H. See Section 5.5.1 for a proof.

To prepare for the proof of Theorem 4.3, The following lemma is similar to Lemma 4.2 in the fixed design, but applicable in the random design with conditions (i)–(iii).

Lemma 4.3 (Concentration of inner products in random design). *For $0 < \epsilon < 1$, let $\Omega_1(\lambda_1)$ be the event that*

$$\max_{j=1,\ldots,p} |\langle \mathbf{Y} - \mathbf{X}\beta^*, \mathbf{X}_j \rangle_n| \leq \lambda_1,$$

where $\mathbf{X}_j$ is the jth column of $\mathbf{X}$. Under conditions (i)–(iii), if $\lambda_1 \geq \sigma C_0 \sqrt{\frac{2\log(p/\epsilon)}{n}}$, then $P(\Omega_1(\lambda_1)) \geq 1 - 2\epsilon$.

Proof. With ε_i mean-zero and sub-Gaussian given X_i for $i = 1, \ldots, n$, a conditional result (which implies the original result) can be obtained: If $\lambda_1 \geq \sigma C_0 \sqrt{\frac{2\log(p/\epsilon)}{n}}$, then $P(\Omega_1(\lambda_1)|\mathbf{X}) \geq 1 - 2\epsilon$. The result can be shown by similar steps as in the proof of Lemma 4.2 (Concentration in fixed design) while conditioning on $\mathbf{X}$. This is left as an exercise.

Alternatively, we provide an unconditional proof, which is important for Lemma 4.7 (Concentration in random design with linear approximation). From the proof of Lemma 4.2, it suffices to show that $\varepsilon_i X_{ij}$ is (marginally) mean-zero and sub-Gaussian with scale parameter σC_0. In fact, by condition (i), $E(\varepsilon_i|X_i) = 0$ and hence $E(\varepsilon_i X_{ij}) = E\{E(\varepsilon_i|X_i)X_{ij}\} = 0$. Moreover, by conditions (ii) and (iii),

$$E(\mathrm{e}^{\lambda\varepsilon_i X_{ij}}|X_i) \le \mathrm{e}^{\frac{\sigma^2(\lambda X_{ij})^2}{2}} \le \mathrm{e}^{\frac{(\sigma C_0)^2\lambda^2}{2}}, \quad \lambda \in \mathbb{R},$$

and hence

$$E(\mathrm{e}^{\lambda\varepsilon_i X_{ij}}) = E\{E(\mathrm{e}^{\lambda\varepsilon_i X_{ij}}|X_i)\} \le \mathrm{e}^{\frac{(\sigma C_0)^2\lambda^2}{2}}, \quad \lambda \in \mathbb{R},$$

indicating that $\varepsilon_i X_{ij}$ is sub-Gaussian with scale parameter σC_0.

[Exercise: Show that Lemma 4.3 can be extended if condition (ii) is weakened such that $\varepsilon_1, \ldots, \varepsilon_n$ are marginally sub-Gaussian with scale parameter σ. Hint: Use $E(\mathrm{e}^{\lambda\varepsilon_i X_{ij}}) \le E\max(\mathrm{e}^{\lambda C_0\varepsilon_i}, \mathrm{e}^{-\lambda C_0\varepsilon_i}) \le E(\mathrm{e}^{\lambda C_0\varepsilon_i} + \mathrm{e}^{-\lambda C_0\varepsilon_i})$.] □

We use the following two lemmas to convert the theoretical compatibility condition to an empirical version with high probability.

Lemma 4.4 (Concentration of Gram matrix). *For $0 < \epsilon < 1$, let $\Omega_2(\lambda_2)$ be the event that*

$$\max_{j,k=1,\ldots,p} |\hat{H}_{jk} - H_{jk}| \le \lambda_2,$$

where $\hat{H}_{jk}$ is the (j,k)th element of $\hat{H}$, and H_{jk} is the (j,k)th element of H. Under independence of $(X_1, \ldots, X_n)$ and condition (iii), if $\lambda_2 \ge 2C_0^2\sqrt{\frac{\log(p/\epsilon)}{n}}$, then $P(\Omega_2(\lambda_2)) \ge 1 - 2\epsilon^2$.

Proof. By definition, $\hat{H}_{jk} = n^{-1}\sum_{i=1}^n X_{ij}X_{ik}$ and $H_{jk} = n^{-1}\sum_{i=1}^n E(X_{ij}X_{ik})$. Let $Z_{i,jk} = X_{ij}X_{ik}$. Then $|Z_{i,jk}| \le C_0^2$. By the Hoeffding inequality (Corollary 3.1),

$$\begin{aligned} P(|\hat{H}_{jk} - H_{ik}| \ge t) &= P\left(\left|\frac{1}{n}\sum_{i=1}^n (Z_{i,jk} - E(Z_{i,jk}))\right| \ge t\right) \\ &\le 2\mathrm{e}^{-\frac{2nt^2}{(2C_0^2)^2}}, \quad t \ge 0. \end{aligned}$$

By the union bound,

$$P(\max_{j,k=1,\ldots,p} |\hat{H}_{jk} - H_{ik}| \ge t) = P\left(\max_{j,k=1,\ldots,p}\left|\frac{1}{n}\sum_{i=1}^n Z_{i,jk} - E(Z_{i,jk})\right| \ge t\right)$$

$$\le 2p^2 \mathrm{e}^{-\frac{2nt^2}{(2C_0^2)^2}}.$$

Take $t = 2C_0^2\sqrt{\frac{\log(p/\epsilon)}{n}}$ such that $2p^2\mathrm{e}^{-\frac{2nt^2}{(2C_0^2)^2}} = 2p^2\mathrm{e}^{-2\log(p/\epsilon)} = 2\epsilon^2$. If $\lambda_2 \ge t$, then the preceding displays gives $P(\Omega_2(\lambda_2)) \ge P(\max_{j,k=1,\ldots,p}|\hat{H}_{jk} - H_{ik}| \le t) \ge 1 - 2\epsilon^2$. □

Lemma 4.5 (From theoretical to empirical compatibility condition). *In the event $\Omega_2(\lambda_2)$, a compatibility condition (S^*, ξ_0^*, ν_0^*) for H implies the compatibility condition (S^*, ξ_0, ν_0) for $\hat{H}$, with $\xi_0 = \xi_0^*$ and $\nu_0 = \nu_0^*(1-\eta_0)^{1/2}$, provided that $(1+\xi_0^*)^2\nu_0^{*-2}|S^*|\lambda_2 \le \eta_0$ for a constant $\eta_0 \in (0,1)$.*

Proof. Suppose that the compatibility condition (S^*, ξ_0^*, ν_0^*) holds for H. Then for any $b \in \mathbb{R}^p$ satisfying $\sum_{j\notin S^*}|b_j| \le \xi_0^*(\sum_{j\in S^*}|b_j|)$, we have

$$\nu_0^{*2}(\sum_{j\in S^*}|b_j|)^2 \le |S^*|(b^{\mathrm{T}}Hb) = |S^*|b^{\mathrm{T}}\hat{H}b + |S^*|b^{\mathrm{T}}(H-\hat{H})b.$$

In the event $\Omega_2(\lambda_2)$,

$$\begin{aligned}
|b^{\mathrm{T}}(H-\hat{H})b| &= |\sum_{j,k=1,\ldots,p}(H_{jk}-\hat{H}_{jk})b_jb_k| \\
&\le \sum_{j,k=1,\ldots,p}|H_{jk}-\hat{H}_{jk}||b_j||b_k| \qquad (4.27)\\
&\le \lambda_2\sum_{j,k=1,\ldots,p}|b_j||b_k| = \lambda_2\|b\|_1^2.
\end{aligned}$$

Combining the preceding two displays shows that for any $b \in \mathbb{R}^p$ satisfying $\sum_{j\notin S^*}|b_j| \le \xi_0^*(\sum_{j\in S^*}|b_j|)$,

$$\begin{aligned}
\nu_0^{*2}(\sum_{j\in S^*}|b_j|)^2 &\le |S^*|b^{\mathrm{T}}\hat{H}b + |S^*|\lambda_2\|b\|_1^2 \\
&\le |S^*|b^{\mathrm{T}}\hat{H}b + |S^*|\lambda_2(1+\xi_0^*)^2(\sum_{j\in S^*}|b_j|)^2,
\end{aligned}$$

where the second inequality follows because $\|b\|_1 \le (1+\xi_0^*)\sum_{j\in S^*}|b_j|$ by the cone condition. If $\nu_0^{*-2}|S^*|\lambda_2(1+\xi_0^*)^2 \le \eta_0$, then the above display gives

$$\nu_0^{*2}(1-\eta_0)(\sum_{j\in S^*}|b_j|)^2 \le |S^*|b^{\mathrm{T}}\hat{H}b,$$

that is, the compatibility condition (S^*, ξ_0, ν_0) holds for $\hat{H}$, with $\xi_0 = \xi_0^*$ and $\nu_0 = \nu_0^*(1-\eta_0)^{1/2}$. □

Proof of Theorem 4.3. We use similar arguments as in the proof of Theorem 4.2. First, inequality (4.23) holds in the event $\Omega_1 = \Omega_1(\lambda_1)$ with $\lambda_1 = \lambda_0$ in

Lemma 4.3. By Lemma 4.5, the empirical compatibility condition $(S_{\beta*}, \xi_0, \nu_0)$ holds for $\hat{H}$, with $\xi_0 = \xi_0^*$ and $\nu_0 = \nu_0^*(1-\eta_0)^{1/2}$, in the event $\Omega_2 = \Omega_2(\lambda_0)$ with $\lambda_2 = \lambda_0$ in Lemma 4.4. Then (4.26) can be deduced from (4.23) in the event $\Omega_1 \cap \Omega_2$. For λ_0 in (4.25), the probability of $\Omega_1 \cap \Omega_2$ is at least $1-2\epsilon-2\epsilon^2$ by Lemmas 4.3 and 4.4. □

The error bound (4.26) with the related probability in Theorem 4.3 is a finite-sample result, valid for any (n, p, ϵ) under the stated conditions. Similarly as in Corollary 4.1 (O_p error rate in fixed design), the result can be converted into O_p statements about the convergence rate of $\hat{\beta}_\lambda$ to β^* (Exercise 4.5). A technical difference, however, is that $n \to \infty$ is required here to ensure the rate condition $|S_{\beta^*}|\sqrt{\frac{\log(n\vee p)}{n}} = o(1)$, whereas either $n \to \infty$ or $p \to \infty$ is allowed in Corollary 4.1.

Corollary 4.2 (O_p error rate in random design)**.** *Assume that conditions (i)–(iv) in Theorem 4.3 hold, and $|S_{\beta^*}|\sqrt{\frac{\log(n\vee p)}{n}} = o(1)$ as $n \to \infty$ and p may increase as n increases. Then the Lasso estimator $\hat{\beta}_\lambda$ in (4.7) with a suitable choice $\lambda = O(1)\sqrt{\frac{\log(n\vee p)}{n}}$ satisfies*

$$\|\mathbf{X}\hat{\beta}_\lambda - \mathbf{X}\beta^*\|_n^2 = O_p(1)|S_{\beta^*}|\frac{\log(n \vee p)}{n},$$

$$\|\hat{\beta}_\lambda - \beta^*\|_1 = O_p(1)|S_{\beta^*}|\sqrt{\frac{\log(n \vee p)}{n}},$$

where the dependency of $|S_{\beta^*}|$ on (n,p) is suppressed in the notation.

Similarly to Corollary 4.1, the simple O_p statements in Corollary 4.2 implicitly assumes that (ξ_0^*, ν_0^*) are independent of (n_k, p_k). Formally, we consider a triangular array of observations $(X_i, Y_i) = (X_i^{(k)}, Y_i^{(k)})$ with $X_i^{(k)} \in \mathbb{R}^{p_k}$, $Y_i^{(k)} \in \mathbb{R}$ for $i = 1, \ldots, n_k$. On each row, the theoretical compatibility condition is imposed on the expected Gram matrix from $X_i^{(k)}$, $i = 1, \ldots, n_k$. In contrast, for Corollary 4.1, the empirical compatibility condition is imposed on the empirical Gram matrix from $X_i^{(k)}$, $i = 1, \ldots, n_k$.

4.3 Lasso estimation with model misspecification

For nonparametric regression, suppose that

$$Y_i = f^*(X_i) + \omega_i, \quad i = 1, \ldots, n, \tag{4.28}$$

where $f^*(\cdot)$ is an unknown function and ω_i is a noise variable with mean zero given X_i for $i = 1, \ldots, n$. Linear regression (4.1) is correctly specified if and only if there exists $\beta^* \in \mathbb{R}^p$ such that $f^*(X_i) = X_i^{\mathrm{T}}\beta^*$ for $i = 1, \ldots, n$. With possible misspecification, linear regression is considered a working model such that for a suitable estimator $\hat{\beta}$, the fitted value $X_i^{\mathrm{T}}\hat{\beta}$ provides a linear approximation to $f^*(X_i)$, $i = 1, \ldots, n$.

4.3.1 Fixed-design linear regression

We study Lasso estimation with possible misspecification of linear regression (4.1) in a fixed design, where $X_1, \ldots, X_n$ are fixed and $Y_1, \ldots, Y_n$ are random (equivalently, $\omega_1, \ldots, \omega_n$ are random).

The target value $\bar{\beta}$ is defined as

$$\bar{\beta} = \mathrm{argmin}_{\beta \in \mathbb{R}^p} E\|\mathbf{Y} - \mathbf{X}^{\mathrm{T}}\beta\|_n^2 \tag{4.29a}$$

$$= \mathrm{argmin}_{\beta \in \mathbb{R}^p} \|\mathbf{f}^* - \mathbf{X}^{\mathrm{T}}\beta\|_n^2, \tag{4.29b}$$

where $\mathbf{f}^* = (f^*(X_1), \ldots, f^*(X_n))^{\mathrm{T}}$, and no expectation is taken for $\|\mathbf{f}^* - \mathbf{X}^{\mathrm{T}}\beta\|_n^2$. The second equality follows from the decomposition, $E\|\mathbf{Y} - \mathbf{X}^{\mathrm{T}}\beta\|_n^2 = E\|\mathbf{Y} - \mathbf{f}^*\|_n^2 + \|\mathbf{f}^* - \mathbf{X}^{\mathrm{T}}\beta\|_n^2$ [Exercise]. If linear regression (4.1) is correctly specified (i.e., $\mathbf{f}^* = \mathbf{X}\beta^*$), then $\bar{\beta}$ coincides with the true value β^* (more precisely, any choice of $\bar{\beta}$ is a valid choice of β^*, and vice versa). Otherwise, $\bar{\beta}$ remains well-defined, even though β^* does not exist. Note that $\bar{\beta}$ depends on $\mathbf{X}$, but the dependency is suppressed in the notation.

For theoretical analysis, consider the following conditions:

(i) [Nonparametric regression in fixed design] $X_1, \ldots, X_n$ are fixed and $\omega_1, \ldots, \omega_n$ are independent with mean zero, where $\omega_i = Y_i - f^*(X_i)$ in (4.28);

(ii) [Sub-Gaussian true noises] $\omega_1, \ldots, \omega_n$ are sub-Gaussian with scale parameter σ;

(iii) [Normalized covariates] $\mathbf{X}_j^{\mathrm{T}}\mathbf{X}_j/n = 1$ for $j = 1, \ldots, p$, where $\mathbf{X}_j = (X_{1j}, \ldots, X_{nj})^{\mathrm{T}}$ is the jth column of $\mathbf{X}$;

(iv) [Empirical compatibility condition] The compatibility condition $(S_{\bar{\beta}}, \xi_0, \nu_0)$ holds for $\hat{H} = \mathbf{X}^{\mathrm{T}}\mathbf{X}/n$, where $S_{\bar{\beta}} = \{1 \le j \le p : \bar{\beta}_j \neq 0\}$ and $\xi_0 > 1$ and $\nu_0 > 0$ are constants.

The choice of target value $\bar{\beta}$ in (4.29) is unique when the compatibility condition $(S_{\bar{\beta}}, \xi_0, \nu_0)$ is assumed to hold with $\xi_0 \ge 1$ for $\hat{H}$ (see Section 5.5.1). Compared with the conditions in Theorem 4.2 (Correct linear regression in fixed design), conditions (i) and (ii) here concern the "true" noises $\omega_1, \ldots, \omega_n$ from nonparametric regression (4.28), and the compatibility condition (iv) on $\hat{H}$ involves the active set $S_{\bar{\beta}}$ of the target value $\bar{\beta}$. If linear regression is correctly specified, then conditions (i)–(iv) here reduce to the corresponding ones in Theorem 4.2.

For any $0 < \epsilon < 1$, suppose that the tuning parameter λ in (4.7) is specified as $\lambda = A_0\lambda_0$ with a constant $A_0 > 1$ and

$$\lambda_0 = \sigma\sqrt{\frac{2\log(p/\epsilon)}{n}},$$

in the same form as in Theorem 4.2. The following result extends the fast-rate error bound in Theorem 4.2 to allow possible model misspecification.

Theorem 4.4 (Fast rate in fixed design with model misspecification). *Assume that conditions (i)–(iv) above are satisfied. For any $0 < \epsilon < 1$, it holds with probability at least $1 - 2\epsilon$ uniformly over $A_0 \ge \frac{\xi_0+1}{\xi_0-1}$ that the Lasso estimator $\hat{\beta}_\lambda$ with $\lambda = A_0\lambda_0$ in (4.7) satisfies*

$$\|\mathbf{X}\hat{\beta}_\lambda - \mathbf{X}\bar{\beta}\|_n^2 + (A_0 - 1)\lambda_0\|\hat{\beta}_\lambda - \bar{\beta}\|_1 \le M_0|S_{\bar{\beta}}|\lambda_0^2, \tag{4.30}$$

or equivalently

$$\begin{aligned} &\frac{1}{2}\|\mathbf{X}\hat{\beta}_\lambda - \mathbf{f}^*\|_n^2 + \frac{1}{2}\|\mathbf{X}\hat{\beta}_\lambda - \mathbf{X}\bar{\beta}\|_n^2 + (A_0 - 1)\lambda_0\|\hat{\beta}_\lambda - \bar{\beta}\|_1 \\ &\le \frac{1}{2}\|\mathbf{X}\bar{\beta} - \mathbf{f}^*\|_n^2 + M_0|S_{\bar{\beta}}|\lambda_0^2, \end{aligned} \tag{4.31}$$

where $M_0 = 4A_0^2\nu_0^{-2}$, depending only on (A_0, ν_0).

The two inequalities (4.30) and (4.31) are equivalent to each other, because algebraically

$$\|\mathbf{X}\tilde{\beta} - \mathbf{X}\bar{\beta}\|_n^2 = \|\mathbf{X}\tilde{\beta} - \mathbf{f}^*\|_n^2 - \|\mathbf{X}\bar{\beta} - \mathbf{f}^*\|_n^2, \tag{4.32}$$

for any $\tilde{\beta} \in \mathbb{R}^p$. In fact, by standard manipulation,

$$\|\mathbf{X}\tilde{\beta} - \mathbf{X}\bar{\beta}\|_n^2 = \|\mathbf{X}\tilde{\beta} - \mathbf{f}^*\|_n^2 - \|\mathbf{X}\bar{\beta} - \mathbf{f}^*\|_n^2 - 2\langle \mathbf{X}\bar{\beta} - \mathbf{f}^*, \mathbf{X}(\tilde{\beta} - \bar{\beta})\rangle_n. \tag{4.33}$$

This also holds with $\bar{\beta}$ replaced by any β and $\mathbf{f}^*$ replaced by any $\mathbf{f}$ (Exercise 4.7). By the definition of $\mathbf{f}^*$ and $\bar{\beta}$, $\langle \mathbf{f}^* - \mathbf{X}\bar{\beta}, \mathbf{X}_j\rangle_n = E\langle \mathbf{Y} - \mathbf{X}\bar{\beta}, \mathbf{X}_j\rangle_n = 0$ for $j = 1, \ldots, p$, and hence $\langle \mathbf{X}\bar{\beta} - \mathbf{f}^*, \mathbf{X}(\tilde{\beta} - \bar{\beta})\rangle_n = 0$. Then (4.32) follows from (4.33).

The error bound (4.30) or (4.31) recovers (4.19) in Theorem 4.2 when linear regression (4.1) is correctly specified, i.e., $\mathbf{f}^* = \mathbf{X}\beta^*$ and $\bar{\beta} = \beta^*$. With possible misspecification of linear regression (4.1), the error bound (4.30) implies that

$$\|\mathbf{X}\hat{\beta}_\lambda - \mathbf{X}\bar{\beta}\|_n^2 \le M_0|S_{\bar{\beta}}|\lambda_0^2, \tag{4.34a}$$

$$\|\hat{\beta}_\lambda - \bar{\beta}\|_1 \le \frac{M_0}{A_0 - 1}|S_{\bar{\beta}}|\lambda_0. \tag{4.34b}$$

which indicate the convergence of $\hat{\beta}_\lambda$ to $\bar{\beta}$ in terms of both the prediction error $\|\mathbf{X}\hat{\beta}_\lambda - \mathbf{X}\bar{\beta}\|_n^2$ and the L_1 estimation error $\|\hat{\beta}_\lambda - \bar{\beta}\|_1$, in a similar manner as in

(4.20a)–(4.20b) on the convergence of $\hat{\beta}_\lambda$ to β^* with correctly specified linear regression. The error bound (4.31) implies that with probability at least $1-2\epsilon$,

$$\|\mathbf{X}\hat{\beta}_\lambda - \mathbf{f}^*\|_n^2 \le \|\mathbf{X}\bar{\beta} - \mathbf{f}^*\|_n^2 + 2M_0|S_{\bar{\beta}}|\lambda_0^2, \tag{4.35}$$

that is, the prediction error $\|\mathbf{X}\hat{\beta}_\lambda - \mathbf{f}^*\|_n^2$ (from comparing $\mathbf{X}\hat{\beta}_\lambda$ with $\mathbf{f}^*$) is upper bounded by $\|\mathbf{X}\bar{\beta}-\mathbf{f}^*\|_n^2$ (from comparing $\mathbf{X}\bar{\beta}$ with $\mathbf{f}^*$), up to a difference of order $|S_{\bar{\beta}}|\frac{\log(ep)}{n}$.

Theorem 4.4 can be shown by carefully extending the proof of Theorem 4.2 (Correct linear regression in fixed design). Taking $\beta = \bar{\beta}$ in the basic inequality (4.11) gives

$$\|\mathbf{X}(\hat{\beta}_\lambda - \bar{\beta})\|_n^2 - \langle \mathbf{Y} - \mathbf{X}\bar{\beta}, \mathbf{X}(\hat{\beta}_\lambda - \bar{\beta})\rangle_n + \lambda\|\hat{\beta}_\lambda\|_1 \le \lambda\|\bar{\beta}\|_1. \tag{4.36}$$

Then the remaining proof of Theorem 4.4 can be completed (Exercise 4.6) by employing similar arguments as in the proof of Theorem 4.2 but using the following lemma instead of Lemma 4.2.

Lemma 4.6 (Concentration of inner products in fixed design with linear approximation). *For $0 < \epsilon < 1$, let $\Omega_1(\lambda_1)$ be the event that*

$$\max_{j=1,\ldots,p} |\langle \mathbf{Y} - \mathbf{X}\bar{\beta}, \mathbf{X}_j\rangle_n| \le \lambda_1,$$

where $\mathbf{X}_j$ is the jth column of $\mathbf{X}$. If $\lambda_1 \ge \sigma\sqrt{\frac{2\log(p/\epsilon)}{n}}$, then $P(\Omega_1(\lambda_1)) \ge 1 - 2\epsilon$.

Proof. By the definition of $\bar{\beta}$ and $\mathbf{f}^*$,

$$E\langle \mathbf{Y} - \mathbf{X}\bar{\beta}, \mathbf{X}_j\rangle_n = \langle \mathbf{f}^* - \mathbf{X}\bar{\beta}, \mathbf{X}_j\rangle_n = 0, \tag{4.37}$$

and hence $\langle \mathbf{Y}-\mathbf{X}\bar{\beta}, \mathbf{X}_j\rangle_n = \langle \mathbf{Y}-\mathbf{f}^*, \mathbf{X}_j\rangle_n$ algebraically for $j = 1, \ldots, p$. Then the event $\Omega_1(\lambda_1)$ is identical to the event that

$$\max_{j=1,\ldots,p} |\langle \mathbf{Y} - \mathbf{f}^*, \mathbf{X}_j\rangle_n| \le \lambda_1.$$

The result follows from the proof of Lemma 4.2 with $\mathbf{X}\beta^*$ replaced by $\mathbf{f}^*$, under the condition that the true noises $\omega_1, \ldots, \omega_n$ are mean-zero sub-Gaussian. □

4.3.2 Random-design linear regression

We study Lasso estimation with possible misspecification of linear regression (4.1) in a random design, where $(X_1, Y_1), \ldots, (X_n, Y_n)$ are random. The target value $\bar{\beta}$ is defined as

$$\bar{\beta} = \text{argmin}_{\beta\in\mathbb{R}^p} E\|\mathbf{Y} - \mathbf{X}^{\mathrm{T}}\beta\|_n^2 \tag{4.38a}$$

$$= \operatorname{argmin}_{\beta \in \mathbb{R}^p} E\|\mathbf{f}^* - \mathbf{X}^{\mathrm{T}}\beta\|_n^2, \tag{4.38b}$$

where $\mathbf{f}^* = (f^*(X_1), \ldots, f^*(X_n))^{\mathrm{T}}$, and the expectation is taken over $(\mathbf{X}, \mathbf{Y})$ jointly in (4.38a) and over $\mathbf{X}$ in (4.38b). The second equality follows from the decomposition, $E\|\mathbf{Y} - \mathbf{X}^{\mathrm{T}}\beta\|_n^2 = E\|\mathbf{Y} - \mathbf{f}^*\|_n^2 + E\|\mathbf{f}^* - \mathbf{X}^{\mathrm{T}}\beta\|_n^2$ [Exercise]. The random-design target value $\bar\beta$ in (4.38a)–(4.38b) does not depend on $\mathbf{X}$ (or the randomness of $\mathbf{X}$), and hence differs from the fixed-design target value $\bar\beta$ in (4.29a)–(4.29b), depending on $\mathbf{X}$.

For theoretical analysis, consider the following conditions:

(i) [Nonparametric regression in random design] $(X_1, Y_1) \ldots, (X_n, Y_n)$ are independent with $E(\omega_i X_i) = 0$, where $\omega_i = Y_i - f^*(X_i)$, for $i = 1, \ldots, n$;

(ii) [Sub-Gaussian true noises] ω_i given X_i is sub-Gaussian with scale parameter σ for $i = 1, \ldots, n$;

(iii) [Bounded covariates] $\max_{j=1,\ldots,p} |X_{ij}| \le C_0$ almost surely for $i = 1, \ldots, n$ with a constant $C_0 > 0$;

(iv) [Theoretical compatibility condition] The compatibility condition $(S_{\bar\beta}, \xi_0^*, \nu_0^*)$ holds for $H = E(\mathbf{X}^{\mathrm{T}}\mathbf{X}/n)$, where $S_{\bar\beta} = \{1 \le j \le p : \bar\beta_j \ne 0\}$ and $\xi_0^* > 1$ and $\nu_0^* > 0$ are constants.

The choice of target value $\bar\beta$ in (4.38) is unique when the compatibility condition $(S_{\bar\beta}, \xi_0^*, \nu_0^*)$ is assumed to hold with $\xi_0^* \ge 1$ for H (see Section 5.5.1). If linear regression is correctly specified, then conditions (i)–(iv) reduce to the corresponding ones in Theorem 4.3 (Correct linear regression in random design).

To establish the convergence of $\hat\beta_\lambda$ to $\bar\beta$ with possible model misspecification in a random design, it requires more careful reasoning than in a fixed design, i.e., in the extension from Theorem 4.2 to Theorem 4.4. The complication can be seen from the fact that Lemma 4.6 (Concentration of the empirical inner products) in a fixed design relies on (4.37) and *algebraically* $\langle \mathbf{Y} - \mathbf{X}\bar\beta, \mathbf{X}_j\rangle_n = \langle \mathbf{Y} - \mathbf{f}^*, \mathbf{X}_j\rangle_n$, which no longer hold in a random design. In fact, by the definition of $\mathbf{f}^*$ and $\bar\beta$, we have for $j = 1, \ldots, p$,

$$E\langle \mathbf{Y} - \mathbf{X}\bar\beta, \mathbf{X}_j\rangle_n = E\langle \mathbf{f}^* - \mathbf{X}\bar\beta, \mathbf{X}_j\rangle_n = 0, \tag{4.39}$$

where $E(\cdot)$ is taken over $\mathbf{X}$ and $\mathbf{Y}$. In contrast, $E(\cdot)$ is taken over only $\mathbf{Y}$ and, without any expectation taken, $\langle \mathbf{f}^* - \mathbf{X}\bar\beta, \mathbf{X}_j\rangle_n = 0$ in (4.37) for a fixed design. From (4.39), $E\langle \mathbf{Y} - \mathbf{X}\bar\beta, \mathbf{X}_j\rangle_n = E\langle \mathbf{Y} - \mathbf{f}^*, \mathbf{X}_j\rangle_n = 0$, i.e., $\langle \mathbf{Y} - \mathbf{X}\bar\beta, \mathbf{X}_j\rangle_n$ and $\langle \mathbf{Y} - \mathbf{f}^*, \mathbf{X}_j\rangle_n$ are mean-zero (*on average* the same), marginally over $(\mathbf{Y}, \mathbf{X})$. But conditionally on $\mathbf{X}$, $\langle \mathbf{Y} - \mathbf{X}\bar\beta, \mathbf{X}_j\rangle_n$ is in general not mean-zero due to model misspecification, whereas $\langle \mathbf{Y} - \mathbf{f}^*, \mathbf{X}_j\rangle_n$ remains mean-zero. Hence in general $\langle \mathbf{Y} - \mathbf{X}\bar\beta, \mathbf{X}_j\rangle_n \ne \langle \mathbf{Y} - \mathbf{f}^*, \mathbf{X}_j\rangle_n$.

For theoretical analysis, consider the following sub-Gaussian condition in place of condition (ii):

(ii†) [Sub-Gaussian pseudo-noises] $Y_i - X_i^{\mathrm{T}}\bar\beta$ given X_i is sub-Gaussian with ψ_2 norm upper bounded by τ, $E(\mathrm{e}^{(Y_i - X_i^{\mathrm{T}}\bar\beta)^2/\tau^2} | X_i) \le 2$, for $i = 1, \ldots, n$.

If linear regression is correctly specified, then condition (ii†) is equivalent to condition (ii) and both reduce to condition (ii) in Theorem 4.3, except that, the sub-Gaussianity of $Y_i - X_i^{\mathrm{T}}\bar\beta$ (which reduces to $\omega_i = Y_i - f^*(X_i) = Y_i - X_i^{\mathrm{T}}\beta^*$ with $f^*(X_i) = X_i^{\mathrm{T}}\beta^*$ and $\bar\beta = \beta^*$) is described in terms of the ψ_2 norm (which is within a constant factor of the scale parameter for mean-zero sub-Gaussian variables). The description by the ψ_2 norm instead of the scale parameter is more convenient with model misspecification, for dealing with the fact that $Y_i - X_i^{\mathrm{T}}\bar\beta$ given X_i is in general not mean-zero, but $(Y_i - X_i^{\mathrm{T}}\bar\beta)X_i$ remains mean-zero marginally by the definition of $\bar\beta$. See Lemma 4.7 for further discussion.

For any $0 < \epsilon < 1$, suppose that the tuning parameter λ in (4.7) is specified as $\lambda = A_0\lambda_0$ with a constant $A_0 > 1$ and

$$\lambda_0 = \max\left\{ \tau C_0\sqrt{\frac{8\log(p/\epsilon)}{n}},\ 2C_0^2\sqrt{\frac{\log(p/\epsilon)}{n}} \right\},$$

depending on τ in condition (ii†) above. The following result extends the fast-rate error bound in Theorem 4.3 to allow possible model misspecification.

Theorem 4.5 (Fast rate in random design with model misspecification). *Assume that conditions (i), (ii†), (iii), and (iv) above are satisfied. For any $0 < \epsilon < 1$, suppose further that*

(v) [Rate condition] $(1+\xi_0^*)^2 \nu_0^{*-2} |S_{\bar\beta}|\lambda_0 \le \eta_0$ for a constant $\eta_0 \in (0,1)$.

Then it holds with probability at least $1 - 2\epsilon - 2\epsilon^2$ uniformly over $A_0 \ge \frac{\xi_0^*+1}{\xi_0^*-1}$ that the Lasso estimator $\hat\beta_\lambda$ with $\lambda = A_0\lambda_0$ in (4.7) satisfies

$$\|\mathbf{X}\hat\beta_\lambda - \mathbf{X}\bar\beta\|_n^2 + (A_0 - 1)\lambda_0 \|\hat\beta_\lambda - \bar\beta\|_1 \le M_0 |S_{\bar\beta}| \lambda_0^2, \tag{4.40}$$

where $M_0 = 4A_0^2\nu_0^{-2}$ with $\nu_0 = \nu_0^*(1-\eta_0)^{1/2}$. Moreover, with probability at least $1 - 4\epsilon - 2\epsilon^2$,

$$\begin{aligned} &\frac{1}{2}\|\mathbf{X}\hat\beta_\lambda - \mathbf{f}^*\|_n^2 + \frac{1}{2}\|\mathbf{X}\hat\beta_\lambda - \mathbf{X}\bar\beta\|_n^2 + (A_0 - 1)\lambda_0\|\hat\beta_\lambda - \bar\beta\|_1 \\ &\le \frac{1}{2}\|\mathbf{X}\bar\beta - \mathbf{f}^*\|_n^2 + \frac{A_0}{A_0 - 1} M_0 |S_{\bar\beta}|\lambda_0^2. \end{aligned} \tag{4.41}$$

The error bound (4.40) recovers (4.26) in Theorem 4.3 when linear regression (4.1) is correctly specified, i.e., $\bar\beta = \beta^*$, in a random design. With possible misspecification of linear regression (4.1), the error bound (4.40) implies the convergence of $\hat\beta_\lambda$ to $\bar\beta$ in terms of both the prediction error $\|\mathbf{X}\hat\beta_\lambda - \mathbf{X}\bar\beta\|_n^2$ and the L_1 estimation error $\|\hat\beta_\lambda - \bar\beta\|_1$, in a similar manner as in (4.35)–(4.34b) with misspecified linear regression for a fixed design. The error bound (4.41) implies that with probability at least $1 - 4\epsilon - 2\epsilon^2$,

$$\|\mathbf{X}\hat\beta_\lambda - \mathbf{f}^*\|_n^2 \le \|\mathbf{X}\bar\beta - \mathbf{f}^*\|_n^2 + \frac{2A_0}{A_0 - 1} M_0 |S_{\bar\beta}|\lambda_0^2,$$

which is comparable to the corresponding bound (4.35) for a fixed design, but with a slightly reduced probability and an additional prefactor $\frac{A_0}{A_0-1}$. This difference stems from the fact that the decomposition (4.32) holds algebraically in a fixed design with the fixed-design target value $\bar{\beta}$, but not in a random design with the random-design target value $\bar{\beta}$.

The error bound (4.40) can be shown by carefully extending the proof of Theorem 4.3 (Correct linear regression in random design) and the proof of Theorem 4.4 (Misspecified linear regression in fixed design). The basic inequality (4.36) can be deduced from (4.11), but with the target value $\bar{\beta}$ defined from (4.38a)–(4.38b) in a random design. The inner-product term $\langle \mathbf{Y} - \mathbf{X}\bar{\beta}, \mathbf{X}(\hat{\beta}_\lambda - \bar{\beta})\rangle_n$ can be controlled in the event $\Omega_1 = \Omega_1(\lambda_0)$ with $\lambda_1 = \lambda_0$ in Lemma 4.7 below. Moreover, the error bound (4.41) can be obtained by combining (4.40) and the following two inequalities. First, from the identity (4.33), we have

$$\begin{aligned}
&\frac{1}{2}\|\mathbf{X}\hat{\beta}_\lambda - \mathbf{f}^*\|_n^2 - \frac{1}{2}\|\mathbf{X}\bar{\beta} - \mathbf{f}^*\|_n^2 \\
&\le \frac{1}{2}\|\mathbf{X}\hat{\beta}_\lambda - \mathbf{X}\bar{\beta}\|_n^2 + |\langle \mathbf{X}\bar{\beta} - \mathbf{f}^*, \mathbf{X}(\hat{\beta}_\lambda - \bar{\beta})\rangle_n|.
\end{aligned}$$

Second, in the intersection of (4.40) and the event $\Omega_3 = \Omega_3(\lambda_0)$ with $\lambda_1 = \lambda_0$ in Lemma 4.7, we have

$$|\langle \mathbf{X}\bar{\beta} - \mathbf{f}^*, \mathbf{X}(\hat{\beta}_\lambda - \bar{\beta})\rangle_n| \le \lambda_0\|\hat{\beta}_\lambda - \bar{\beta}\|_1 \le \frac{1}{A_0 - 1} M_0 |S_{\bar{\beta}}| \lambda_0^2.$$

Proof details for Theorem 4.5 are left as an exercise.

Lemma 4.7 (Concentration of inner products in random design with linear approximation). *For $0 < \epsilon < 1$, let $\Omega_1(\lambda_1)$ be the event that*

$$\max_{j=1,\ldots,p} |\langle \mathbf{Y} - \mathbf{X}\bar{\beta}, \mathbf{X}_j\rangle_n| \le \lambda_1,$$

and let $\Omega_3(\lambda_1)$ be the event that

$$\max_{j=1,\ldots,p} |\langle \mathbf{f}^* - \mathbf{X}\bar{\beta}, \mathbf{X}_j\rangle_n| \le \lambda_1.$$

Under conditions (i), (ii†) and (iii), if $\lambda_1 \ge \tau C_0\sqrt{\frac{8\log(p/\epsilon)}{n}}$, then $P(\Omega_1(\lambda_1)) \ge 1 - 2\epsilon$ and $P(\Omega_3(\lambda_1)) \ge 1 - 2\epsilon$.

Proof. For $j = 1, \ldots, p$, denote $Z_j = \langle \mathbf{Y} - \mathbf{X}\bar{\beta}, \mathbf{X}_j\rangle_n = n^{-1}\sum_{i=1}^n (Y_i - X_i^{\mathrm{T}}\bar{\beta})X_{ij}$. First, we show that $(Y_i - X_i^{\mathrm{T}}\bar{\beta})X_{ij}$ is sub-Gaussian with ψ_2 norm upper-bounded by τC_0. In fact, by conditions (ii†) and (iii),

$$\begin{aligned}
E(\mathrm{e}^{(Y_i - X_i^{\mathrm{T}}\bar{\beta})^2 X_{ij}^2/(\tau^2 C_0^2)}) &\le E(\mathrm{e}^{(Y_i - X_i^{\mathrm{T}}\bar{\beta})^2/\tau^2}) \\
&= E\{E(\mathrm{e}^{(Y_i - X_i^{\mathrm{T}}\bar{\beta})^2/\tau^2}|X_i)\} \le 2. \qquad (4.42)
\end{aligned}$$

Note that $(Y_i - X_i^{\mathrm{T}}\bar{\beta})X_{ij}$ is mean-zero by the definition of $\bar{\beta}$. Therefore, $(Y_i - X_i^{\mathrm{T}}\bar{\beta})X_{ij}$, $i = 1, \ldots, n$, are independent, mean-zero, and sub-Gaussian with ψ_2 norms upper-bounded by τC_0 for each j. Next, by the Hoeffding inequality with ψ_2 norms (Corollary 3.2),

$$P\left(\left|\sum_{i=1}^{n}(Y_i - X_i^{\mathrm{T}}\bar{\beta})X_{ij}\right| \geq t\right) \leq \mathrm{e}^{-\frac{t^2}{8n(\tau^2 C_0^2)}}, \quad t \geq 0.$$

Taking $t = ns$ in the preceding inequality gives

$$P(|Z_j| \geq s) \leq 2\mathrm{e}^{-\frac{ns^2}{8(\tau^2 C_0^2)}}, \quad s \geq 0.$$

By the union bound,

$$\begin{aligned} P(\max_{j=1,\ldots,p} |Z_j| \geq s) &= P(\cup_{j=1}^{p}\{|Z_j \geq s\}) \\ &\leq \sum_{j=1}^{p} P(|Z_j| \geq s) \leq 2p\mathrm{e}^{-\frac{ns^2}{8(\tau^2 C_0^2)}}, \quad s \geq 0. \end{aligned}$$

Take $s = \tau C_0 \sqrt{\frac{8\log(p/\epsilon)}{n}} (= \lambda_0)$ such that $2p\mathrm{e}^{-\frac{ns^2}{8(\tau^2 C_0^2)}} = 2p\mathrm{e}^{-\log(p/\epsilon)} = 2\epsilon$. If $\lambda_1 \geq s$, then the preceding display gives $P(\Omega_1(\lambda_1)) \geq P(\max_{j=1,\ldots,p}|Z_j| \leq s) \geq 1 - 2\epsilon$.

Similarly, it can be shown that $P(\Omega_3(\lambda_1)) \geq 1 - 2\epsilon$ for the stated range of λ_1. First, $(f^*(X_i) - X_i^{\mathrm{T}}\bar{\beta})X_{ij}$ is also sub-Gaussian with ψ_2 norm upper-bounded by τC_0, because

$$\begin{aligned} E(\mathrm{e}^{(f^*(X_i) - X_i^{\mathrm{T}}\bar{\beta})^2 X_{ij}^2/(\tau^2 C_0^2)}) &\leq E\{E(\mathrm{e}^{(Y_i - X_i^{\mathrm{T}}\bar{\beta})^2 X_{ij}^2/(\tau^2 C_0^2)}|X_i)\} \\ &= E(\mathrm{e}^{(Y_i - X_i^{\mathrm{T}}\bar{\beta})^2 X_{ij}^2/(\tau^2 C_0^2)}) \leq 2, \end{aligned} \qquad (4.43)$$

using the definition $f^*(X_i) = E(Y_i|X_i)$ and Jensen's inequality with the convex function $\mathrm{e}^{(y-b)^2/c}$ in y,

$$\mathrm{e}^{(f^*(X_i) - X_i^{\mathrm{T}}\bar{\beta})^2 X_{ij}^2/(\tau^2 C_0^2)} \leq E(\mathrm{e}^{(Y_i - X_i^{\mathrm{T}}\bar{\beta})^2 X_{ij}^2/(\tau^2 C_0^2)}|X_i).$$

Moreover, the variable $(f^*(X_i) - X_i^{\mathrm{T}}\bar{\beta})X_{ij}$ is also mean-zero: $E\{(f^*(X_i) - X_i^{\mathrm{T}}\bar{\beta})X_{ij}\} = E\{(Y_i - X_i^{\mathrm{T}}\bar{\beta})X_{ij}\} = 0$. Then the result follows from the Hoeffding inequality and the union bound. □

We provide additional comments about condition (ii†) and Lemma 4.7. First, the pseudo-noise $Y_i - X_i^{\mathrm{T}}\bar{\beta}$ can be decomposed as

$$Y_i - X_i^{\mathrm{T}}\bar{\beta} = \omega_i + \{f^*(X_i) - X_i^{\mathrm{T}}\bar{\beta}\},$$

where $E(\omega_i|X_i) = 0$ and $f^*(X_i) - X_i^{\mathrm{T}}\bar{\beta}$ is the conditional mean of $Y_i - X_i^{\mathrm{T}}\bar{\beta}$ given X_i. By Theorem 3.7(i), if ω_i given X_i is sub-Gaussian with scale parameter σ and $|f^*(X_i) - X_i^{\mathrm{T}}\bar{\beta}|$ is bounded by a constant D_0 for $i = 1, \ldots, n$, then condition (ii†) is satisfied, with τ within an absolute constant factor of $\sigma + D_0$. This suggests that condition (ii†) tends to be more restrictive than the conditional sub-Gaussianity of $\omega_1, \ldots, \omega_n$.

Second, to alleviate the limitation noted above, Lemma 4.7 (hence Theorem 4.5) remains valid if condition (ii†) is weakened to

(ii‡) [Marginally sub-Gaussian pseudo-noises] $Y_i - X_i^{\mathrm{T}}\bar\beta$ is marginally sub-Gaussian with ψ_2 norm upper bounded by τ, $E(\mathrm{e}^{(Y_i - X_i^{\mathrm{T}}\bar\beta)^2/\tau^2}) \le 2$, for $i = 1, \ldots, n$.

This extension of Lemma 4.7 can be easily shown by modifying (4.42) as

$$E(\mathrm{e}^{(Y_i - X_i^{\mathrm{T}}\bar\beta)^2 X_{ij}^2/(\tau^2 C_0^2)}) \le E(\mathrm{e}^{(Y_i - X_i^{\mathrm{T}}\bar\beta)^2/\tau^2}) \le 2,$$

where the two inequalities follow from conditions (iii) and (ii†) respectively. Then inequality (4.43) holds without any change. For condition (ii‡) to hold, it is sufficient that ω_i and $f^*(X_i) - X_i^{\mathrm{T}}\bar\beta$ are marginally sub-Gaussian with ψ_2 norms upper bounded by a constant for $i = 1, \ldots, n$. This sufficient condition is weaker than requiring boundedness of $|f^*(X_i) - X_i^{\mathrm{T}}\bar\beta|$ for $i = 1, \ldots, n$.

Finally, Lemma 4.7 and accordingly Theorem 4.5 can also be extended by relaxing condition (iii), for example, allowing individual covariates X_{ij} to be sub-Gaussian with bounded ψ_2 norms. In that case, with condition (ii†) or (ii‡) used, the product $(Y_i - X_i^{\mathrm{T}}\bar\beta)X_{ij}$ is sub-Exponential by Theorem 3.10 (Product of Sub-Gaussian variables). Hence additional arguments will be needed, related to sub-Exponential variables.

4.4 Dantzig selector

For linear regression (4.1), the Dantzig selector is defined as

$$\hat\beta_\lambda = \mathrm{argmin}_{\beta \in \mathbb{R}^p} \|\beta\|_1 \tag{4.44a}$$

$$\text{subject to} \quad \|n^{-1}\mathbf{X}^{\mathrm{T}}(\mathbf{Y} - \mathbf{X}\beta)\|_\infty \le \lambda, \tag{4.44b}$$

where $\lambda > 0$ is a tuning parameter and $\|\cdot\|_\infty$ denotes the L_∞ norm. The Dantzig selector can be shown to achieve error bounds similar to those of Lasso estimation.

We mainly discuss the setting of linear regression correctly specified in a fixed design. The case of a random design is left as Exercise 4.8. For any $0 < \epsilon < 1$, let

$$\lambda_0 = \sigma\sqrt{\frac{2\log(p/\epsilon)}{n}},$$

as in Lasso estimation in Section 4.2.1 (Correct linear regression in fixed design). The following result, comparable to Theorem 4.2, gives fast-rate error bounds for the Dantzig selector.

Theorem 4.6 (Dantzig selector in fixed design). *Assume that the following conditions are satisfied:*

(i)–(iii) Same as in Theorem 4.1;

(iv) [Empirical compatibility condition] the compatibility condition $(S_{\beta^*}, \xi_0, \nu_0)$ *holds for* $\hat{H} = \mathbf{X}^{\mathrm{T}}\mathbf{X}/n$, *where* $S_{\beta^*} = \{1 \le j \le p : \beta_j^* \ne 0\}$ *and* $\xi_0 = 1$ *and* $\nu_0 > 0$ *are constants. Note that the compatibility condition* $(S_{\beta^*}, \xi_0 = 1, \nu_0)$ *is implied by (hence weaker than) the compatibility condition* $(S_{\beta^*}, \xi_0 = c, \nu_0)$ *for any* $c \ge 1$.

For any $0 < \epsilon < 1$, *the Dantzig selector* $\hat{\beta}_\lambda$ *in* (4.44) *with* $\lambda = \lambda_0$ *satisfies that with probability at least* $1 - 2\epsilon$,

$$\|\mathbf{X}\hat{\beta}_\lambda - \mathbf{X}\beta^*\|_n^2 \le (16\nu_0^{-2})|S_{\beta^*}|\lambda_0^2, \tag{4.45a}$$

$$\|\hat{\beta}_\lambda - \beta^*\|_1 \le (8\nu_0^{-2})|S_{\beta^*}|\lambda_0. \tag{4.45b}$$

To prove Theorem 4.6, we use Lemma 4.2 (Concentration of inner products) and the following consequence of the definition (4.44), in place of Lemma 4.1 (Basic inequality).

Lemma 4.8. *For any* $\beta = (\beta_1, \ldots, \beta_p)^{\mathrm{T}}$ *satisfying the constraint* (4.44b),

$$\sum_{j \notin S_\beta} |\hat{\beta}_{\lambda,j} - \beta_j| \le \sum_{j \in S_\beta} |\hat{\beta}_{\lambda,j} - \beta_j|,$$

where $S_\beta = \{1 \le j \le p : \beta_j \ne 0\}$. *In other words,* $\hat{\beta}_\lambda - \beta$ *satisfies the cone condition associated with the compatibility condition in (iv).*

Proof. For any β satisfying the constraint (4.44b), by the definition of $\hat{\beta}_\lambda$, we have $\|\hat{\beta}_\lambda\|_1 \le \|\beta\|_1$, which by the definition of S_β can be stated as

$$\sum_{j \in S_\beta} |\hat{\beta}_{\lambda,j}| + \sum_{j \notin S_\beta} |\hat{\beta}_{\lambda,j}| \le \sum_{j \in S_\beta} |\beta_j|.$$

The desired result follows from the triangle inequality:

$$\begin{aligned}\sum_{j \notin S_\beta} |\hat{\beta}_{\lambda,j} - \beta_j| &= \sum_{j \notin S_\beta} |\hat{\beta}_{\lambda,j}| \\ &\le \sum_{j \in S_\beta} |\beta_j| - \sum_{j \in S_\beta} |\hat{\beta}_{\lambda,j}| \le \sum_{j \in S_\beta} |\hat{\beta}_{\lambda,j} - \beta_j|.\end{aligned}$$

□

Proof of Theorem 4.6. In the event $\Omega_1 = \Omega_1(\lambda_0)$ with $\lambda_1 = \lambda_0$ in Lemma 4.2,

$$\|n^{-1}\mathbf{X}^{\mathrm{T}}(\mathbf{Y} - \mathbf{X}\beta^*)\|_\infty = \max_{j=1,\ldots,p} |\langle \mathbf{Y} - \mathbf{X}\beta^*, \mathbf{X}_j\rangle_n| \le \lambda_0,$$

i.e., β^* satisfies the constraint (4.44b) with $\lambda = \lambda_0$. Taking $\beta = \beta^*$ in Lemma 4.8 shows that

$$\sum_{j\notin S_{\beta^*}} |\hat\beta_{\lambda,j} - \beta_j^*| \le \sum_{j\in S_{\beta^*}} |\hat\beta_{\lambda,j} - \beta_j^*|,$$

i.e., $\hat\beta_\lambda - \beta^*$ satisfies the cone condition with $\xi_0 = 1$. Then by the compatibility condition,

$$\begin{aligned}
&\nu_0^2\Big(\sum_{j\in S_{\beta^*}} |\hat\beta_{\lambda,j} - \beta_j^*|\Big)^2 \\
&\le |S_{\beta^*}|(\hat\beta_\lambda - \beta^*)^{\mathrm{T}}\hat H(\hat\beta_\lambda - \beta^*) = |S_{\beta^*}| \cdot \|\mathbf{X}(\hat\beta_\lambda - \beta^*)\|_n^2.
\end{aligned} \tag{4.46}$$

By the Hölder inequality and the triangle inequality,

$$\begin{aligned}
&\|\mathbf{X}(\hat\beta_\lambda - \beta^*)\|_n^2 \\
&= n^{-1}(\hat\beta_\lambda - \beta^*)^{\mathrm{T}}\mathbf{X}^{\mathrm{T}}\mathbf{X}(\hat\beta_\lambda - \beta^*) \\
&\le n^{-1}\|\mathbf{X}^{\mathrm{T}}\mathbf{X}(\hat\beta_\lambda - \beta^*)\|_\infty \cdot \|\hat\beta_\lambda - \beta^*\|_1 \\
&\le \Big\{n^{-1}\|\mathbf{X}^{\mathrm{T}}(\mathbf{Y} - \mathbf{X}\hat\beta)\|_\infty + n^{-1}\|\mathbf{X}^{\mathrm{T}}(\mathbf{Y} - \mathbf{X}\beta^*)\|_\infty\Big\} \cdot \|\hat\beta_\lambda - \beta^*\|_1,
\end{aligned}$$

which in the event Ω_1 leads to

$$\|\mathbf{X}(\hat\beta_\lambda - \beta^*)\|_n^2 \le 2\lambda_0\|\hat\beta_\lambda - \beta^*\|_1 \le 4\lambda_0 \sum_{j\in S_{\beta^*}} |\hat\beta_{\lambda,j} - \beta_j^*|. \tag{4.47}$$

The first inequality (4.47) holds because $\|n^{-1}\mathbf{X}^{\mathrm{T}}(\mathbf{Y} - \mathbf{X}\beta^*)\|_\infty \le \lambda_0$ and, by the definition of $\hat\beta_\lambda$, $\|n^{-1}\mathbf{X}^{\mathrm{T}}(\mathbf{Y} - \mathbf{X}\hat\beta_\lambda)\|_\infty \le \lambda_0$. The second inequality holds because $\|\hat\beta_\lambda - \beta^*\|_1 \le 2\sum_{j\in S_{\beta^*}} |\hat\beta_{\lambda,j} - \beta_j^*|$. Combining (4.46) and (4.47) yields

$$\|\mathbf{X}(\hat\beta_\lambda - \beta^*)\|_n^2 \le 4\lambda_0 \sum_{j\in S_{\beta^*}} |\hat\beta_{\lambda,j} - \beta_j^*| \le 4\lambda_0 \cdot \nu_0^{-1}|S_{\beta^*}|^{1/2}\|\mathbf{X}(\hat\beta_\lambda - \beta^*)\|_n,$$

which implies that $\|\mathbf{X}(\hat\beta_\lambda - \beta^*)\|_n \le 4\lambda_0\nu_0^{-1}|S_{\beta^*}|^{1/2}$, that is, inequality (4.45a). Moreover, returning to $\|\hat\beta_\lambda - \beta^*\|_1 \le 2\sum_{j\in S_{\beta^*}} |\hat\beta_{\lambda,j} - \beta_j^*|$ and using (4.46) and (4.45a) gives

$$\begin{aligned}
\|\hat\beta_\lambda - \beta^*\|_1 &\le 2\sum_{j\in S_{\beta^*}} |\hat\beta_{\lambda,j} - \beta_j^*| \\
&\le 2\nu_0^{-1}|S_{\beta^*}|^{1/2}\|\mathbf{X}(\hat\beta_\lambda - \beta^*)\|_n \le 8\nu_0^{-2}|S_{\beta^*}|\lambda_0,
\end{aligned}$$

that is, inequality (4.45b). To conclude, inequalities (4.45a) and (4.45b) hold in the event Ω_1, which has probability at least $1 - 2\epsilon$ by Lemma 4.2. □

With possible misspecification of linear regression, the Dantzig selector $\hat{\beta}_\lambda$ can be shown to satisfy similar bounds on the prediction error $\|\mathbf{X}\hat{\beta}_\lambda - \mathbf{X}\bar{\beta}\|_n^2$ and L_1 estimation error $\|\hat{\beta}_\lambda - \bar{\beta}\|_1$ as in Theorems 4.4 and 4.5 for Lasso estimation (Exercise 4.8).

Exercises

Ex 4.1. Lemma 4.1 (Basic inequality) is proved in Section 4.2.1 using the stationary condition for $\hat{\beta}_\lambda$. Use the following idea to give an alternative proof of Lemma 4.1: Apply the weaker inequality (4.12) to compare the Lasso objective function at $\hat{\beta}_\lambda$ and $(1-t)\hat{\beta}_\lambda + t\beta$ for $0 \le t \le 1$.

Ex 4.2. For the quadratic function (4.17) and the cone $\{b \in \mathbb{R}^2 : |b_2| \le 1.5|b_1|\}$, calculate the minimum values stated in (4.18).

Ex 4.3. Complete the proof of Corollary 4.1 (O_p error rate in fixed design).

Ex 4.4. Consider fixed-design linear regression (4.1), correctly specified, as in Section 4.2.1. For any $0 < q \le 1$, denote $\|\beta\|_q = \sum_{j=1}^p |\beta_j|^q$ for $\beta = (\beta_1, \ldots, \beta_p)^{\mathrm{T}}$. For the Lasso estimator $\hat{\beta}_\lambda$, show that under some conditions,

$$\|\mathbf{X}\hat{\beta}_\lambda - \mathbf{X}\beta^*\|_n^2 = O_p(1)\|\beta^*\|_q \lambda_0^{2-q},$$

where $\lambda_0 = \sqrt{\log(n \vee p)/n}$. Present your conditions and proof of the result. [Hint: Choose the subset $S = \{1 \le j \le p : |\beta_j^*| > \lambda_{0,\epsilon}\}$ with $\lambda_{0,\epsilon} = c\sqrt{\log(p/\epsilon)/n}$ for some constant c.]

Ex 4.5. Provide a proof of Corollary 4.2 (O_p error rate in random design).

Ex 4.6. Complete the proof of Theorem 4.4 (Fixed-design linear regression with model misspecification). [Hint: The proof involves similar steps as in the proof of Theorem 4.2. The essence of this problem is to present those steps in a clear and simple manner.]

Ex 4.7. For any $\mathbf{X} \in \mathbb{R}^{n\times p}$, $\beta, \tilde{\beta} \in \mathbb{R}^p$, and $\mathbf{f} \in \mathbb{R}^n$, show that

$$\|\mathbf{X}\tilde{\beta} - \mathbf{X}\beta\|_n^2 = \|\mathbf{X}\tilde{\beta} - \mathbf{f}\|_n^2 - \|\mathbf{X}\beta - \mathbf{f}\|_n^2 - 2\langle \mathbf{X}\beta - \mathbf{f}, \mathbf{X}(\tilde{\beta} - \beta)\rangle_n.$$

Ex 4.8. Let $\hat{\beta}_\lambda$ be the Dantzig estimator in (4.44).

(i) Consider linear regression correctly specified in a random design, as in Section 4.2.1. Show that under suitable conditions, we have with probability at least $1 - c\,\epsilon$,

$$\|\mathbf{X}\hat{\beta}_\lambda - \mathbf{X}\beta^*\|_n^2 \le M_0|S_{\beta^*}|\lambda_0^2,$$
$$\|\hat{\beta}_\lambda - \beta^*\|_1 \le M_1|S_{\beta^*}|\lambda_0,$$

for some constants (c, M_0, M_1), where λ_0 is defined as in Theorem 4.6 (Dantzig selector in fixed design). Present your technical conditions and proof of the result, including explicit (c, M_0, M_1).

(ii) Suppose that the linear regression may be misspecified. Can a similar result to Theorem 4.6 be obtained in a fixed design, with β^* replaced by the target value $\bar{\beta}$ in (4.29)? Can a similar result to that in (i) be obtained in a random design, with β^* replaced by the target value $\bar{\beta}$ in (4.38)? Give sufficient reasoning to justify your answer.

Ex 4.9. Suppose that the training data (X_i, Y_i), $i = 1, \ldots, n$, are IID, satisfying linear regression (4.1). Let (X_0, Y_0) be a new observation IID as the training data. For the Lasso estimator $\hat{\beta}_\lambda$ in (4.7), the problem of interest is to study the out-of-sample prediction MSE:

$$E\{(Y_0 - X_0^{\mathrm{T}}\hat{\beta}_\lambda)^2|\mathbf{X}, \mathbf{Y}\}, \tag{4.48}$$

where $\mathbf{X} = (X_1, \ldots, X_n)^{\mathrm{T}}$ and $\mathbf{Y} = (Y_1, \ldots, Y_n)^{\mathrm{T}}$, and the expectation is taken over (X_0, Y_0) conditionally on $(\mathbf{X}, \mathbf{Y})$. Hence (4.48) is a function of the training data. For $0 < \epsilon < 1$, denote $\lambda_0 = \sqrt{\frac{log(p/\epsilon)}{n}}$, with dependency on ϵ suppressed in the notation.

(i) Show that

$$\begin{aligned} E\{(Y_0 - X_0^{\mathrm{T}}\hat{\beta}_\lambda)^2|\mathbf{X}, \mathbf{Y}\} &= E\{(Y_0 - X_0^{\mathrm{T}}\beta^*)^2\} + E\{(X_0^{\mathrm{T}}(\hat{\beta}_\lambda - \beta^*))^2|\mathbf{X}, \mathbf{Y}\} \\ &= E\{(Y_0 - X_0^{\mathrm{T}}\beta^*)^2\} + (\hat{\beta}_\lambda - \beta^*)^{\mathrm{T}}\Sigma(\hat{\beta}_\lambda - \beta^*), \end{aligned}$$

where $\Sigma = E(X_0 X_0^{\mathrm{T}})$.

(ii) Assume that conditions (i)–(v) are satisfied as in Theorem 4.3 (Fast rate in random design). For $\hat{\beta}_\lambda$ with $\lambda = A_0\lambda_0$ and $A_0 \ge \frac{\xi_0^*+1}{\xi_0^*-1}$, show that in the event $\Omega_1 \cap \Omega_2$ as in the proof of Theorem 4.3,

$$E\{(Y_0 - X_0^{\mathrm{T}}\hat{\beta}_\lambda)^2|\mathbf{X}, \mathbf{Y}\} \le E\{(Y_0 - X_0^{\mathrm{T}}\beta^*)^2\} + M_1|S_{\beta^*}|\lambda_0^2,$$

where $M_1 > 0$ is some constant. State your M_1 explicitly.

Ex 4.10. Suppose that $X_1, \ldots, X_n$ are IID from a p-variate normal distribution with mean vector 0 and unknown variance matrix Σ. Assume that the smallest eigenvalue of Σ is lower bounded by a constant $\sigma_0 > 0$ (hence Σ is non-singular), and the diagonal elements of Σ (i.e., the marginal variances of the components of X_i) are upper bounded by a constant $\sigma_1 > 0$.

Given a vector $x^* \in \mathbb{R}^p$, it is of interest to estimate $u^* = \Gamma x^*$, where $\Gamma = \Sigma^{-1}$ is called the precision matrix. For p much less than n, a simple estimator of u^* is $\hat{u} = \hat{\Sigma}^{-1}x^*$, where $\hat{\Sigma} = \mathbf{X}^{\mathrm{T}}\mathbf{X}/n$, and $\mathbf{X} = (X_1, \ldots, X_n)^{\mathrm{T}}$. Equivalently, $\hat{u}$ can be determined as

$$\hat{u} = \mathrm{argmin}_{u \in \mathbb{R}^p} \left\{ \frac{1}{2} u^{\mathrm{T}} \hat{\Sigma} u - u^{\mathrm{T}} x^* \right\}.$$

For p close to or larger than n, we consider a Lasso penalized estimator

$$\hat{u}_\lambda = \mathrm{argmin}_{u \in \mathbb{R}^p} \left\{ \frac{1}{2} u^{\mathrm{T}} \hat{\Sigma} u - u^{\mathrm{T}} x^* + \lambda \|u\|_1 \right\}, \tag{4.49}$$

where $\lambda > 0$ is a tuning parameter. The Lasso penalized estimator is expected to perform well in the setting where the true value u^* is sparse even with p larger than n. In this problem, you are asked to study convergence of $\hat{u}_\lambda$ to u^*.

(i) Show that $u^{*\mathrm{T}} X_i$ is sub-Gaussian with scale parameter $\sigma_1 \|u^*\|_2$.

(ii) Let $\Omega_0 = \Omega_0(\lambda_0)$ be the event that

$$\|\hat{\Sigma} u^* - x^*\|_\infty \leq \lambda_0.$$

For any $0 < \epsilon < 1$, find λ_0, depending on (n, p, ϵ), such that $P(\Omega_0(\lambda_0)) \geq 1 - 2\epsilon$.

(iii) Let Ω_{Comp} be the event that empirical compatibility condition (S_{u^*}, ξ_0, ν_0) holds for $\hat{\Sigma}$ for some constants $\xi_0 > 1$ and $\nu_0 > 0$. Take $\lambda = A_0 \lambda_0$ in (4.49) for a constant $A_0 > 1$. Show that in the event $\Omega_0(\lambda_0) \cap \Omega_{\mathrm{Comp}}$, for a suitable choice of A_0,

$$\begin{aligned} \|\mathbf{X}(\hat{u}_\lambda - u^*)\|_n^2 &\leq M_0 |S_{u^*}| \lambda_0^2, \\ \|\hat{u}_\lambda - u^*\|_1 &\leq M_1 |S_{u^*}| \lambda_0, \end{aligned} \tag{4.50}$$

where $M_0, M_1 > 0$ are some constants, depending on (A_0, ν_0, η_0) but not on (n, p, ϵ), and $S_{u^*} = \{1 \leq j \leq p : u_j^* \neq 0\}$.

(iv) Let $\Omega_1 = \Omega_1(\lambda_1)$ be the event that

$$\max_{j,k=1,\ldots,p} |\hat{\Sigma}_{jk} - \Sigma_{jk}| \leq \lambda_1,$$

where $\hat{\Sigma}_{jk}$ or Σ_{jk} denotes the (j, k)th element of $\hat{\Sigma}$ or Σ respectively. For any $0 < \epsilon < 1$, find λ_1, depending on (n, p, ϵ), such that $P(\Omega_1(\lambda_1)) \geq 1 - 2\epsilon$.

(v) Show that if $|S_{u^*}|\lambda_1$ is sufficiently small, then for $\xi_0 = 2$ and $\nu_0 = \sigma_0/2$, $\Omega_{\mathrm{Comp}} \supset \Omega_1(\lambda_1)$, and hence $P(\Omega_{\mathrm{Comp}}) \geq 1 - 2\epsilon$.

To conclude, combining the preceding results shows that the error bounds in (4.50) hold with probability at least $1 - 4\epsilon$.

5

High-dimensional generalized linear regression

In this chapter, we study regularized convex M-estimation in high-dimensional, sparse generalized linear regression, where the number of covariates is close to or greater than the sample size, and a small but unknown subset of covariates are associated with nonzero coefficients. We focus on Lasso-penalized estimation and discuss both slow and fast rates for error bounds, in fixed or random design, with correct model specification or with possible misspecification. Furthermore, we discuss error bounds for Lasso-penalized M-estimation with a non-smooth loss function such as in quantile regression.

We study statistical properties of regularized convex M-estimation in high-dimensional generalized linear regression, where the number of covariates (or associated coefficients) is close to or greater than the sample size. We use the phrase "generalized linear regression" interchangeably with perhaps the more common "generalized linear models" (e.g., McCullagh & Nelder, 1989). We focus on Lasso estimation (Tibshirani, 1996; Hastie et al., 2015), and discuss finite-sample (or non-asymptotic) error bounds (including slow and fast rates), in fixed or random design, with correct model specification or with possible misspecification. As mentioned in Chapter 4, the broader subject of high-dimensional statistics is covered by many books (e.g., Bühlmann & van de Geer, 2011; Wainwright, 2019).

Compared with linear regression, a technical complication is that the (non-penalized) objective function, also called the loss function, is non-quadratic and hence additional arguments are required. We mainly present theoretical analysis, based on Tan (2020b), Supplement Section II, for convex loss functions in a specific form that accommodates both likelihood estimation and non-likelihood estimation in generalized linear regression. The analysis can be readily extended to M-estimation with twice differentiable loss functions. In Section 5.5.2, we discuss theoretical analysis for Lasso-penalized M-estimation with a non-smooth loss function such as in quantile regression.

DOI: 10.1201/9781003680734-5

5.1 Introduction

Let (X_i, Y_i), $i = 1, \ldots, n$, be a collection of n observations, where $X_i = (X_{i1}, \ldots, X_{ip})^{\mathrm{T}}$ is a $p \times 1$ covariate vector and Y_i is a response variable. Consider a generalized linear regression model, with mean restriction

$$E(Y_i|X_i) = g^{-}(X_i^{\mathrm{T}}\beta^*), \quad i = 1, \ldots, n, \tag{5.1}$$

where $\beta^* = (\beta_1^*, \ldots, \beta_p^*)^{\mathrm{T}}$ is an unknown coefficient vector, $g(\cdot)$ is a *known*, increasing differentiable function, called the link function, and $g^{-}(\cdot)$ is the inverse function of $g(\cdot)$. An exponential-family assumption can also be introduced, indicating that the conditional density of Y_i given X_i (or with fixed X_i) is of the form

$$f(Y_i|X_i) \propto \exp\left\{\frac{Y_i\theta_i^* - a(\theta_i^*)}{\phi^*} + c(Y_i, \phi^*)\right\}, \tag{5.2}$$

where $\theta_i^* = \theta(X_i; \beta^*)$ is the canonical parameter, $\phi^* > 0$ is the dispersion parameter (assumed to be constant in X_i), $a(\cdot)$ is a *known*, strictly convex, twice differentiable function, and $c(\cdot)$ is a *known* function. For a continuous random variable, the density function is defined with respect to the Lebesgue measure and corresponds to the usual probability density function. For a discrete random variable, the density function is defined with respect to the counting measure, and corresponds to the usual probability mass function.

The exponential family (5.2) has two simple but important properties:

$$E(Y_i|X_i) = a'(\theta_i^*), \tag{5.3a}$$
$$\mathrm{var}(Y_i|X_i) = \phi^* a''(\theta_i^*). \tag{5.3b}$$

The variance identity (5.3b) underlies the convexity of $a(\cdot)$. From the mean identity (5.3a), the canonical parameter $\theta_i^* = \theta(X_i; \beta^*)$ is linked to the mean parameter $\mu_i^* = E(Y_i|X_i)$ by $\mu_i^* = a'(\theta_i^*)$ or equivalently $\theta_i^* = (a')^{-}(\mu_i^*)$. Combining this and the mean restriction (5.1), $\mu_i^* = g^{-}(X_i^{\mathrm{T}}\beta^*)$, indicates that the canonical parameter $\theta_i^* = \theta(X_i; \beta^*)$ is linked to the linear predictor $X_i^{\mathrm{T}}\beta^*$ as follows:

$$\theta_i^* = (a')^{-}(g^{-}(X_i^{\mathrm{T}}\beta^*)).$$

Note that $g(\cdot)$ and $a(\cdot)$ are two separate choices made by users (or data analysts): $g(\cdot)$ is a choice for directly indicating how the mean $E(Y_i|X_i)$ is linked to the linear predictor $X_i^{\mathrm{T}}\beta^*$, whereas $a(\cdot)$ is determined by a choice of the exponential family (5.2) for the distribution of Y_i given X_i. See McCullagh and Nelder (1989) and Tan (2025), Section 2.1, for further information.

For simplicity, we consider the canonical link such that

$$g^{-}(\cdot) = a'(\cdot) \quad \Longleftrightarrow \quad \theta_i^* = X_i^{\mathrm{T}}\beta^*.$$

For maximum likelihood estimation (with known ϕ^*), the log-likelihood function of β is

$$\ell_n(\beta) = \sum_{i=1}^{n} \frac{Y_i X_i^{\mathrm{T}}\beta - a(X_i^{\mathrm{T}}\beta)}{\phi^*} + \text{constant}.$$

After dropping the constant factor ϕ^* and the additive constant, the objective function becomes

$$K_n(\beta) = \frac{-1}{n}\sum_{i=1}^{n} \left\{Y_i X_i^{\mathrm{T}}\beta - a(X_i^{\mathrm{T}}\beta)\right\}, \tag{5.4}$$

which is convex in β by the convexity of $a(\cdot)$. The maximum likelihood estimator of β is defined as a minimizer of $K_n(\beta)$, even with ϕ^* unknown.

Example 5.1 (Logistic regression)**.** For binary responses, consider logistic regression

$$P(Y_i = 1|X_i) = \text{expit}(X_i^{\mathrm{T}}\beta) = \frac{\mathrm{e}^{X_i^{\mathrm{T}}\beta^*}}{1+\mathrm{e}^{X_i^{\mathrm{T}}\beta^*}}, \quad i = 1, \ldots, n. \tag{5.5}$$

The link function $g(p) = \log \frac{p}{1-p}$ is the canonical link for the Bernoulli family,

$$f(y; p) = p^y(1-p)^{1-y} = \exp\left\{y \log \frac{p}{1-p} + \log(1-p)\right\}, \quad y = 0, 1,$$

where the canonical parameter is $\theta = \log \frac{p}{1-p}$ and $a(\theta) = -\log(1-p) = \log(1+\mathrm{e}^{\theta})$. The average negative log-likelihood function of β is

$$K_n(\beta) = \frac{-1}{n}\sum_{i=1}^{n} \left\{Y_i X_i^{\mathrm{T}}\beta - \log(1+\mathrm{e}^{X_i^{\mathrm{T}}\beta})\right\}, \tag{5.6}$$

which can be directly shown to be convex in β. The maximum likelihood estimator of β, defined as a minimizer of $K_n(\beta)$, satisfies the score equation setting $\frac{\partial}{\partial\beta}K_n(\beta)$ to 0, i.e.,

$$0 = \frac{-1}{n}\sum_{i=1}^{n} \left\{Y_i - \text{expit}(X_i^{\mathrm{T}}\beta)\right\} X_i.$$

Example 5.2 (Log-linear regression)**.** For nonnegative responses, consider log-linear regression

$$P(Y_i = 1|X_i) = \mathrm{e}^{X_i^{\mathrm{T}}\beta^*}, \quad i = 1, \ldots, n. \tag{5.7}$$

The link function $g(p) = \log\mu$ is the canonical link for the Poisson family,

$$f(y; \mu) = \frac{\mu^y \mathrm{e}^{-\mu}}{y!} = \exp\left\{y\log\mu - \mu + \log(y!)\right\}, \quad y = 0, 1, \ldots,$$

where the canonical parameter is $\theta = \log \mu$ and $a(\theta) = \mu = e^{\theta}$. The average negative log-likelihood function of β is

$$K_n(\beta) = \frac{-1}{n} \sum_{i=1}^{n} \left(Y_i X_i^{\mathrm{T}} \beta - e^{X_i^{\mathrm{T}} \beta} \right), \tag{5.8}$$

which can be directly shown to be convex in β. The maximum likelihood estimator of β, defined as a minimizer of $K_n(\beta)$, satisfies the score equation setting $\frac{\partial}{\partial \beta} K_n(\beta)$ to 0, i.e.,

$$0 = \frac{-1}{n} \sum_{i=1}^{n} \left(Y_i - e^{X_i^{\mathrm{T}} \beta} \right) X_i.$$

In high-dimensional generalized linear regression (with p close to or even greater than n), Lasso estimation can be introduced similarly as in linear regression. For an (non-penalized) objective function $K_n(\beta)$, we define the Lasso estimator as

$$\hat{\beta}_\lambda = \operatorname{argmin}_{\beta \in \mathbb{R}^p} \left\{ K_n(\beta) + \lambda \|\beta\|_1 \right\}, \tag{5.9}$$

where $\lambda > 0$ is a tuning parameter. For the objective function (5.4) associated with the log-likelihood function in generalized linear regression with a canonical link, the resulting estimator $\hat{\beta}_\lambda$ is called a Lasso-penalized likelihood estimator. Nevertheless, Lasso estimation can also be combined with an objective function which is distinct from a log-likelihood function, as shown in the following two examples. Such an estimator is called a Lasso-penalized M-estimator.

Example 5.3 (Probit regression). For binary responses, consider probit regression

$$P(Y_i = 1 | X_i) = \Phi(X_i^{\mathrm{T}} \beta^*), \quad i = 1, \ldots, n.$$

where $\Phi(\cdot)$ is the cumulative distribution function of $\mathrm{N}(0,1)$. The link function $g(p) = \Phi^{-}(p)$ is not the canonical link for the Bernoulli family. The average negative log-likelihood function of β is

$$\frac{-1}{n} \ell_n(\theta) = \frac{-1}{n} \sum_{i=1}^{n} \left\{ Y_i \log \Phi(X_i^{\mathrm{T}} \beta) + (1 - Y_i) \log \Phi(-X_i^{\mathrm{T}} \beta) \right\},$$

which may not be convex in β. The maximum likelihood estimator of β, defined as a maximizer of $\ell_n(\beta)$, satisfies the score equation

$$0 = \frac{-1}{n} \sum_{i=1}^{n} \left\{ Y_i - \Phi(X_i^{\mathrm{T}} \beta) \right\} \frac{\Phi'(X^{\mathrm{T}} \beta)}{\Phi(X^{\mathrm{T}} \beta)(1 - \Phi(X^{\mathrm{T}} \beta))} X_i.$$

Alternatively, consider an objective function in the form of (5.4):

$$K_n(\beta) = \frac{-1}{n}\sum_{i=1}^{n}\{Y_i X_i^{\mathrm{T}}\beta - a(X_i^{\mathrm{T}}\beta)\}, \tag{5.10}$$

where $a(\eta) = \int_0^\eta \Phi(u)\,du$ such that $a'(\eta) = g^{-}(\eta) = \Phi(\eta)$. Note that $a'(\eta)$ is increasing in η and hence $a(\eta)$ is convex in η. Then the objective function $K(\beta)$ above is convex in β. The estimator defined as a minimizer of $K_n(\beta)$ satisfies the estimating equation

$$0 = \frac{-1}{n}\sum_{i=1}^{n}\{Y_i - \Phi(X_i^{\mathrm{T}}\beta)\}\,X_i,$$

which differs from the score equation by maximum likelihood estimation.

Example 5.4 (Calibrated estimation for logistic regression). For logistic regression (5.5), non-likelihood objective functions can also be considered. For example, consider the objective function, called the calibration loss (Tan, 2020a, 2020b),

$$K_n(\beta) = \frac{1}{n}\sum_{i=1}^{n}\{Y_i \exp(-X_i^{\mathrm{T}}\beta) + (1-Y_i)X_i^{\mathrm{T}}\beta\}, \tag{5.11}$$

which can be directly shown to be convex in β. The calibrated estimator of β, defined as a minimizer of $K_n(\beta)$, satisfies the estimating equation setting $\frac{\partial}{\partial\beta}K_n(\beta)$ to 0, i.e.,

$$\begin{aligned} 0 &= \frac{-1}{n}\sum_{i=1}^{n}\left\{Y_i \mathrm{e}^{-X_i^{\mathrm{T}}\beta} - (1-Y_i)\right\}X_i \\ &= \frac{-1}{n}\sum_{i=1}^{n}\left\{\frac{Y_i}{\mathrm{expit}(X_i^{\mathrm{T}}\beta)} - 1\right\}X_i, \end{aligned} \tag{5.12}$$

which is called calibration equation. Equivalently, (5.12) can be written as

$$\frac{1}{n}\sum_{1\le i\le n: Y_i=1} w_i(\beta)X_i = \frac{1}{n}\sum_{i=1}^{n} X_i.$$

where $w_i(\beta) = \{\mathrm{expit}(X_i^{\mathrm{T}}\beta)\}^{-1}$ is an inverse probability weight, in being the inverse of the probability, $\mathrm{expit}(X_i^{\mathrm{T}}\beta)$, of $Y_i = 1$ given X_i (with β treated as the true value). Assuming that 1 is included in X_i, the preceding equation implies that $\sum_{1\le i\le n: Y_i=1} w_i(\beta) = n$ and hence

$$\frac{\sum_{1\le i\le n: Y_i=1} w_i(\beta)X_i}{\sum_{1\le i\le n: Y_i=1} w_i(\beta)} = \frac{1}{n}\sum_{i=1}^{n} X_i,$$

i.e., the weighted average of X_i with weight $w_i(\beta)$ in the subsample $\{1 \le i \le n : Y_i = 1\}$ matches the simple average of X_i in the overall sample. By this design, calibrated estimation provides an appealing alternative to maximum likelihood estimation when the fitted probabilities from logistic regression are used for inverse probability weighting.

In the remainder of this chapter, for concreteness, we consider generalized linear regression with mean restriction in the form

$$E(Y_i|X_i) = a'(X_i^{\mathrm{T}}\beta^*), \quad i = 1, \dots, n, \tag{5.13}$$

where $\beta^* = (\beta_1^*, \dots, \beta_p^*)^{\mathrm{T}}$ is an unknown coefficient vector, and $a(\cdot)$ is a *known*, strictly convex, twice differentiable function. Taking $a(\eta) = \log(1 + \mathrm{e}^{\eta})$ or $a(\eta) = \int_0^{\eta} \Phi(u)\, du$ leads to logistic regression or probit regression respectively. The exponential-family assumption (5.2) may be removed. We study Lasso-penalized M-estimator (5.9) with the objective function

$$K_n(\beta) = \frac{-1}{n} \sum_{i=1}^{n} \{Y_i X_i^{\mathrm{T}}\beta - a(X_i^{\mathrm{T}}\beta)\}, \tag{5.14}$$

which has the same form as (5.4) for maximum likelihood estimation in generalized linear regression with a canonical link (i.e., (5.1) and (5.2) with $g^{-} = a'$). However, the choice of $a(\cdot)$ is flexible, not constrained by the exponential-family assumption (5.2). An example is the non-likelihood objective function (5.10) for probit regression.

The objective function (5.14) is convex in β by the convexity of $a(\cdot)$. The associated stationary condition for $\hat\beta_\lambda$ in (5.9) gives

$$\langle \mathbf{Y} - a'(\mathbf{X}\hat\beta_\lambda), \mathbf{X}_j \rangle_n = \begin{cases} \lambda, & \text{if } \hat\beta_{\lambda,j} > 0, \\ -\lambda, & \text{if } \hat\beta_{\lambda,j} < 0, \\ c_j, & \text{if } \hat\beta_{\lambda,j} = 0, \end{cases} \tag{5.15}$$

where c_j is some value in $[-\lambda, \lambda]$.

Objective functions in the form (5.14) are representative, although not the only choices allowed in M-estimation for generalized linear regression. This setup accommodates likelihood estimation in generalized linear regression with a canonical link, such as likelihood estimation in logistic regression (Example 5.1) and log-linear, Poisson regression (Example 5.2). This setup also includes non-likelihood estimation in generalized linear regression with a non-canonical link, such as non-likelihood estimation in probit regression (Example 5.3). But the setup does not include calibrated estimation for logistic regression in Example 5.4. Nevertheless, the development can be extended to such choices, as presented in Tan (2020b), Supplement Section II.

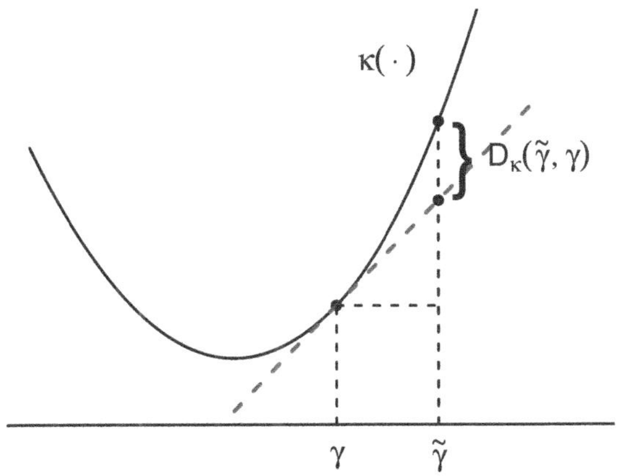

FIGURE 5.1
Illustration of the Bregman divergence, where a tangent line passes $(\gamma, \kappa(\gamma))$.

5.2 Bregman divergences

For linear regression, the prediction performance using an estimator $\hat{\beta}_\lambda$ can be directly measured by the squared L_2 distance (normalized by n), $\|\mathbf{X}\hat{\beta}_\lambda - \mathbf{X}\beta^*\|_n^2$, which is the excess in-sample prediction error by a similar identity as (4.3) for the least-squares estimator. For generalized linear regression, we use a generalization of the L_2 distance, called Bregman divergences, to measure prediction performances.

Definition 5.1. Let $\kappa(\cdot)$ be a convex, differentiable (scalar) function on $\mathbb{R}^p$. The Bregman divergence at two points, $(\tilde{\gamma}, \gamma)$, is defined as

$$D_\kappa(\tilde{\gamma}, \gamma) = \kappa(\tilde{\gamma}) - \kappa(\gamma) - \langle \nabla\kappa(\gamma), \tilde{\gamma} - \gamma \rangle,$$

where $\nabla\kappa = \frac{\partial\kappa}{\partial\tilde{\gamma}}$ is the gradient of $\kappa(\cdot)$, and $\langle u, v \rangle = u^{\mathrm{T}} v$ for two (column) vectors $u, v \in \mathbb{R}^p$.

By the convexity of $\kappa(\cdot)$, $D_\kappa(\tilde{\gamma}, \gamma) \geq 0$ for any $\tilde{\gamma}, \gamma \in \mathbb{R}^p$. For strictly convex $\kappa(\cdot)$, $D_\kappa(\tilde{\gamma}, \gamma) = 0$ if and only if $\tilde{\gamma} = \gamma$. See Figure 5.1 for an illustration with a univariate convex function $\kappa(\cdot)$.

Example 5.5. Let $\kappa(\tilde{\gamma}) = \frac{1}{2}\tilde{\gamma}^{\mathrm{T}}\tilde{\gamma}$. Then $\nabla\kappa(\tilde{\gamma}) = \tilde{\gamma}$ and

$$\begin{aligned} D_\kappa(\tilde{\gamma}, \gamma) &= \frac{1}{2}\tilde{\gamma}^{\mathrm{T}}\tilde{\gamma} - \frac{1}{2}\gamma\gamma^{\mathrm{T}} - \langle \gamma, \tilde{\gamma} - \gamma \rangle \\ &= \frac{1}{2}\tilde{\gamma}^{\mathrm{T}}\tilde{\gamma} + \frac{1}{2}\gamma\gamma^{\mathrm{T}} - \gamma^{\mathrm{T}}\tilde{\gamma} = \frac{1}{2}\|\tilde{\gamma} - \gamma\|_2^2, \end{aligned}$$

that is, the Bregman divergence recovers the squared L_2 distance.

Example 5.6. For linear regression, let $\kappa(\beta) = \frac{1}{2n}\sum_{i=1}^n (Y_i - X_i^{\mathrm{T}}\beta)^2$. Then by direct calculation as above,

$$\begin{aligned} D_\kappa(\tilde\beta, \beta) &= \kappa(\tilde\beta) - \kappa(\beta) - \langle \nabla\kappa(\beta), \tilde\beta - \beta\rangle \\ &= \frac{1}{2n}\|\mathbf{X}\tilde\beta - \mathbf{X}\beta\|_2^2 = \frac{1}{2}\|\mathbf{X}\tilde\beta - \mathbf{X}\beta\|_n^2, \end{aligned}$$

which is independent of $(Y_1, \ldots, Y_n)$.

Example 5.7. For logistic regression, let $\kappa(\cdot)$ be the likelihood objective function in (5.6), i.e.,

$$\kappa(\beta) = \frac{-1}{n}\sum_{i=1}^n \left\{ Y_i X_i^{\mathrm{T}}\beta - \log(1 + \mathrm{e}^{X_i^{\mathrm{T}}\beta}) \right\}.$$

Then $\nabla\kappa(\beta) = \frac{-1}{n}\sum_{i=1}^n \{Y_i - \mu_i(\beta)\}X_i$, where $\mu_i(\beta) = \mathrm{expit}(X_i^{\mathrm{T}}\beta)$, and

$$\begin{aligned} D_\kappa(\tilde\beta, \beta) &= \kappa(\tilde\beta) - \kappa(\beta) - \langle \nabla\kappa(\beta), \tilde\beta - \beta\rangle \\ &= \frac{1}{n}\sum_{i=1}^n \left\{ \log(1 + \mathrm{e}^{X_i^{\mathrm{T}}\tilde\beta}) - \log(1 + \mathrm{e}^{X_i^{\mathrm{T}}\beta}) - \mu_i(\beta) X_i^{\mathrm{T}}(\tilde\beta - \beta) \right\}, \end{aligned}$$

which is independent of $(Y_1, \ldots, Y_n)$. In fact, $\kappa(\beta)$ can be decomposed as

$$\kappa(\beta) = \underbrace{\frac{-1}{n}\sum_{i=1}^n Y_i X_i^{\mathrm{T}}\beta}_{\text{linear in } \beta} + \underbrace{\frac{1}{n}\sum_{i=1}^n \log(1 + \mathrm{e}^{X_i^{\mathrm{T}}\beta})}_{\substack{\text{nonlinear in } \beta \text{ and} \\ \text{independent of } Y_i\text{'s}}}.$$

Hence $D_\kappa(\tilde\beta, \beta)$ is identical to the Bregman divergence associated with the nonlinear term above, $\frac{1}{n}\sum_{i=1}^n \log(1 + \mathrm{e}^{X_i^{\mathrm{T}}\beta})$, independent of $(Y_1, \ldots, Y_n)$. Similarly, an interesting property of any objective function $K_n(\beta)$ in the form (5.14) is that the associated Bregman divergence is independent of $(Y_1, \ldots, Y_n)$. See Section 5.4 for related discussion.

We show that the Bregman divergence can be expressed in terms of the KL divergences:

$$D_\kappa(\tilde\beta, \beta) = \frac{1}{n}\sum_{i=1}^n \mathrm{KL}(\mathbf{p}_i(\beta), \mathbf{p}_i(\tilde\beta)), \tag{5.16}$$

where $\mathbf{p}_i(\beta) = (1 - \mu_i(\beta), \mu_i(\beta))^{\mathrm{T}}$ and $\mathrm{KL}(\mathbf{q}, \tilde{\mathbf{q}}) = \sum_{j=1}^m q_j \log(q_j/\tilde q_j)$ is the KL divergence between two probability vectors $\mathbf{q} = (q_1, \ldots, q_m)^{\mathrm{T}}$ and $\tilde{\mathbf{q}} = (\tilde q_1, \ldots, \tilde q_m)^{\mathrm{T}}$. Note that both $D_\kappa(\tilde\beta, \beta)$ and $\mathrm{KL}(\mathbf{p}_i(\beta), \mathbf{p}_i(\tilde\beta))$ are asymmetric in $\tilde\beta$ and β. In fact, the KL divergence between $\mathbf{p}_i(\beta)$ and $\mathbf{p}_i(\tilde\beta)$ can be calculated as

$$\mathrm{KL}(\mathbf{p}_i(\beta), \mathbf{p}_i(\tilde\beta)) = \mu_i(\beta)\log\frac{\mu_i(\beta)}{\mu_i(\tilde\beta)} + (1 - \mu_i(\beta))\log\frac{1 - \mu_i(\beta)}{1 - \mu_i(\tilde\beta)}$$

$$= \mu_i(\beta)\log\frac{(1-\mu_i(\beta))e^{X_i^{\mathrm{T}}\beta}}{(1-\mu_i(\tilde\beta))e^{X_i^{\mathrm{T}}\tilde\beta}} + (1-\mu_i(\beta))\log\frac{1-\mu_i(\beta)}{1-\mu_i(\tilde\beta)}$$
$$= \log\frac{1-\mu_i(\beta)}{1-\mu_i(\tilde\beta)} + \mu_i(\beta)X_i^{\mathrm{T}}(\beta-\tilde\beta)$$
$$= \log(1+e^{X_i^{\mathrm{T}}\tilde\beta}) - \log(1+e^{X_i^{\mathrm{T}}\beta}) - \mu_i(\beta)X_i^{\mathrm{T}}(\tilde\beta-\beta),$$

using the expression $1-\mu_i(\beta) = (1+e^{X_i^{\mathrm{T}}\beta})^{-1}$.

A relationship similar to (5.16), connecting the Bregman divergence with the KL divergence, holds in general for the likelihood objective function (5.4) in generalized linear regression with a canonical link based on exponential family (5.2) (Exercise 5.1).

5.3 Lasso estimation with correct model specification

5.3.1 Fixed-design generalized linear regression

We study Lasso estimation for generalized linear regression, assumed to be correctly specified in a fixed design. For theoretical analysis, consider the following conditions:

(i) [Generalized linear regression in fixed design] Generalized linear regression (5.13) is correctly specified in a fixed design, i.e., $X_1,\ldots,X_n$ are fixed and $\varepsilon_1,\ldots,\varepsilon_n$ are independent, where $\varepsilon_i = Y_i - a'(X_i^{\mathrm{T}}\beta^*)$;

(ii) [Sub-Gaussian noises] $\varepsilon_1,\ldots,\varepsilon_n$ are sub-Gaussian with scale parameter σ;

(iii) [Bounded covariates] $\max_{j=1,\ldots,p}|X_{ij}| \le C_0$ almost surely for $i = 1,\ldots,n$ with a constant $C_0 > 0$;

Conditions (i)–(ii), in the case of $a(u) = \frac{u^2}{2}$, reduce to those for linear regression in Section 4.2.1. Condition (iii) is used to facilitate a lower bound on the Bregman divergence for a fast-rate error bound (see Lemma 5.3). For the slow-rate error bound in Theorem 5.1, condition (iii) can be replaced by the condition of normalized covariates as in Section 4.2.1.

For any $0 < \epsilon < 1$, suppose that the tuning parameter λ in (5.9) is specified as $\lambda = A_0\lambda_0$ with a constant $A_0 > 1$ and

$$\lambda_0 = \sigma C_0\sqrt{\frac{2\log(p/\epsilon)}{n}}.$$

The following result gives a slow-rate error bound for Lasso-penalized M-estimation.

Theorem 5.1 (Slow rate in fixed design). *Assume that conditions (i)–(iii) are satisfied. For any $0 < \epsilon < 1$, it holds with probability at least $1 - 2\epsilon$ uniformly over $A_0 > 1$ that the Lasso estimator $\hat{\beta}_\lambda$ with $\lambda = A_0\lambda_0$ in (5.9) satisfies*

$$D_{K_n}(\hat{\beta}_\lambda, \beta^*) + D_{K_n}(\beta^*, \hat{\beta}_\lambda) + (A_0 - 1)\lambda_0\|\hat{\beta}_\lambda - \beta^*\|_1 \le 2A_0\lambda_0\|\beta^*\|_1. \quad (5.17)$$

The result holds for any choice of β^* in regression equation (5.13), which is not unique in the case of $p > n$.

Theorem 5.1 extends the slow-rate error bound for linear regression (Theorem 4.1). Commonly in this chapter, the prediction error is measured by the symmetrized Bregman divergence $D_{K_n}(\hat{\beta}_\lambda, \beta^*) + D_{K_n}(\beta^*, \hat{\beta}_\lambda)$. By direct calculation,

$$\begin{aligned} D_{K_n}(\hat{\beta}_\lambda, \beta) + D_{K_n}(\beta, \hat{\beta}_\lambda) &= \langle \nabla K_n(\hat{\beta}_\lambda) - \nabla K_n(\beta), \hat{\beta}_\lambda - \beta \rangle \\ &= \langle a'(\mathbf{X}\hat{\beta}_\lambda) - a'(\mathbf{X}\beta), \mathbf{X}(\hat{\beta}_\lambda - \beta)\rangle_n, \end{aligned} \quad (5.18)$$

which is independent of $(Y_1, \ldots, Y_n)$. Throughout, $a'(\cdot)$ is applied componentwise. The first line follows from Definition 5.1, and the second line uses the expression of $K_n(\beta)$ in (5.14) . For linear regression with $a(u) = \frac{u^2}{2}$, by Example 5.6,

$$D_{K_n}(\hat{\beta}_\lambda, \beta) = D_{K_n}(\beta, \hat{\beta}_\lambda) = \frac{1}{2}\|\mathbf{X}\hat{\beta}_\lambda - \mathbf{X}\beta\|_n^2,$$

and hence $D_{K_n}(\hat{\beta}_\lambda, \beta) + D_{K_n}(\beta, \hat{\beta}_\lambda)$ reduces to $\|\mathbf{X}\hat{\beta}_\lambda - \mathbf{X}\beta\|_n^2$. Theorem 5.1 can be shown similarly as Theorem 4.1, but using the following two lemmas which extend Lemma 4.1 (Basic inequality) and Lemma 4.2 (Concentration of inner products). Proof details for Theorem 5.1 are left as an exercise.

Lemma 5.1 (Basic inequality). *For any $\beta \in \mathbb{R}^p$,*

$$D_{K_n}(\hat{\beta}_\lambda, \beta) + D_{K_n}(\beta, \hat{\beta}_\lambda) + \langle \nabla K_n(\beta), \hat{\beta}_\lambda - \beta \rangle + \lambda\|\hat{\beta}_\lambda\|_1 \le \lambda\|\beta\|_1, \quad (5.19)$$

where $\hat{\beta}_\lambda$ is defined as (5.9) for an objective function $K_n(\beta)$.
[From the proof, the result holds for general $K_n(\beta)$, including $K_n(\beta)$ nonconvex in β. But in that case, $D_{K_n}(\hat{\beta}_\lambda, \beta)$ and $D_{K_n}(\beta, \hat{\beta}_\lambda)$ may not be nonnegative.]

Proof. We first show a weaker version that for any $\beta \in \mathbb{R}^p$,

$$D_{K_n}(\hat{\beta}_\lambda, \beta) + \langle \nabla K_n(\beta), \hat{\beta}_\lambda - \beta \rangle + \lambda\|\hat{\beta}_\lambda\|_1 \le \lambda\|\beta\|_1. \quad (5.20)$$

By the definition of $\hat{\beta}_\lambda$ in (5.9), for any $\beta \in \mathbb{R}^p$,

$$K_n(\hat{\beta}_\lambda) + \lambda\|\hat{\beta}_\lambda\|_1 \le K_n(\beta) + \lambda\|\beta\|_1.$$

which equivalently gives (5.20) because $K_n(\hat\beta_\lambda) - K_n(\beta) = D_{K_n}(\hat\beta_\lambda, \beta) + \langle \nabla K_n(\beta), \hat\beta_\lambda - \beta\rangle$.

Next, we show the refined version (5.19). By the definition of $\hat\beta_\lambda$ in (5.9), for any $\beta \in \mathbb{R}^p$ and $t \in [0,1]$,

$$\begin{aligned} K_n(\hat\beta_\lambda) + \lambda\|\hat\beta_\lambda\|_1 &\le K_n((1-t)\hat\beta_\lambda + t\beta) + \lambda\|(1-t)\hat\beta_\lambda + t\beta\|_1 \\ &\le K_n((1-t)\hat\beta_\lambda + t\beta) + \lambda\{(1-t)\|\hat\beta_\lambda\|_1 + t\|\beta\|_1\}, \end{aligned}$$

where the second step uses the convexity of $\|\cdot\|_1$. Rearranging the inequality gives for $t \in (0,1]$,

$$\frac{K_n(\hat\beta_\lambda) - K_n((1-t)\hat\beta_\lambda + t\beta)}{t} + \lambda\|\hat\beta_\lambda\|_1 \le \lambda\|\beta\|_1.$$

Letting $t \to 0+$ yields

$$\langle \nabla K_n(\hat\beta_\lambda), \hat\beta_\lambda - \beta\rangle + \lambda\|\hat\beta_\lambda\|_1 \le \lambda\|\beta\|_1, \tag{5.21}$$

which equivalently gives (5.19) because by (5.18),

$$\langle \nabla K_n(\hat\beta_\lambda), \hat\beta_\lambda - \beta\rangle = D_{K_n}(\hat\beta_\lambda, \beta) + D_{K_n}(\beta, \hat\beta_\lambda) + \langle \nabla K_n(\beta), \hat\beta_\lambda - \beta\rangle.$$

Alternatively, (5.19), which is equivalent to (5.21), can be shown by using the stationary condition for $\hat\beta_\lambda$, similarly as in the proof of Lemma 4.1 (Exercise 5.2). □

For the objective function $K_n(\beta)$ in (5.14),

$$\langle \nabla K_n(\beta), \hat\beta_\lambda - \beta\rangle = -\langle \mathbf{Y} - a'(\mathbf{X}\beta), \mathbf{X}(\hat\beta_\lambda - \beta)\rangle_n.$$

Then the basic inequality (5.19) reduces to

$$D_{K_n}(\hat\beta_\lambda, \beta) + D_{K_n}(\beta, \hat\beta_\lambda) - \langle \mathbf{Y} - a'(\mathbf{X}\beta), \mathbf{X}(\hat\beta_\lambda - \beta)\rangle_n + \lambda\|\hat\beta_\lambda\|_1 \le \lambda\|\beta\|_1, \tag{5.22}$$

The inner-product term $\langle \mathbf{Y} - a'(\mathbf{X}\beta), \mathbf{X}(\hat\beta_\lambda - \beta)\rangle_n$ can be controlled using the Hölder inequality and the following lemma.

Lemma 5.2 (Concentration of inner products in fixed design). *For $0 < \epsilon < 1$, let $\Omega_1(\lambda_1)$ be the event that*

$$\max_{j=1,\ldots,p} |\langle \mathbf{Y} - a'(\mathbf{X}\beta^*), \mathbf{X}_j\rangle_n| \le \lambda_1,$$

where $\mathbf{X}_j$ is the jth column of $\mathbf{X}$. Under conditions (i)–(iii), if $\lambda_1 \ge \sigma C_0\sqrt{\frac{2\log(p/\epsilon)}{n}}$, then $P(\Omega_1(\lambda_1)) \ge 1 - 2\epsilon$.

Proof. The result follows by similar steps in the proof of Lemma 4.2 with $\mathbf{X}\beta^*$ replaced by $a'(\mathbf{X}\beta^*)$. □

Similarly as in linear regression, a fast-rate error bound can be deduced by invoking an empirical compatibility condition on $\hat{H}(\beta^*)$, the Hessian of the objective function $K_n(\cdot)$ at β^* (hence depending on β^*), where $\hat{H}(\beta)$ denotes the Hessian of $K_n(\cdot)$ at β,

$$\hat{H}(\beta) = \frac{\partial^2}{\partial\beta\partial\beta^{\mathrm{T}}} K_n(\beta) = \frac{1}{n}\sum_{i=1}^n a''(X_i^{\mathrm{T}}\beta)X_iX_i^{\mathrm{T}}.$$

A compatibility condition on $\hat{H}(\beta^*)$ is implied by a compatibility condition on the (scaled) Gram matrix $\hat{H} = \mathbf{X}^{\mathrm{T}}\mathbf{X}/n$ (as used in Chapter 4) if $a''(X_i^{\mathrm{T}}\beta^*)$ is lower bounded by a positive constant for $i = 1, \ldots, n$.

Another related complication is that the objective function $K_n(\beta)$ is non-quadratic in β and hence the symmetrized Bregman divergence $D_{K_n}(\beta, \tilde{\beta}) + D_{K_n}(\tilde{\beta}, \beta)$ is also non-quadratic in $\beta - \tilde{\beta}$. The following lemma can be used to lower bound the symmetrized Bregman divergence in terms of a quadratic form of $\beta - \tilde{\beta}$ with matrix $\hat{H}(\beta)$.

Lemma 5.3 (Lower bound of Bregman divergence). *Suppose that condition (iii) holds, i.e.,* $\max_{j=1,\ldots,p}|X_{ij}| \le C_0$ *for* $i = 1, \ldots, n$ *with a constant* $C_0 > 0$, *and that [condition (iv)]*

$$a''(\tilde{u}) \le a''(u)\mathrm{e}^{C_1|\tilde{u}-u|}, \tag{5.23}$$

for any $u, \tilde{u} \in \mathbb{R}$ *with a constant* $C_1 > 0$. *Then for any* $\beta, \tilde{\beta} \in \mathbb{R}^p$,

$$D_{K_n}(\tilde{\beta}, \beta) + D_{K_n}(\beta, \tilde{\beta}) \ge \int_0^1 \mathrm{e}^{-\tilde{C}_1\|b\|_1 t}\,\mathrm{d}t \cdot \{b^{\mathrm{T}}\hat{H}(\beta)b\} \tag{5.24a}$$

$$= \frac{1-\mathrm{e}^{-\tilde{C}_1\|b\|_1}}{\tilde{C}_1\|b\|_1}\{b^{\mathrm{T}}\hat{H}(\beta)b\}, \tag{5.24b}$$

where $b = \tilde{\beta} - \beta$ *and* $\tilde{C}_1 = C_0C_1$. *If* $b = 0$, *then redefine* $\frac{1-\mathrm{e}^{-\tilde{C}_1\|b\|_1}}{\tilde{C}_1\|b\|_1} = 1$.

Proof. From the expression (5.18). we proceed as follows:

$$\begin{aligned}
&D_{K_n}(\tilde{\beta}, \beta) + D_{K_n}(\beta, \tilde{\beta}) \\
&= \frac{1}{n}\sum_{i=1}^n \left\{a'(X_i^{\mathrm{T}}\tilde{\beta}) - a'(X_i^{\mathrm{T}}\beta)\right\} X_i^{\mathrm{T}}(\tilde{\beta} - \beta) \\
&= \frac{1}{n}\sum_{i=1}^n \int_0^1 a''(X_i^{\mathrm{T}}(\beta + t(\tilde{\beta} - \beta)))\,\mathrm{d}t \cdot (X_i^{\mathrm{T}}(\tilde{\beta} - \beta))^2 \qquad (5.25) \\
&\ge \frac{1}{n}\sum_{i=1}^n \int_0^1 a''(X_i^{\mathrm{T}}\beta)\mathrm{e}^{-C_1 t|X_i^{\mathrm{T}}(\tilde{\beta}-\beta)|}\,\mathrm{d}t \cdot (X_i^{\mathrm{T}}(\tilde{\beta} - \beta))^2
\end{aligned}$$

$$
\begin{aligned}
&= \frac{1}{n}\sum_{i=1}^{n} a''(X_i^{\mathrm{T}}\beta)(X_i^{\mathrm{T}}(\tilde{\beta}-\beta))^2 \cdot \int_0^1 \mathrm{e}^{-C_1 t|X_i^{\mathrm{T}}(\tilde{\beta}-\beta)|}\,\mathrm{d}t \\
&\geq \frac{1}{n}\sum_{i=1}^{n} a''(X_i^{\mathrm{T}}\beta)(X_i^{\mathrm{T}}(\tilde{\beta}-\beta))^2 \cdot \int_0^1 \mathrm{e}^{-tC_1C_0\|\tilde{\beta}-\beta\|_1}\,\mathrm{d}t \\
&= \{b^{\mathrm{T}}\hat{H}(\beta)b\}\frac{1-\mathrm{e}^{-\tilde{C}_1\|b\|_1}}{\tilde{C}_1\|b\|_1}.
\end{aligned}
$$

The third line uses the fundamental theorem of calculus (Exercise 5.3), the fourth line uses the condition on $a''(\cdot)$, and the sixth line uses the boundedness of the covariates. □

In a related manner, the modified self-concordance condition (Bach, 2010) is defined as for any $u \in \mathbb{R}$,

$$|a'''(u)| \leq B a''(u), \tag{5.26}$$

with a constant $B > 0$, provided that $a(\cdot)$ is strictly convex and three-times differentiable. We note that condition (5.26) implies (hence is more restrictive than) condition (5.23) with $C_1 = B$. In fact, condition (5.26) is equivalent to $|\frac{\mathrm{d}}{\mathrm{d}u}\log a''(u)| \leq B$, and hence implies by the mean value theorem that for any $u, \tilde{u} \in \mathbb{R}$,

$$\log a''(\tilde{u}) \leq \log a''(u) + B|\tilde{u} - u|.$$

Taking exponentials on both side gives

$$a''(\tilde{u}) \leq a''(u)\mathrm{e}^{B|\tilde{u}-u|},$$

that is, condition (5.23) holds with $C_1 = B$.

Example 5.8. For logistic regression (5.5), consider the likelihood objective function in (5.6), $K_n(\beta) = \frac{-1}{n}\sum_{i=1}^{n}\{Y_i X_i^{\mathrm{T}}\beta - \log(1+\mathrm{e}^{X_i^{\mathrm{T}}\beta})\}$. Direct calculation gives

$$a(u) = \log(1+\mathrm{e}^u), \quad a'(u) = \frac{\mathrm{e}^u}{1+\mathrm{e}^u}, \quad a''(u) = \frac{\mathrm{e}^u}{(1+\mathrm{e}^u)^2}.$$

For any $\tilde{u}, u \in \mathbb{R}$, if $\tilde{u} \geq u$, then

$$\frac{a''(\tilde{u})}{a''(u)} = \mathrm{e}^{\tilde{u}-u}\frac{(1+\mathrm{e}^u)^2}{(1+\mathrm{e}^{\tilde{u}})^2} \leq \mathrm{e}^{\tilde{u}-u} = \mathrm{e}^{|\tilde{u}-u|}.$$

If $u \geq \tilde{u}$, then

$$\frac{a''(\tilde{u})}{a''(u)} = \mathrm{e}^{u-\tilde{u}}\frac{(1+\mathrm{e}^{-u})^2}{(1+\mathrm{e}^{-\tilde{u}})^2} \leq \mathrm{e}^{u-\tilde{u}} = \mathrm{e}^{|\tilde{u}-u|}.$$

Hence condition (5.23) is satisfied with $C_1 = 1$. Alternatively, this can be proved through condition (5.26). In fact, $a'' = a'(1-a')$ and $a''' = (1-2a')a''$. Then $|a'''| \leq a''$ because $a' \in (0,1)$. Hence condition (5.26) holds with $B = 1$, which implies condition (5.23) with $C_1 = 1$.

Example 5.9. For log-linear regression (5.7), consider the likelihood objective function in (5.8), $K_n(\beta) = \frac{-1}{n}\sum_{i=1}^n (Y_i X_i^{\mathrm{T}}\beta - \mathrm{e}^{X_i^{\mathrm{T}}\beta})$. Direct calculation gives

$$a(u) = \mathrm{e}^u, \quad a'(u) = \mathrm{e}^u, \quad a''(u) = \mathrm{e}^u.$$

For any $\tilde{u}, u \in \mathbb{R}$,

$$\frac{a''(\tilde{u})}{a''(u)} = \mathrm{e}^{\tilde{u}-u} \leq \mathrm{e}^{|\tilde{u}-u|}.$$

Hence condition (5.23) is satisfied with $C_1 = 1$.

Example 5.10. For probit regression (5.3), consider the non-likelihood objective function in (5.10), $K_n(\beta) = \frac{-1}{n}\sum_{i=1}^n \{Y_i X_i^{\mathrm{T}}\beta - a(X_i^{\mathrm{T}}\beta)\}$, where $a(\eta) = \int_0^\eta \Phi(u)\,du$. Direct calculation gives

$$a'(u) = \Phi(u), \quad a''(u) = \frac{1}{\sqrt{2\pi}}\mathrm{e}^{-\frac{u^2}{2}}.$$

For any $\tilde{u}, u \in \mathbb{R}$,

$$\frac{a''(\tilde{u})}{a''(u)} = \mathrm{e}^{\frac{u^2-\tilde{u}^2}{2}}.$$

In general, condition (5.23) is not satisfied. Nevertheless, a fast-rate error bound can be obtained by introducing some additional condition and generalizing the arguments here. See Exercise 5.7.

The following result gives a fast-rate error bound for Lasso-penalized M-estimation. The choice of true value β^* in regression equation (5.13) is unique when the compatibility condition $(S_{\beta^*}, \xi_0, \nu_0)$ is assumed to hold with $\xi_0 \geq 1$ for $\hat{H}(\beta^*)$ (see Section 5.5.1).

Theorem 5.2 (Fast rate in fixed design). *Assume that the following conditions are satisfied:*

(i)–(iii) Same as in Theorem 5.1;

(iv) [Hessian condition] $a''(\tilde{u}) \leq a''(u)\mathrm{e}^{C_1|\tilde{u}-u|}$ *for any* $u, \tilde{u} \in \mathbb{R}$ *where* $C_1 > 0$ *is a constant;*

(v) [Empirical compatibility condition] The compatibility condition $(S_{\beta^*}, \xi_0, \nu_0)$ *holds for* $\hat{H}(\beta^*) = \frac{1}{n}\sum_{i=1}^n a''(X_i^{\mathrm{T}}\beta^*)X_i X_i^{\mathrm{T}}$*, where* $S_{\beta^*} = \{1 \leq j \leq p : \beta_j^* \neq 0\}$ *and* $\xi_0 > 1$ *and* $\nu_0 > 0$ *are constants;*

For any $0 < \epsilon < 1$ *and* $A_0 \geq \frac{\xi_0+1}{\xi_0-1}$*, suppose further that*

(vi) [Rate condition] $\frac{4A_0^2}{A_0-1}C_0 C_1 \nu_0^{-2}|S_{\beta^*}|\lambda_0 \leq \eta_1$ *for a constant* $\eta_1 \in (0,1)$*.*

Then it holds with probability at least $1 - 2\epsilon$ *uniformly over* A_0 *subject to condition (vi) that the Lasso estimator* $\hat{\beta}_\lambda$ *with* $\lambda = A_0\lambda_0$ *in* (5.9) *satisfies*

$$D_{K_n}(\hat{\beta}_\lambda, \beta^*) + D_{K_n}(\beta^*, \hat{\beta}_\lambda) + (A_0 - 1)\lambda_0\|\hat{\beta}_\lambda - \beta^*\|_1 \le M_0|S_{\beta^*}|\lambda_0^2, \quad (5.27)$$

where $M_0 = 4A_0^2(1-\eta_1)^{-1}\nu_0^{-2}$, *depending only on* (A_0, ν_0, η_1).

Rate condition (vi) requires that $|S_{\beta^*}|\lambda_0$ is sufficiently small as $n \to \infty$. This condition is similar to condition (v) in Theorem 4.3 (Linear regression in random design), but is imposed to localize $\hat{\beta}_\lambda$ near β^*, as shown in (5.34) later, using Lemma 5.3 (Lower bound of Bregman divergence) with non-quadratic objective function $K_n(\beta)$.

Theorem 5.2 extends the fast-rate error bound for linear regression (Theorem 4.2). The error bound (5.27) implies the following bounds: with probability at least $1 - 2\epsilon$,

$$D_{K_n}(\hat{\beta}_\lambda, \beta^*) + D_{K_n}(\beta^*, \hat{\beta}_\lambda) \le M_0|S_{\beta^*}|\lambda_0^2, \quad (5.28a)$$

$$\|\hat{\beta}_\lambda - \beta^*\|_1 \le \frac{M_0}{A_0 - 1}|S_{\beta^*}|\lambda_0. \quad (5.28b)$$

The symmetrized Bregman divergence $D_{K_n}(\hat{\beta}_\lambda, \beta^*) + D_{K_n}(\beta^*, \hat{\beta}_\lambda)$ serves as a measure of the in-sample prediction error, i.e., the difference between $\mathbf{X}\hat{\beta}_\lambda$ and $\mathbf{X}\beta^*$. Hence the prediction error is controlled in (5.28a). The L_1 estimation error is controlled in (5.28b).

Proof. By similar steps as leading to (4.13) and (4.22) in the proof of Theorem 4.2, we use the basic inequality (5.22) and the Hölder inequality to obtain that in the event $\Omega_1 = \Omega_1(\lambda_0)$ with $\lambda_1 = \lambda_0$ in Lemma 5.2,

$$D_{K_n}(\hat{\beta}_\lambda, \beta^*) + D_{K_n}(\beta^*, \hat{\beta}_\lambda) + A_0\lambda_0\|\hat{\beta}_\lambda\|_1 \le \lambda_0\|\hat{\beta}_\lambda - \beta^*\|_1 + A_0\lambda_0\|\beta^*\|_1.$$

and then use the triangle inequality to obtain in the event Ω_1,

$$\begin{aligned} &D_{K_n}(\hat{\beta}_\lambda, \beta^*) + D_{K_n}(\beta^*, \hat{\beta}_\lambda) + (A_0 - 1)\lambda_0\|\hat{\beta}_\lambda - \beta^*\|_1 \\ &\le 2A_0\lambda_0\left\{\sum_{j\in S}|\hat{\beta}_{\lambda,j} - \beta_j^*| + \sum_{j\notin S}|\beta_j^*|\right\}, \end{aligned}$$

for any subset $S \subset \{1, \ldots, p\}$. Taking $S = \emptyset$ yields the slow-rate error bound (5.17). Alternatively, taking $S = S_{\beta^*}$ shows that in the event Ω_1,

$$\begin{aligned} &D_{K_n}(\hat{\beta}_\lambda, \beta^*) + D_{K_n}(\beta^*, \hat{\beta}_\lambda) + (A_0 - 1)\lambda_0\|\hat{\beta}_\lambda - \beta^*\|_1 \\ &\le 2A_0\lambda_0 \sum_{j\in S_{\beta^*}}|\hat{\beta}_{\lambda,j} - \beta_j^*|, \end{aligned} \quad (5.29)$$

which corresponds to (4.23) in the proof of Theorem 4.19. Denote $\Delta_n = D_{K_n}(\hat{\beta}_\lambda, \beta^*) + D_{K_n}(\beta^*, \hat{\beta}_\lambda) + (A_0 - 1)\lambda_0\|\hat{\beta}_\lambda - \beta^*\|_1$, the left-hand side of (5.29).

It suffices to deduce (5.27) from (5.29). In fact, (5.29) implies that

$$(A_0 - 1)\lambda_0\|\hat{\beta}_\lambda - \beta^*\|_1 \le 2A_0\lambda_0 \sum_{j\in S_{\beta^*}} |\hat{\beta}_{\lambda,j} - \beta_j^*|,$$

or equivalently

$$\sum_{j\notin S_{\beta^*}} |\hat{\beta}_{\lambda,j} - \beta_j^*| \le \frac{A_0+1}{A_0-1} \sum_{j\in S_{\beta^*}} |\hat{\beta}_{\lambda,j} - \beta_j^*|. \tag{5.30}$$

Note that $\frac{A_0+1}{A_0-1} \le \xi_0$ for $A_0 \ge \frac{\xi_0+1}{\xi_0-1}$. Then $\hat{\beta}_\lambda - \beta^*$ satisfies the cone condition, $\sum_{j\notin S_{\beta^*}} |\hat{\beta}_{\lambda,j} - \beta_j^*| \le \xi_0 \sum_{j\in S_{\beta^*}} |\hat{\beta}_{\lambda,j} - \beta_j^*|$, and by the compatibility condition,

$$\nu_0^2\Big(\sum_{j\in S_{\beta^*}} |\hat{\beta}_{\lambda,j} - \beta_j^*|\Big)^2 \le |S_{\beta^*}| \underbrace{(\hat{\beta}_\lambda - \beta^*)^{\mathrm{T}}\hat{H}(\beta^*)(\hat{\beta}_\lambda - \beta^*)}_{Q_n(\hat{\beta}_\lambda,\beta^*)}. \tag{5.31}$$

To derive a quadratic inequality from (5.29) and (5.31), we need to localize $\hat{\beta}_\lambda$ near β^*, i.e., showing that $\hat{\beta}_\lambda - \beta^*$ is upper bounded in a certain sense. By Lemma 5.3 with conditions (iii) and (iv),

$$D_{K_n}(\hat{\beta}_\lambda, \beta^*) + D_{K_n}(\beta^*, \hat{\beta}_\lambda) \ge \frac{1 - \mathrm{e}^{-\tilde{C}_1\|\hat{\beta}_\lambda - \beta^*\|_1}}{\tilde{C}_1\|\hat{\beta}_\lambda - \beta^*\|_1} Q_n(\hat{\beta}_\lambda, \beta^*), \tag{5.32}$$

where $\tilde{C}_1 = C_0C_1$. Combining (5.29), (5.31), and (5.32) yields a quadratic inequality in $Q_n(\hat{\beta}_\lambda, \beta^*)^{1/2}$:

$$\begin{aligned}
&\frac{1 - \mathrm{e}^{-\tilde{C}_1\|\hat{\beta}_\lambda - \beta^*\|_1}}{\tilde{C}_1\|\hat{\beta}_\lambda - \beta^*\|_1} Q_n(\hat{\beta}_\lambda, \beta^*) \le D_{K_n}(\hat{\beta}_\lambda, \beta^*) + D_{K_n}(\beta^*, \hat{\beta}_\lambda)\\
&\le 2A_0\lambda_0 \sum_{j\in S_{\beta^*}} |\hat{\beta}_{\lambda,j} - \beta_j^*|\\
&\le 2A_0\lambda_0 \cdot \nu_0^{-1}|S_{\beta^*}|^{1/2} Q_n(\hat{\beta}_\lambda, \beta^*)^{1/2},
\end{aligned}$$

and hence ("inverting a quadratic inequality")

$$Q_n(\hat{\beta}_\lambda, \beta^*) \le (2A_0\lambda_0)^2\nu_0^{-2}|S_{\beta^*}| \left(\frac{\tilde{C}_1\|\hat{\beta}_\lambda - \beta^*\|_1}{1 - \mathrm{e}^{-\tilde{C}_1\|\hat{\beta}_\lambda - \beta^*\|_1}}\right)^2. \tag{5.33}$$

Returning to (5.31) with (5.33) gives

$$\begin{aligned}
\sum_{j\in S_{\beta^*}} |\hat{\beta}_{\lambda,j} - \beta_j^*| &\le \nu_0^{-1}|S_{\beta^*}|^{1/2} Q_n(\hat{\beta}_\lambda, \beta^*)^{1/2}\\
&\le \nu_0^{-1}|S_{\beta^*}|^{1/2}(2A_0\lambda_0)\nu_0^{-1}|S_{\beta^*}|^{1/2} \frac{\tilde{C}_1\|\hat{\beta}_\lambda - \beta^*\|_1}{1 - \mathrm{e}^{-\tilde{C}_1\|\hat{\beta}_\lambda - \beta^*\|_1}},
\end{aligned}$$

$$= \nu_0^{-2}|S_{\beta^*}|(2A_0\lambda_0)\frac{\tilde{C}_1\|\hat{\beta}_\lambda - \beta^*\|_1}{1 - \mathrm{e}^{-\tilde{C}_1\|\hat{\beta}_\lambda - \beta^*\|_1}},$$

which, in conjunction with (5.30), gives

$$\begin{aligned}\|\hat{\beta}_\lambda - \beta^*\|_1 &\le \frac{2A_0}{A_0 - 1}\sum_{j\in S_{\beta^*}} |\hat{\beta}_{\lambda,j} - \beta_j^*| \\ &\le \frac{2A_0}{A_0 - 1}\nu_0^{-2}|S_{\beta^*}|(2A_0\lambda_0)\frac{\tilde{C}_1\|\hat{\beta}_\lambda - \beta^*\|_1}{1 - \mathrm{e}^{-\tilde{C}_1\|\hat{\beta}_\lambda - \beta^*\|_1}}.\end{aligned}$$

Hence by condition (vi),

$$1 - \mathrm{e}^{-\tilde{C}_1\|\hat{\beta}_\lambda - \beta^*\|_1} \le \frac{2A_0}{A_0 - 1}\nu_0^{-2}|S_{\beta^*}|(2A_0\lambda_0)\tilde{C}_1 \le \eta_1, \tag{5.34}$$

which implies that $\|\hat{\beta}_\lambda - \beta^*\|_1$ is upper bounded by $C_1^{-1}\log\frac{1}{1-\eta_1}$ and

$$\frac{1 - \mathrm{e}^{-\tilde{C}_1\|\hat{\beta}_\lambda - \beta^*\|_1}}{\tilde{C}_1\|\hat{\beta}_\lambda - \beta^*\|_1} = \int_0^1 \mathrm{e}^{-\tilde{C}_1\|\hat{\beta}_\lambda - \beta^*\|_1 t}\,\mathrm{d}t \ge \mathrm{e}^{-\tilde{C}_1\|\hat{\beta}_\lambda - \beta^*\|_1} \ge 1 - \eta_1. \tag{5.35}$$

[Note: By a second look, the derivation from (5.32) to (5.35) can be done using the integral expression of the Bregman divergence lower bound (5.24a), but not directly the closed-form expression (5.24b) (Exercise 5.4).]

Finally, returning to (5.33) with (5.35) gives

$$Q_n(\hat{\beta}_\lambda, \beta^*) \le (1 - \eta_1)^{-2}(2A_0\lambda_0)^2\nu_0^{-2}|S_{\beta^*}|, \tag{5.36}$$

which, in conjunction with (5.29) and (5.31), yields

$$\begin{aligned}\Delta_n &\le 2A_0\lambda_0 \sum_{j\in S_{\beta^*}} |\hat{\beta}_{\lambda,j} - \beta_j^*| \\ &\le 2A_0\lambda_0 \cdot \nu_0^{-1}|S_{\beta^*}|^{1/2}\{(\hat{\beta}_\lambda - \beta^*)^{\mathrm{T}}\hat{H}(\beta^*)(\hat{\beta}_\lambda - \beta^*)\}^{1/2} \\ &\le 4A_0^2(1 - \eta_1)^{-1}\nu_0^{-2}|S_{\beta^*}|\lambda_0^2,\end{aligned}$$

that is, $\Delta_n \le M_0|S_{\beta^*}|\lambda_0^2$ as stated in (5.27). In summary, (5.27) holds in the event Ω_1, with probability at least $1 - 2\epsilon$ by Lemma 5.2. □

Like the Bregman divergence, the quadratic form

$$\begin{aligned}Q_n(\hat{\beta}_\lambda, \beta^*) &= (\hat{\beta}_\lambda - \beta^*)^{\mathrm{T}}\hat{H}(\beta^*)(\hat{\beta}_\lambda - \beta^*) \\ &= \frac{1}{n}\sum_{i=1}^n a''(X_i^{\mathrm{T}}\beta^*)(X_i^{\mathrm{T}}\hat{\beta}_\lambda - X_i^{\mathrm{T}}\bar{\beta})^2\end{aligned}$$

is an in-sample weighted MSE of the linear predictors $X_i^{\mathrm{T}}\hat{\beta}_\lambda$ against $X_i^{\mathrm{T}}\beta^*$. From (5.36) in the proof, it holds that with probability at least $1 - 2\epsilon$,

$$Q_n(\hat{\beta}_\lambda, \beta^*) \le (1 - \eta_1)^{-1}M_0|S_{\beta^*}|\lambda_0^2,$$

in addition to the inequality (5.27) in Theorem 5.2.

5.3.2 Random-design generalized linear regression

We study Lasso estimation for generalized linear regression, assumed to be correctly specified in a random design. Similarly as in linear regression, a central issue is to show that the empirical compatibility condition can be satisfied with a high probability under suitable conditions.

For theoretical analysis, consider the following conditions:

(i) [Generalized linear regression in random design] Generalized linear regression (5.13) is correctly specified in a random design, i.e., $(X_1, Y_1), \ldots, (X_n, Y_n)$ are independent with $E(\varepsilon_i|X_i) = 0$, where $\varepsilon_i = Y_i - a'(X_i^{\mathrm{T}}\beta^*)$, for $i = 1, \ldots, n$;

(ii) [Sub-Gaussian noises] ε_i given X_i is sub-Gaussian with scale parameter σ for $i = 1, \ldots, n$;

(iii) [Bounded covariates] $\max_{j=1,\ldots,p} |X_{ij}| \le C_0$ almost surely for $i = 1, \ldots, n$ with a constant $C_0 > 0$;

(iv) [Hessian condition] $a''(\tilde{u}) \le a''(u)\mathrm{e}^{C_1|\tilde{u}-u|}$ for any $u, \tilde{u} \in \mathbb{R}$ with a constant $C_1 > 0$, and $a''(X_i^{\mathrm{T}}\beta^*) \le C_2$ almost surely for $i = 1, \ldots, n$ with a constant $C_2 > 0$;

(v) [Theoretical compatibility condition] The compatibility condition $(S_{\beta^*}, \xi_0^*, \nu_0^*)$ holds for $H(\beta^*) = E\{\frac{1}{n}\sum_{i=1}^n a''(X_i^{\mathrm{T}}\beta^*)X_i X_i^{\mathrm{T}}\}$, where $S_{\beta^*} = \{1 \le j \le p : \beta_j^* \ne 0\}$ and $\xi_0^* > 1$ and $\nu_0^* > 0$ are constants.

The upper boundedness of $a''(X_i^{\mathrm{T}}\beta^*)$, $i = 1, \ldots, n$, is used to show the concentration of the empirical Hessian $\hat{H}(\beta^*)$ about the theoretical Hessian $H(\beta^*)$. A compatibility condition on $H(\beta^*)$ is implied by a compatibility condition on the expected (scaled) Gram matrix $H = E(\mathbf{X}^{\mathrm{T}}\mathbf{X}/n)$ if $a''(X_i^{\mathrm{T}}\beta^*)$ is lower bounded almost surely by a positive constant for $i = 1, \ldots, n$. For logistic regression, $a''(X_i^{\mathrm{T}}\beta^*)$ is automatically upper bounded by $1/4$, whereas is lower bounded by a positive constant if and only if $|X_i^{\mathrm{T}}\beta^*|$ is bounded almost surely for $i = 1, \ldots, n$. For log-linear regression, $a''(X_i^{\mathrm{T}}\beta^*)$ is upper or lower bounded by a positive constant if and only if $X_i^{\mathrm{T}}\beta^*$ is, respectively, upper or lower bounded almost surely for $i = 1, \ldots, n$. In general, conditions (iv) and (v) are implied by jointly the following two conditions (Exercise 5.8):

(iv†) [Hessian and bounded mean condition] $a''(\tilde{u}) \le a''(u)\mathrm{e}^{C_1|\tilde{u}-u|}$ for any $u, \tilde{u} \in \mathbb{R}$ with a constant $C_1 > 0$, and $|X_i^{\mathrm{T}}\beta^*| \le B_0$ almost surely for $i = 1, \ldots, n$ with a constant $B_0 > 0$;

(v†) [Theoretical compatibility condition] The compatibility condition $(S_{\beta^*}, \xi_0^*, \nu_0^*)$ holds for $H = E(\mathbf{X}^{\mathrm{T}}\mathbf{X}/n)$, where $S_{\beta^*} = \{1 \le j \le p : \beta_j^* \ne 0\}$ and $\xi_0^* > 1$ and $\nu_0^* > 0$ are constants.

These two conditions may be easier to interpret than (iv) and (v).

For any $0 < \epsilon < 1$, suppose that the tuning parameter λ in (5.9) is specified as $\lambda = A_0\lambda_0$ with a constant $A_0 > 1$ and

$$\lambda_0 = \max\left\{\sigma C_0\sqrt{\frac{2\log(p/\epsilon)}{n}},\ 2C_0^2C_2\sqrt{\frac{\log(p/\epsilon)}{n}}\right\},$$

where C_0 and C_2 are constants from condition (iii) and (iv). The following result gives a fast-rate error bound for Lasso-penalized M-estimation in the random design. The choice of true value β^* in regression equation (5.13) is unique when the compatibility condition $(S_{\beta^*}, \xi_0^*, \nu_0^*)$ is assumed to hold with $\xi_0^* \geq 1$ for $H(\beta^*)$ (see Section 5.5.1).

Theorem 5.3 (Fast rate in random design). *Assume that conditions (i)–(v) above are satisfied. For any* $0 < \epsilon < 1$ *and* $A_0 \geq \frac{\xi_0^*+1}{\xi_0^*-1}$, *suppose further that*

(vi) [Rate condition] $(1+\xi_0^*)^2\nu_0^{*-2}|S_{\beta^*}|\lambda_0 \leq \eta_0$ *for a constant* $\eta_0 \in (0,1)$, *and* $\frac{4A_0^2}{A_0-1}C_0C_1(1-\eta_0)^{-1}\nu_0^{*-2}|S_{\beta^*}|\lambda_0 \leq \eta_1$ *for a constant* $\eta_1 \in (0,1)$.

Then it holds with probability at least $1 - 2\epsilon - 2\epsilon^2$ *uniformly over* A_0 *subject to condition (vi) that the Lasso estimator* $\hat\beta_\lambda$ *with* $\lambda = A_0\lambda_0$ *in* (5.9) *satisfies*

$$D_{K_n}(\hat\beta_\lambda, \beta^*) + D_{K_n}(\beta^*, \hat\beta_\lambda) + (A_0-1)\lambda_0\|\hat\beta_\lambda - \beta^*\|_1 \leq M_0|S_{\beta^*}|\lambda_0^2, \quad (5.37a)$$

$$(\hat\beta_\lambda - \beta^*)^{\mathrm{T}}\hat H(\beta^*)(\hat\beta_\lambda - \beta^*) \leq (1-\eta_1)^{-1}M_0|S_{\beta^*}|\lambda_0^2, \quad (5.37b)$$

where $M_0 = 4A_0^2(1-\eta_1)^{-1}(1-\eta_0)^{-1}\nu_0^{*-2}$, *depending only on* $(A_0, \nu_0^*, \eta_0, \eta_1)$.

Rate condition (vi) is obtained by merging condition (v) in Theorem 4.3 (Linear regression in random design) and condition (vi) in Theorem 5.2 (Generalized linear regression in fixed design), which are used to convert the theoretical compatibility condition to an empirical version and to localize $\hat\beta_\lambda$ near β^* with non-quadratic objective function $K_n(\beta)$.

Theorem 5.3 can be shown by properly combining the related arguments in the proof of Theorem 4.3 (Linear regression in random design) and the proof of Theorem 5.2 (Generalized linear regression in fixed design). In particular, Lemma 4.4 (Concentration of Hessian matrix) can be extended as follows, and Lemma 4.5 (From theoretical to empirical compatibility condition) can be extended with $\hat H$ and H replaced by $\hat H(\beta^*)$ and $H(\beta^*)$. Lemma 5.2 (Concentration of inner products in fixed design) remains valid in a random design. Proof details for Theorem 5.3 are left as an exercise.

Lemma 5.4 (Concentration of Hessian matrix). *For* $0 < \epsilon < 1$, *let* $\Omega_2(\lambda_2)$ *be the event that*

$$\max_{j,k=1,\ldots,p} |\hat H_{jk}(\beta^*) - H_{jk}(\beta^*)| \leq \lambda_2,$$

where $\hat H_{jk}(\beta^*)$ *is the* (j,k)*th element of* $\hat H(\beta^*)$, *and* $H_{jk}(\beta^*)$ *is the* (j,k)*th element of* $H(\beta^*)$. *Under independence of* $(X_1, \ldots, X_n)$ *and condition (iii), if* $\lambda_2 \geq 2C_0^2C_2\sqrt{\frac{\log(p/\epsilon)}{n}}$, *then* $P(\Omega_2(\lambda_2)) \geq 1 - 2\epsilon^2$.

5.4 Lasso estimation with model misspecification

For a nonparametric version of mean restriction (5.13), suppose that

$$E(Y_i|X_i) = a'(f^*(X_i)), \quad i = 1, \ldots, n, \tag{5.38}$$

where $f^*(\cdot)$ is an unknown function and $a(\cdot)$ is a *known*, convex differentiable function (hence $a'(\cdot)$ is non-decreasing). The function $a'(\cdot)$ can, in principle, be absorbed into $f^*(\cdot)$, leading to nonparametric regression (4.28), but is used as an inverse link function to ensure that the range of $f^*(\cdot)$ is unrestricted. Generalized linear regression (5.13) is correctly specified if and only if there exists $\beta^* \in \mathbb{R}^p$ such that $f^*(X_i) = X_i^{\mathrm{T}}\beta^*$ for $i = 1, \ldots, n$. With possible misspecification, linear regression is considered a working model such that for a suitable estimator $\hat{\beta}$, the fitted value $X_i^{\mathrm{T}}\hat{\beta}$ provides a linear approximation to $f^*(X_i)$, $i = 1, \ldots, n$.

For a possibly nonlinear function $f(x)$ as a prediction for $f^*(x)$, the objective function in (5.14) can be extended as

$$\kappa_n(\mathbf{f}) = \frac{-1}{n} \sum_{i=1}^n \{Y_i f(X_i) - a(f(X_i))\},$$

where $\mathbf{f} = (f_1, \ldots, f_n)^{\mathrm{T}}$ with $f_i = f(X_i)$. For a linear function $f_\beta(x) = x^{\mathrm{T}}\beta$, the objective function $K_n(\beta)$ in (5.14) can be rewritten as $K_n(\beta) = \kappa_n(\mathbf{f}_\beta)$, where $\mathbf{f}_\beta = (X_1^{\mathrm{T}}\beta, \ldots, X_n^{\mathrm{T}}\beta)^{\mathrm{T}}$. For two functions $\tilde{f}(x)$ and $f(x)$, the Bregman divergence associated with $\kappa_n(\cdot)$ is

$$D_{\kappa_n}(\tilde{\mathbf{f}}, \mathbf{f}) = \kappa_n(\tilde{\mathbf{f}}) - \kappa_n(\mathbf{f}) - \langle \nabla \kappa_n(\mathbf{f}), \tilde{\mathbf{f}} - \mathbf{f} \rangle,$$

where $\nabla \kappa_n(\mathbf{f})$ is the gradient of $\kappa_n(\mathbf{f})$ with respect to $\mathbf{f}$ as a vector in $\mathbb{R}^n$, such that $\langle \nabla \kappa_n(\mathbf{f}), \tilde{\mathbf{f}} - \mathbf{f} \rangle = -\langle \mathbf{Y} - a'(\mathbf{f}), \tilde{\mathbf{f}} - \mathbf{f} \rangle_n$. For $\tilde{\beta}, \beta \in \mathbb{R}^p$, the Bregman divergence associated with $K_n(\cdot)$ can be rewritten as $D_{K_n}(\tilde{\beta}, \beta) = D_{\kappa_n}(\mathbf{f}_{\tilde{\beta}}, \mathbf{f}_\beta)$. However, $\kappa_n(\cdot)$ and the associated Bregman divergence can also be applied to nonlinear functions such as $f^*(\cdot)$.

Similarly as discussed in Example 5.7, $\kappa_n(\mathbf{f})$ can be decomposed as

$$\kappa_n(\mathbf{f}) = \underbrace{\frac{-1}{n} \sum_{i=1}^n Y_i f(X_i)}_{\text{linear in } \mathbf{f}} + \underbrace{\frac{1}{n} \sum_{i=1}^n a(f(X_i))}_{\substack{\text{nonlinear in } \mathbf{f} \text{ and} \\ \text{independent of } \mathbf{Y}}}.$$

Hence $D_{\kappa_n}(\tilde{\mathbf{f}}, \mathbf{f})$ is identical to the Bregman divergence associated with the nonlinear term above, $\frac{1}{n}\sum_{i=1}^n a(f(X_i))$, independent of $(Y_1, \ldots, Y_n)$.

There is an interesting extension of the algebraic identity (4.33) in linear regression with $a(u) = \frac{u^2}{2}$. For any $\tilde{\beta}, \beta \in \mathbb{R}^p$ and $\mathbf{f} \in \mathbb{R}^n$,

$$D_{\kappa_n}(\mathbf{f}_{\tilde{\beta}}, \mathbf{f}_\beta)$$

$$= D_{\kappa_n}(\mathbf{f}_{\tilde\beta}, \mathbf{f}) - D_{\kappa_n}(\mathbf{f}_{\beta}, \mathbf{f}) + \langle \nabla\kappa_n(\mathbf{f}), \mathbf{f}_{\tilde\beta} - \mathbf{f}_\beta\rangle - \langle \nabla\kappa_n(\mathbf{f}_\beta), \mathbf{f}_{\tilde\beta} - \mathbf{f}_\beta\rangle$$
$$= D_{\kappa_n}(\mathbf{f}_{\tilde\beta}, \mathbf{f}) - D_{\kappa_n}(\mathbf{f}_{\beta}, \mathbf{f}) + \langle a'(\mathbf{X}\beta) - a'(\mathbf{f}), \mathbf{X}(\tilde\beta - \beta)\rangle. \tag{5.39}$$

The first line follows from the definition of D_{κ_n} and direct calculation, and the second line uses the expression of $\kappa_n(\cdot)$. For a fixed design, the relationship (4.32) in linear regression can be extended as [Exercise]

$$D_{\kappa_n}(\mathbf{f}_{\tilde\beta}, \mathbf{f}_{\bar\beta}) = D_{\kappa_n}(\mathbf{f}_{\tilde\beta}, \mathbf{f}^*) - D_{\kappa_n}(\mathbf{f}_{\bar\beta}, \mathbf{f}^*), \tag{5.40}$$

for any vector $\tilde\beta \in \mathbb{R}^p$, where the target value $\bar\beta$ is defined later as (5.41), but the expectation is taken over $(Y_1, \ldots, Y_n)$ with $(X_1, \ldots, X_n)$ fixed. Equivalently, $\bar\beta$ can be obtained as (5.42), but no expectation is taken for $D_{\kappa_n}(\mathbf{f}_\beta, \mathbf{f}^*)$, which is independent of $(Y_1, \ldots, Y_n)$.

5.4.1 Random-design generalized linear regression

We study Lasso estimation with possible misspecification of generalized linear regression (5.13) in a random design, where $(X_1, Y_1), \ldots, (X_n, Y_n)$ are random. The analysis in a fixed design is similar to that of fixed-design linear regression with misspecification (Section 4.3.2), and is left as exercises.

The target value $\bar\beta$ is defined as

$$\bar\beta = \text{argmin}_{\beta\in\mathbb{R}^p}\ E\{K_n(\beta)\}$$
$$= \text{argmin}_{\beta\in\mathbb{R}^p}\ \frac{-1}{n}\sum_{i=1}^n E\left\{Y_i X_i^{\mathrm{T}}\beta - a(X_i^{\mathrm{T}}\beta)\right\}, \tag{5.41}$$

where the expectation is taken over (X_i, Y_i), $i = 1, \ldots, n$, jointly. Equivalently, the target value $\bar\beta$ can be obtained as

$$\bar\beta = \text{argmin}_{\beta\in\mathbb{R}^p}\ E\{\kappa_n(\mathbf{f}_\beta)\}$$
$$= \text{argmin}_{\beta\in\mathbb{R}^p}\ E\{D_{\kappa_n}(\mathbf{f}_\beta, \mathbf{f}^*)\}, \tag{5.42}$$

by the decomposition $E\{\kappa_n(\mathbf{f}_\beta)\} = E\{\kappa_n(\mathbf{f}^*)\} + E\{D_{\kappa_n}(\mathbf{f}_\beta, \mathbf{f}^*)\}$ [Exercise]. In parallel to (5.40) in a fixed design, the following identity holds in a random design [Exercise]:

$$E\{D_{\kappa_n}(\mathbf{f}_{\tilde\beta}, \mathbf{f}_{\bar\beta})\} = E\{D_{\kappa_n}(\mathbf{f}_{\tilde\beta}, \mathbf{f}^*)\} - E\{D_{\kappa_n}(\mathbf{f}_{\bar\beta}, \mathbf{f}^*)\}, \tag{5.43}$$

for any vector $\tilde\beta \in \mathbb{R}^p$ and the target value $\bar\beta$ in (5.41) or (5.42).

For theoretical analysis, consider the following conditions:

(i) [Generalized nonparametric regression in random design] $(X_1, Y_1), \ldots, (X_n, Y_n)$ are independent with $E(\omega_i|X_i) = 0$, where $\omega_i = Y_i - a'(f^*(X_i))$, for $i = 1, \ldots, n$;

(ii) [Sub-Gaussian true noises] ω_i given X_i is sub-Gaussian with scale parameter σ for $i = 1, \ldots, n$;

(iii) [Bounded covariates] $\max_{j=1,\ldots,p} |X_{ij}| \leq C_0$ almost surely for $i = 1,\ldots,n$ with a constant $C_0 > 0$;

(iv) [Hessian condition] $a''(\tilde{u}) \leq a''(u)\mathrm{e}^{C_1|\tilde{u}-u|}$ for any $u, \tilde{u} \in \mathbb{R}$ with a constant $C_1 > 0$, and $a''(X_i^{\mathrm{T}}\bar{\beta}) \leq C_2$ almost surely for $i = 1,\ldots,n$ with a constant $C_2 > 0$;

(v) [Theoretical compatibility condition] The compatibility condition $(S_{\bar{\beta}}, \xi_0^*, \nu_0^*)$ holds for $H(\bar{\beta}) = E\{\frac{1}{n}\sum_{i=1}^n X_i a''(X_i^{\mathrm{T}}\bar{\beta})X_i^{\mathrm{T}}\}$, where $S_{\bar{\beta}} = \{1 \leq j \leq p : \bar{\beta}_j \neq 0\}$ and $\xi_0^* > 1$ and $\nu_0^* > 0$ are constants.

The choice of target value $\bar{\beta}$ in (5.41) is unique when the compatibility condition $(S_{\bar{\beta}}, \xi_0^*, \nu_0^*)$ is assumed to hold with $\xi_0^* \geq 1$ for $H(\bar{\beta})$ (see Section 5.5.1). If generalized linear regression is correctly specified, then conditions (i)–(v) reduce to the corresponding ones in Theorem 5.3 (Correct generalized linear regression in random design). Moreover, conditions (iv) and (v) are implied by jointly conditions (iv†) and (v†) in Section 5.3.2 (Correct generalized linear regression in random design), with β^* replaced by $\bar{\beta}$.

For similar reasons as discussed in Section 4.3.2 (Misspecified linear regression in random design), we replace sub-Gaussian condition (ii) by either of the following two conditions:

(ii†) [Sub-Gaussian pseudo-noises] $Y_i - a'(X_i^{\mathrm{T}}\bar{\beta})$ given X_i is sub-Gaussian with ψ_2 norm upper bounded by τ, $E(\mathrm{e}^{(Y_i - a'(X_i^{\mathrm{T}}\bar{\beta}))^2/\tau^2}|X_i) \leq 2$, for $i = 1,\ldots,n$;

(ii‡) [Marginally sub-Gaussian pseudo-noises] $Y_i - a'(X_i^{\mathrm{T}}\bar{\beta})$ is sub-Gaussian with ψ_2 norm upper bounded by τ, $E(\mathrm{e}^{(Y_i - a'(X_i^{\mathrm{T}}\bar{\beta}))^2/\tau^2}) \leq 2$, for $i = 1,\ldots,n$.

Condition (ii†) involves conditioning on the covariates as in condition (ii), but condition (ii‡) is weaker than condition (ii†).

For any $0 < \epsilon < 1$, suppose that the tuning parameter λ in (5.9) is specified as $\lambda = A_0\lambda_0$ with a constant $A_0 > 1$ and

$$\lambda_0 = \max\left\{\tau C_0\sqrt{\frac{8\log(p/\epsilon)}{n}},\ 2C_0^2C_2\sqrt{\frac{\log(p/\epsilon)}{n}}\right\},$$

depending on τ in condition (ii†) or (ii‡). The following result extends the fast-rate error bound in Theorem 5.3 to allow possible model misspecification.

Theorem 5.4 (Fast rate in random design with model misspecification). *Assume that conditions (i), (ii†) or (ii‡), and (iii)–(v) above are satisfied. For any $0 < \epsilon < 1$ and $A_0 \geq \frac{\xi_0^*+1}{\xi_0^*-1}$, suppose further that*

(vi) [Rate condition] $(1+\xi_0^*)^2\nu_0^{*-2}|S_{\bar{\beta}}|\lambda_0 \leq \eta_0$ for a constant $\eta_0 \in (0,1)$, and $\frac{4A_0^2}{A_0-1}C_0C_1(1-\eta_0)^{-1}\nu_0^{*-2}|S_{\bar{\beta}}|\lambda_0 \leq \eta_1$ for a constant $\eta_1 \in (0,1)$.

Then it holds with probability at least $1 - 2\epsilon - 2\epsilon^2$ *uniformly over* A_0 *subject to condition (vi) that the Lasso estimator* $\hat{\beta}_\lambda$ *with* $\lambda = A_0\lambda_0$ *in* (5.9) *satisfies*

$$D_{\kappa_n}(\mathbf{f}_{\hat{\beta}_\lambda}, \mathbf{f}_{\bar{\beta}}) + D_{\kappa_n}(\mathbf{f}_{\bar{\beta}}, \mathbf{f}_{\hat{\beta}_\lambda}) + (A_0 - 1)\lambda_0\|\hat{\beta}_\lambda - \bar{\beta}\|_1 \leq M_0|S_{\bar{\beta}}|\lambda_0^2, \quad (5.44a)$$

$$(\hat{\beta}_\lambda - \bar{\beta})^{\mathrm{T}}\hat{H}(\bar{\beta})(\hat{\beta}_\lambda - \bar{\beta}) \leq (1 - \eta_1)^{-1}M_0|S_{\bar{\beta}}|\lambda_0^2, \quad (5.44b)$$

where $M_0 = 4A_0^2(1-\eta_1)^{-1}(1-\eta_0)^{-1}\nu_0^{*-2}$, *depending only on* $(A_0, \nu_0^*, \eta_0, \eta_1)$. *Moreover, with probability at least* $1 - 4\epsilon - 2\epsilon^2$,

$$\begin{aligned} &D_{\kappa_n}(\mathbf{f}_{\hat{\beta}_\lambda}, \mathbf{f}^*) + D_{\kappa_n}(\mathbf{f}_{\bar{\beta}}, \mathbf{f}_{\hat{\beta}_\lambda}) + (A_0 - 1)\lambda_0\|\hat{\beta}_\lambda - \bar{\beta}\|_1 \\ &\leq D_{\kappa_n}(\mathbf{f}_{\bar{\beta}}, \mathbf{f}^*) + \frac{A_0}{A_0 - 1}M_0|S_{\bar{\beta}}|\lambda_0^2, \end{aligned} \quad (5.45)$$

where $\mathbf{f}^* = (f^*(X_1), \ldots, f^*(X_n))^{\mathrm{T}}$ *and* $\mathbf{f}_\beta = (X_1^{\mathrm{T}}\beta, \ldots, X_n^{\mathrm{T}}\beta)^{\mathrm{T}}$.

The error bound (5.44) recovers (5.37) in Theorem 5.3 when generalized linear regression (5.13) is correctly specified, i.e., $\bar{\beta} = \beta^*$, in a random design. With possible misspecification of generalized linear regression (5.13), the error bound (5.44) implies the convergence of $\hat{\beta}_\lambda$ to $\bar{\beta}$ in terms of both the prediction error and the L_1 estimation error, whereas the error bound (5.45) implies that the prediction error $D_{\kappa_n}(\mathbf{f}_{\hat{\beta}_\lambda}, \mathbf{f}^*)$ (from comparing $\mathbf{f}_{\hat{\beta}_\lambda}$ with $\mathbf{f}^*$) converges to $D_{\kappa_n}(\mathbf{f}_{\bar{\beta}}, \mathbf{f}^*)$ (from comparing $\mathbf{f}_{\bar{\beta}}$ with $\mathbf{f}^*$) at rate $|S_{\bar{\beta}}|\frac{\log(n \vee p)}{n}$.

The error bound (5.44) can be shown by carefully extending the proof of Theorem 5.3 (Correct generalized linear regression in random design) and the proof of Theorem 4.5 (Misspecified linear regression in random design). Taking $\beta = \bar{\beta}$ from (5.41) or (5.42) in the basic inequality (5.22) gives

$$D_{\kappa_n}(\mathbf{f}_{\hat{\beta}_\lambda}, \mathbf{f}_{\bar{\beta}}) + D_{\kappa_n}(\mathbf{f}_{\bar{\beta}}, \mathbf{f}_{\hat{\beta}_\lambda}) - \langle \mathbf{Y} - a'(\mathbf{X}\bar{\beta}), \mathbf{X}(\hat{\beta}_\lambda - \bar{\beta})\rangle_n + \lambda\|\hat{\beta}_\lambda\|_1 \leq \lambda\|\bar{\beta}\|_1.$$

The inner-product term $\langle \mathbf{Y} - a'(\mathbf{X}\bar{\beta}), \mathbf{X}(\hat{\beta}_\lambda - \bar{\beta})\rangle_n$ can be controlled in the event $\Omega_1 = \Omega_1(\lambda_0)$ with $\lambda_1 = \lambda_0$ in Lemma 5.5 below. Moreover, the error bound (5.45) can be obtained by combining (5.44a) and the following two inequalities. First, from the identity (5.39), we have

$$D_{\kappa_n}(\mathbf{f}_{\hat{\beta}_\lambda}, \mathbf{f}^*) - D_{\kappa_n}(\mathbf{f}_{\bar{\beta}}, \mathbf{f}^*) \leq D_{\kappa_n}(\mathbf{f}_{\bar{\beta}}, \mathbf{f}_{\hat{\beta}_\lambda}) + |\langle a'(\mathbf{X}\bar{\beta}) - \mathbf{f}^*, \mathbf{X}(\hat{\beta}_\lambda - \bar{\beta})\rangle_n|.$$

Second, in the intersection of (5.44a) and the event $\Omega_3 = \Omega_3(\lambda_0)$ with $\lambda_1 = \lambda_0$ in Lemma 5.5, we have

$$|\langle a'(\mathbf{X}\bar{\beta}) - a'(\mathbf{f}^*), \mathbf{X}(\hat{\beta}_\lambda - \bar{\beta})\rangle_n| \leq \lambda_0\|\hat{\beta}_\lambda - \bar{\beta}\|_1 \leq \frac{1}{A_0 - 1}M_0|S_{\bar{\beta}}|\lambda_0^2.$$

Lemma 5.5 can be shown similarly as Lemma 4.7 with linear approximation. Proof details for Theorem 5.4 are left as an exercise.

Lemma 5.5 (Concentration of inner products in random design with generalized linear approximation). *For $0 < \epsilon < 1$, let $\Omega_1(\lambda_1)$ be the event that*

$$\max_{j=1,\ldots,p} |\langle \mathbf{Y} - a'(\mathbf{X}\bar{\beta}), \mathbf{X}_j \rangle_n| \leq \lambda_1,$$

and let $\Omega_3(\lambda_1)$ be the event that

$$\max_{j=1,\ldots,p} |\langle a'(\mathbf{f}^*) - a'(\mathbf{X}\bar{\beta}), \mathbf{X}_j \rangle_n| \leq \lambda_1.$$

Under conditions (i), (ii†) or (ii‡), and (iii), if $\lambda_1 \geq \tau C_0 \sqrt{\frac{8\log(p/\epsilon)}{n}}$, then $P(\Omega_1(\lambda_1)) \geq 1 - 2\epsilon$ and $P(\Omega_3(\lambda_1)) \geq 1 - 2\epsilon$.

5.5 Appendices

5.5.1 Uniqueness of compatibility conditions

We show the uniqueness in the choice of target value $\bar{\beta}$, including true value β^* under correct model specification, when a compatibility condition is assumed to hold with the associated active set $S_{\bar{\beta}}$ for the empirical Hessian in a fixed design or the theoretical Hessian in a random design. Generalized linear regression is considered, which includes linear regression as a special case with $a(u) = \frac{u^2}{2}$.

For a fixed design, the target value $\bar{\beta}$ is defined as (5.41), but the expectation is taken over $(Y_1, \ldots, Y_n)$ with $(X_1, \ldots, X_n)$ fixed. Equivalently, $\bar{\beta}$ can be obtained as (5.42), but no expectation is taken for $D_{\kappa_n}(\mathbf{f}_\beta, \mathbf{f}^*)$, which is independent of $(Y_1, \ldots, Y_n)$.

Proposition 5.1.

(i) Suppose that $\bar{\beta}^{(1)}$ and $\bar{\beta}^{(2)}$ are two choices of target value $\bar{\beta}$, that is,

$$\bar{\beta}^{(1)} = \text{argmin}_{\beta \in \mathbb{R}^p} D_{\kappa_n}(\mathbf{f}_\beta, \mathbf{f}^*),$$
$$\bar{\beta}^{(2)} = \text{argmin}_{\beta \in \mathbb{R}^p} D_{\kappa_n}(\mathbf{f}_\beta, \mathbf{f}^*).$$

Then $\mathbf{X}\bar{\beta}^{(1)} = \mathbf{X}\bar{\beta}^{(2)}$, denoted as $\mathbf{X}\bar{\beta}$, and $\hat{H}(\bar{\beta}^{(1)}) = \hat{H}(\bar{\beta}^{(2)})$, denoted as $\hat{H}(\bar{\beta})$.

(ii) In addition, suppose that for the matrix $\hat{H}(\bar{\beta})$, the compatibility condition is satisfied by two choices $(S_{\bar{\beta}^{(1)}}, \xi_0^{(1)}, \nu_0^{(1)})$ and $(S_{\bar{\beta}^{(2)}}, \xi_0^{(2)}, \nu_0^{(2)})$, with $\xi_0^{(1)} \geq 1$, $\nu_0^{(1)} > 0$, $\xi_0^{(2)} \geq 1$, $\nu_0^{(2)} > 0$. Then $\bar{\beta}^{(1)} = \bar{\beta}^{(2)}$.

Proof. (i) By definition, $D_{\kappa_n}(\mathbf{f}_{\bar{\beta}^{(1)}}, \mathbf{f}^*) = D_{\kappa_n}(\mathbf{f}_{\bar{\beta}^{(2)}}, \mathbf{f}^*)$. By the identity (5.40),

$$D_{\kappa_n}(\mathbf{f}_{\bar{\beta}^{(1)}}, \mathbf{f}_{\bar{\beta}^{(2)}}) = D_{\kappa_n}(\mathbf{f}_{\bar{\beta}^{(1)}}, \mathbf{f}^*) - D_{\kappa_n}(\mathbf{f}_{\bar{\beta}^{(2)}}, \mathbf{f}^*) = 0.$$

This implies that $\mathbf{f}_{\bar\beta^{(1)}} = \mathbf{f}_{\bar\beta^{(2)}}$, because $a(\cdot)$ and hence $\kappa_n(\cdot)$ are strictly convex.

(ii) Let $b = \bar\beta^{(1)} - \bar\beta^{(2)}$. Then $\mathbf{X}b = 0$ and

$$b^{\mathrm{T}}\hat{H}(\bar\beta)b = \langle \mathbf{X}b, a''(\mathbf{X}\bar\beta)\mathbf{X}b\rangle_n = 0.$$

Consider the following decomposition:

$$\sum_{j\in S^{(1)}} |\bar\beta_j^{(1)} - \bar\beta_j^{(2)}| = \sum_{j\in S^{(1)}\cap S^{(2)}} |\bar\beta_j^{(1)} - \bar\beta_j^{(2)}| + \sum_{j\in S^{(1)}\cap S^{(2)c}} |\bar\beta_j^{(1)}|,$$

$$\sum_{j\in S^{(1)c}} |\bar\beta_j^{(1)} - \bar\beta_j^{(2)}| = \sum_{j\in S^{(1)c}\cap S^{(2)}} |\bar\beta_j^{(2)}|,$$

where $S^{(1)} = S_{\bar\beta^{(1)}}$ and $S^{(2)} = S_{\bar\beta^{(2)}}$. Similarly,

$$\sum_{j\in S^{(2)}} |\bar\beta_j^{(1)} - \bar\beta_j^{(2)}| = \sum_{j\in S^{(1)}\cap S^{(2)}} |\bar\beta_j^{(1)} - \bar\beta_j^{(2)}| + \sum_{j\in S^{(1)c}\cap S^{(2)}} |\bar\beta_j^{(2)}|,$$

$$\sum_{j\in S^{(2)c}} |\bar\beta_j^{(1)} - \bar\beta_j^{(2)}| = \sum_{j\in S^{(1)}\cap S^{(2)c}} |\bar\beta_j^{(1)}|.$$

If $\sum_{j\in S^{(1)}\cap S^{(2)c}} |\bar\beta_j^{(1)}| \ge \sum_{j\in S^{(1)c}\cap S^{(2)}} |\bar\beta_j^{(2)}|$, then $\sum_{j\in S^{(1)}} |\bar\beta_j^{(1)} - \bar\beta_j^{(2)}| \ge \sum_{j\in S^{(1)c}} |\bar\beta_j^{(1)} - \bar\beta_j^{(2)}|$ from the first decomposition. Otherwise, $\sum_{j\in S^{(2)}} |\bar\beta_j^{(1)} - \bar\beta_j^{(2)}| \ge \sum_{j\in S^{(2)c}} |\bar\beta_j^{(1)} - \bar\beta_j^{(2)}|$ from the second decomposition. Hence $b = \bar\beta^{(1)} - \bar\beta^{(2)}$ satisfies the cone condition for either the compatibility condition $(S^{(1)}, \xi_0^{(1)}, \nu_0^{(1)})$ or the compatibility condition $(S^{(2)}, \xi_0^{(2)}, \nu_0^{(2)})$. By symmetry, assume that the first case occurs. If $\|b\|_1 \ne 0$, then $2\sum_{j\in S^{(1)}} |b_j| \ge \|b\|_1 > 0$ and

$$|S^{(1)}| \cdot b^{\mathrm{T}}\hat{H}(\bar\beta)b \ge \nu_0^{(1)2}(\sum_{j\in S^{(1)}} |b_j|)^2 > 0,$$

which contradicts $b^{\mathrm{T}}\hat{H}(\bar\beta)b = 0$. Therefore, $\|b\|_1 = 0$, i.e., $b = 0$ and $\bar\beta^{(1)} = \bar\beta^{(2)}$. □

For a random design, the target value $\bar\beta$ is defined as (5.41) or equivalently (5.42), where the expectation is taken over (X_i, Y_i), $i = 1, \ldots, n$, jointly.

Proposition 5.2.

(i) Suppose that $\bar\beta^{(1)}$ and $\bar\beta^{(2)}$ are two choices of target value $\bar\beta$, that is,

$$\bar\beta^{(1)} = \mathrm{argmin}_{\beta\in\mathbb{R}^p} E\{D_{\kappa_n}(\mathbf{f}_\beta, \mathbf{f}^*)\},$$
$$\bar\beta^{(2)} = \mathrm{argmin}_{\beta\in\mathbb{R}^p} E\{D_{\kappa_n}(\mathbf{f}_\beta, \mathbf{f}^*)\}.$$

Then $\mathbf{X}\bar\beta^{(1)} = \mathbf{X}\bar\beta^{(2)}$ almost surely, and $H(\bar\beta^{(1)}) = H(\bar\beta^{(2)})$, denoted as $H(\bar\beta)$.

(ii) In addition, suppose that for the matrix $H(\bar\beta)$, the compatibility condition is satisfied by two choices $(S_{\bar\beta^{(1)}}, \xi_0^{(1)}, \nu_0^{(1)})$ and $(S_{\bar\beta^{(2)}}, \xi_0^{(2)}, \nu_0^{(2)})$, with $\xi_0^{(1)} \ge 1$, $\nu_0^{(1)} > 0$, $\xi_0^{(2)} \ge 1$, $\nu_0^{(2)} > 0$. Then $\bar\beta^{(1)} = \bar\beta^{(2)}$.

Proof. The result can be shown similarly as in the proof of Proposition 5.1. In particular, the proof of result (i) uses the identity (5.43) in place of (5.40). □

5.5.2 Penalized non-smooth M-estimation

We study high-dimensional Lasso-penalized estimation with a non-smooth loss function. Our presentation draws on Bühlmann and van de Geer (2011), Chapter 6, and Tan (2024), Supplement Section IV. The approach of theoretical analysis in previous sections, based on Tan (2020b), Supplement Section II, assumes that the loss function is twice differentiable and hence is not applicable to non-smooth loss functions.

Let $(X_1, Y_1), \ldots, (X_n, Y_n)$ be a collection of n observations, where $X_i = (X_{i1}, \ldots, X_{ip})^{\mathrm{T}} \in \mathbb{R}^p$ is a covariate vector and $Y_i \in \mathbb{R}$ is a response variable. Consider a sample objective (or loss) function in the form

$$K_n(\beta) = \frac{1}{n}\sum_{i=1}^{n} \rho(Y_i, X_i^{\mathrm{T}}\beta), \tag{5.46}$$

where $\rho(y, \eta)$ is a measure of discrepancy between y and η. The Lasso-penalized estimator is defined as

$$\hat\beta_\lambda = \mathrm{argmin}_{\beta\in\mathbb{R}^p} \left\{K_n(\beta) + \lambda\|\beta\|_1\right\}, \tag{5.47}$$

where $\lambda > 0$ is a tuning parameter. The target value $\bar\beta$ is defined as

$$\bar\beta = \mathrm{argmin}_{\beta\in\mathbb{R}^p} K(\beta),$$

where $K(\beta) = E\{K_n(\beta)\}$.

For generalized linear regression, the objective function (5.14) is of the form (5.46), where $\rho(y, \eta) = -y\eta + a(\eta)$ is twice differentiable for a strictly convex, twice differentiable function $a(\cdot)$. As mentioned above, our primary interest is to deal with $\eta(y, \eta)$ which may not be twice differentiable in η. A motivating example is quantile regression.

Example 5.11 (Quantile regression). For $0 < \tau < 1$, a τ-quantile regression model postulates

$$q^*_{Y_i|X_i}(\tau) = X_i^{\mathrm{T}}\beta^*, \quad i = 1, \ldots, n, \tag{5.48}$$

where $q^*_{Y_i|X_i}(\tau)$ denotes the τ-quantile of the conditional distribution of Y_i given X_i. By the standard characterization of quantiles (Example 2.3), an

M-estimator for β^* is $\hat{\beta} = \mathrm{argmin}_{\beta \in \mathbb{R}^p} K_n(\beta)$ with

$$K_n(\beta) = \frac{1}{n} \sum_{i=1}^{n} \{\tau(Y_i - X_i^{\mathrm{T}}\beta)_+ + (1-\tau)(X_i^{\mathrm{T}}\beta - Y_i)_+\},$$

which is of the form (5.46) with $\rho(y, \eta) = \tau(y-\eta)_+ + (1-\tau)(\eta - y)_+$. The function $\rho(y, \eta)$ is not differentiable everywhere in η. In high-dimensional settings, a Lasso-penalized estimator $\hat{\beta}_\lambda$ is defined as in (5.47). If quantile regression (5.48) is correctly specified with true value β^*, then the target value $\bar{\beta}$ coincides with β^* and hence $X_i^{\mathrm{T}}\bar{\beta} = X_i^{\mathrm{T}}\beta^*$ gives the conditional quantile $q^*_{Y_i|X_i}(\tau)$. Otherwise, $\bar{\beta}$ remains well-defined but $X_i^{\mathrm{T}}\bar{\beta}$ does not give the conditional quantile $q^*_{Y_i|X_i}(\tau)$ for all $i = 1, \ldots, n$.

For theoretical analysis of Lasso-penalized non-smooth M-estimation, consider the following conditions in a random design. The analysis in a fixed design can be similarly conducted and is left as an exercise.

(i) [Random design] $(X_1, Y_1), \ldots, (X_n, Y_n)$ are independent observations.

(ii) [Convex and Lipschitz loss] $\rho(y, \eta)$ is convex and Lipschitz-L in η for any fixed y, where $L > 0$ is a constant.

(iii) [Bounded covariates] $\max_{j=1,\ldots,p} |X_{ij}| \le C_0$ almost surely for $i = 1, \ldots, n$ with a constant $C_0 > 0$;

(iv) [Margin condition] The quadratic margin condition holds for some constants $\zeta_0^* > 0$ and $\kappa_0^* > 0$: $K(\beta) - K(\bar{\beta}) \ge \kappa_0^* (\beta - \bar{\beta})^{\mathrm{T}} H (\beta - \bar{\beta})$ for any β satisfying $\|\beta - \bar{\beta}\|_1 \le \zeta_0^*$, where $H = E(\mathbf{X}\mathbf{X}^{\mathrm{T}}/n)$.

(v) [Theoretical compatibility condition] The compatibility condition $(S_{\bar{\beta}}, \xi_0^*, \nu_0^*)$ holds for H, where $S_{\bar{\beta}} = \{1 \le j \le p : \bar{\beta}_j \ne 0\}$ and $\xi_0^* > 1$ and $\nu_0^* > 0$ are constants.

The Lipschitz property of $\rho(y, \eta)$ in η is exploited to establish concentration of the sample loss difference $K_n(\beta) - K_n(\bar{\beta})$, as shown in Proposition 3.1 (Section 3.4.3), whereas the convexity of $\rho(Y, \eta)$ in η is used to localize $\hat{\beta}_\lambda$ near $\bar{\beta}$ (i.e., eliminate the possibility of $\hat{\beta}_\lambda$ far away from $\bar{\beta}$). For τ-quantile regression, $\rho(y, \eta)$ is convex and Lipschitz-L in η with $L = \max(\tau, 1-\tau)$. For logistic regression with maximum likelihood estimation, $\rho(y, \eta) = -y\eta + \log(1 + \mathrm{e}^\eta)$ is convex and Lipschitz-1 in η.

The margin condition (iv) requires a lower bound of the population loss difference $K(\beta) - K(\bar{\beta})$ by a quadratic form with matrix H for β near $\bar{\beta}$. For τ-quantile regression, from Example 2.7,

$$\frac{\partial^2}{\partial \beta \partial \beta^{\mathrm{T}}} K(\beta) = E\left\{ \sum_{i=1}^{n} f_{Y_i|X_i}(X_i^{\mathrm{T}}\beta) X_i X_i^{\mathrm{T}}/n \right\},$$

provided that the CDF of Y_i given X_i is differentiable at $X_i^{\mathrm{T}}\beta$ with derivative $f_{Y_i|X_i}(X_i^{\mathrm{T}}\beta)$, i.e., the conditional distribution of Y_i given X_i admits density

$f_{Y_i|X_i}(X_i^{\mathrm{T}}\beta)$ at $X_i^{\mathrm{T}}\beta$. For logistic regression with maximum likelihood estimation, from Example 2.6,

$$\frac{\partial^2}{\partial\beta\partial\beta^{\mathrm{T}}}K(\beta) = E\left\{\sum_{i=1}^n \mu_i(\beta)(1-\mu_i(\beta))X_iX_i^{\mathrm{T}}/n\right\},$$

where $\mu_i(\beta) = \mathrm{e}^{X_i^{\mathrm{T}}\beta}/(1+\mathrm{e}^{X_i^{\mathrm{T}}\beta})$. Based on these observations, the margin condition (iv) can be justified from primitive conditions as follows. The proof is left as Exercise 5.11.

Lemma 5.6 (Sufficient conditions for the margin condition). *Suppose that condition (iii) is satisfied.*

(i) For τ-quantile regression, if there exist constants $B_1, B_2 > 0$ such that the conditional distribution of Y_i given X_i admits density $f_{Y_i|X_i}(\cdot)$ satisfying $f_{Y_i|X_i}(u) \ge B_2$ for $|u - X_i^{\mathrm{T}}\bar\beta| \le B_1$ for $i = 1,\ldots,n$, then the margin condition (iv) holds with $\zeta_0^* = B_1/C_0$ and $\kappa_0^* = B_2/2$.

(ii) For logistic regression with maximum likelihood estimation, if $|X_i^{\mathrm{T}}\bar\beta| \le B_0$ almost surely for $i = 1,\ldots,n$ with a constant $B_0 > 0$, then the margin condition (iv) holds with $\zeta_0^* = 1/C_0$ and $\kappa_0^* = \{\mathrm{e}^{B_0+1}/(1+\mathrm{e}^{B_0+1})^2\}/2$. [Question: Can you find a larger κ_0^* given the stated ζ_0^*?]

For any $0 < \epsilon < 1$, suppose that the tuning parameter λ in (5.47) is specified as $\lambda = A_0\lambda_0$ with a constant $A_0 > 1$ and

$$\lambda_0 = 8LC_0\sqrt{\frac{2\log(p/\epsilon)}{n}}.$$

The following result gives a fast-rate error bound for Lasso-penalized M-estimation. A slow-rate error bound can also be established under conditions (i)–(iii) only, and is left as Exercise 5.10.

Theorem 5.5 (Fast rate in random design). *Assume that conditions (i)–(iv) above are satisfied. For any $0 < \epsilon < 1$ and $A_0 \ge \frac{\xi_0^*+1}{\xi_0^*-1}$, suppose further that*

(v) [Rate condition] $\frac{4A_0^2}{A_0-1}\nu_0^{*-2}\kappa_0^{*-1}|S_{\bar\beta}|\lambda_0 \le \zeta_0^*$.

Then it holds with probability at least $1-2\epsilon$ uniformly over A_0 subject to condition (v) that the Lasso estimator $\hat\beta_\lambda$ with $\lambda = A_0\lambda_0$ in (5.47) satisfies

$$\|\hat\beta_\lambda - \bar\beta\|_1 \le \frac{M_0}{A_0-1}|S_{\bar\beta}|\lambda_0, \tag{5.49a}$$

$$K(\hat\beta_\lambda) - K(\bar\beta) + (A_0-1)\lambda_0\|\hat\beta_\lambda - \bar\beta\|_1 \le M_1|S_{\bar\beta}|\lambda_0^2, \tag{5.49b}$$

where $M_0 = 4A_0^2\nu_0^{*-2}\kappa_0^{*-1}$ and $M_1 = \frac{4A_0^2(2A_0+1)}{A_0-1}\nu_0^{*-2}\kappa_0^{*-1} = \frac{2A_0+1}{A_0-1}M_0$, depending only on $(A_0, \nu_0^*, \kappa_0^*)$.

The proof of Theorem 5.5 involves similar high-level ideas such as using a basic inequality and concentration of sample quantities and inverting a quadratic inequality as in the proofs of Theorems 5.3 and 5.4 for Lasso estimation in generalized linear regression. However, the details of applying such ideas differ from the previous proof. First, we recall Proposition 3.1 about the concentration of the sample loss difference $K_n(\beta) - K_n(\bar{\beta})$. Let

$$D_n(\beta, \bar{\beta}) = K_n(\beta) - K_n(\bar{\beta}) - \{K(\beta) - K(\bar{\beta})\}.$$

For fixed $r > 0$ and $0 < \epsilon < 1$, let $\Omega_1(r)$ be the event that

$$\sup_{\|\beta - \bar{\beta}\|_1 \le r} |D_n(\beta, \bar{\beta})| \le 8LC_0 r \cdot \sqrt{\frac{\log(p/\epsilon)}{n}} = \lambda_0 r.$$

Then $P(\Omega_1(r)) \ge 1 - 2\epsilon$ by Proposition 3.1. In contrast, Lemmas 5.2 and 5.5 deal with concentration of empirical inner products.

Next, the following result gives a basic inequality concerning the population loss function $K(\cdot)$. In contrast, Lemma 5.1 (Basic inequality) for Lasso estimation in generalized linear regression is obtained in terms of the Bregman divergence associated with the sample loss function $K_n(\cdot)$.

Lemma 5.7 (Basic inequality). *For any $\beta \in \mathbb{R}^p$ satisfying*

$$K_n(\beta) + A_0\lambda_0\|\beta\|_1 \le K_n(\bar{\beta}) + A_0\lambda_0\|\bar{\beta}\|_1, \tag{5.50}$$

we have

$$K(\beta) - K(\bar{\beta}) + A_0\lambda_0\|\beta - \bar{\beta}\|_1 \le 2A_0\lambda_0 \sum_{j \in S_{\bar{\beta}}} |\beta_j - \bar{\beta}| + |D_n(\beta, \bar{\beta})|.$$

Proof. By simple manipulation from (5.50),

$$\begin{aligned} K_n(\beta) - K_n(\bar{\beta}) + A_0\lambda_0\|\beta\|_1 &\le A_0\lambda_0\|\bar{\beta}\|_1 + D_n(\beta, \bar{\beta}) \\ &\le A_0\lambda_0\|\bar{\beta}\|_1 + |D_n(\beta, \bar{\beta})|. \end{aligned}$$

Applying the following identity and inequality

$$\begin{aligned} &|\beta_j| = |\beta_j - \bar{\beta}_j| - |\bar{\beta}_j|, \quad j \notin S_{\bar{\beta}} \text{ (i.e., } \bar{\beta}_j = 0), \\ &|\beta_j| \ge |\bar{\beta}_j| - |\beta_j - \bar{\beta}_j|, \quad j \in S_{\bar{\beta}}, \end{aligned}$$

leads to

$$K_n(\beta) - K_n(\bar{\beta}) + A_0\lambda_0 \left\{ \sum_{j \notin S_{\bar{\beta}}} (|\beta_j - \bar{\beta}_j| - |\bar{\beta}_j|) + \sum_{j \notin S_{\bar{\beta}}} (|\bar{\beta}_j| - |\beta_j - \bar{\beta}_j|) \right\}$$
$$\le A_0\lambda_0\|\bar{\beta}\|_1 + |D_n(\beta, \bar{\beta})|,$$

which gives the desired inequality after rearrangement. □

The following result exploits the margin condition and theoretical compatibility condition to obtain an error bound, for β achieving a penalized loss no greater than the target value $\bar{\beta}$ but on a fixed ball around $\bar{\beta}$.

Lemma 5.8 (Error bound on a fixed parameter ball). *Fix* $0 < r \leq \zeta_0^*$. *In the event* $\Omega_1(r)$, *for any* $\beta \in \mathbb{R}^p$ *satisfying* (5.50) *and* $\|\beta - \bar{\beta}\|_1 = r$, *we have*

$$K(\beta) - K(\bar{\beta}) > 0, \tag{5.51}$$

and

$$K(\beta) - K(\bar{\beta}) + (A_0 - 1)\lambda_0\|\beta - \bar{\beta}\|_1 \leq M_0|S_{\bar{\beta}}|\lambda_0^2, \tag{5.52}$$

where $M_0 = 4A_0^2\nu_0^{*-2}\kappa_0^{*-1}$.

Proof. With $\|\beta - \bar{\beta}\|_1 = r$, we have, by definition, in the event $\Omega_1(r)$,

$$|D_n(\beta, \bar{\beta})| \leq r\lambda_0 = \lambda_0\|\beta - \bar{\beta}\|_1.$$

Moreover, with β satisfying (5.50), we apply Lemma 5.7 and obtain that in the event $\Omega_1(r)$,

$$K(\beta) - K(\bar{\beta}) + A_0\lambda_0\|\beta - \bar{\beta}\|_1 \leq 2A_0\lambda_0 \sum_{j \in S_{\bar{\beta}}} |\beta_j - \bar{\beta}| + \lambda_0\|\beta - \bar{\beta}\|_1,$$

or equivalently

$$K(\beta) - K(\bar{\beta}) + (A_0 - 1)\lambda_0\|\beta - \bar{\beta}\|_1 \leq 2A_0\lambda_0 \sum_{j \in S_{\bar{\beta}}} |\beta_j - \bar{\beta}|. \tag{5.53}$$

Denote $\Delta = K(\beta) - K(\bar{\beta}) + (A_0 - 1)\lambda_0\|\beta - \bar{\beta}\|_1$, the left-hand side of (5.53).

Next, we deduce (5.52) from (5.53). In fact, (5.53) implies that

$$(A_0 - 1)\lambda_0\|\beta - \bar{\beta}\|_1 \leq 2A_0\lambda_0 \sum_{j \in S_{\bar{\beta}}} |\beta_j - \bar{\beta}_j|,$$

or equivalently

$$\sum_{j \notin S_{\bar{\beta}}} |\beta_j - \bar{\beta}_j| \leq \frac{A_0 + 1}{A_0 - 1} \sum_{j \in S_{\bar{\beta}}} |\beta_j - \bar{\beta}_j|. \tag{5.54}$$

Note that $\frac{A_0+1}{A_0-1} \leq \xi_0^*$ for $A_0 \geq \frac{\xi_0^*+1}{\xi_0^*-1}$. Then $\beta - \bar{\beta}$ satisfies the cone condition, $\sum_{j \notin S_{\bar{\beta}}} |\beta_j - \bar{\beta}_j| \leq \xi_0^* \sum_{j \in S_{\bar{\beta}}} |\beta_j - \bar{\beta}_j|$, and by the theoretical compatibility condition,

$$\nu_0^{*2}\Big(\sum_{j \in S_{\bar{\beta}}} |\beta_j - \bar{\beta}_j|\Big)^2 \leq |S_{\bar{\beta}}| \underbrace{(\beta - \bar{\beta})^{\mathrm{T}} H (\beta - \bar{\beta})}_{Q(\beta, \bar{\beta})}. \tag{5.55}$$

Combining (5.53) and (5.55) yields

$$K(\beta) - K(\bar{\beta}) \leq \Delta \leq 2A_0\lambda_0 \sum_{j \in S_{\bar{\beta}}} |\beta_j - \bar{\beta}_j|$$

$$\leq 2A_0\lambda_0 \cdot \nu_0^{-1}|S_{\bar\beta}|^{1/2}Q(\beta,\bar\beta)^{1/2}. \qquad (5.56)$$

By the margin condition, with $\|\beta-\bar\beta\|_1 = r \leq \zeta_0^*$, we have

$$K(\beta) - K(\bar\beta) \geq \kappa_0^* Q(\beta,\bar\beta).$$

Combining the preceding two displays gives a quadratic inequality in $Q(\beta,\bar\beta)^{1/2}$:

$$\kappa_0^* Q(\beta,\bar\beta) \leq 2A_0\lambda_0 \cdot \nu_0^{-1}|S_{\bar\beta}|^{1/2}Q(\beta,\bar\beta)^{1/2},$$

and hence ("inverting a quadratic inequality")

$$Q(\beta,\bar\beta) \leq (2A_0\nu_0^{-1}\kappa_0^{*-1})^2|S_{\bar\beta}|\lambda_0^2.$$

Returning to (5.56) yields

$$\begin{aligned}\Delta &\leq 2A_0\lambda_0 \cdot \nu_0^{-1}|S_{\bar\beta}|^{1/2}Q(\beta,\bar\beta)^{1/2}\\ &\leq (2A_0\nu_0^{-1})^2\kappa_0^{*-1}|S_{\bar\beta}|\lambda_0^2,\end{aligned}$$

that is, $\Delta \leq M_0|S_{\bar\beta}|\lambda_0^2$ as stated in (5.52).

To deduce (5.51), suppose that $K(\beta)-K(\bar\beta)=0$. Then inequality (5.55) implies that $\sum_{j\in S_{\bar\beta}}|\beta_j-\bar\beta_j| = 0$, which in conjunction with (5.54) implies that $\sum_{j\notin S_{\bar\beta}}|\beta_j-\bar\beta_j| = 0$. Hence $\|\beta-\bar\beta\|_1 = 0$, a contradiction. Therefore, $K(\beta)-K(\bar\beta) > 0$ as stated in (5.51). □

Finally, we complete the proof of Theorem 5.5 through a convexity argument. Note that the convexity of $\rho(y,\eta)$ is used in this step to localize $\hat\beta_\lambda$ near $\bar\beta$, whereas, separately, the Lipschitz property of $\rho(y,\eta)$ is used through the concentration result (Proposition 3.1).

Proof of Theorem 5.5. Take $r = \frac{M_0}{A_0-1}|S_{\bar\beta}|\lambda_0$. Consider two possible cases, $\|\hat\beta_\lambda-\bar\beta\|_1 > r$ or $\leq r$, both in the event $\Omega_1(r)$.

In the case of $\|\hat\beta_\lambda-\bar\beta\|_1 > r$, let $\tilde\beta_\lambda = \bar\beta + \{r/\|\hat\beta_\lambda-\bar\beta\|_1\}(\hat\beta_\lambda-\bar\beta)$, satisfying $\|\tilde\beta_\lambda-\bar\beta\|_1 = r$. By definition, $\hat\beta_\lambda$ satisfies (5.50), which, by the convexity of $K_n(\beta)+A_0\lambda_0\|\beta\|_1$, implies that $\tilde\beta_\lambda$ also satisfies (5.50). [Exercise: Prove this statement and explain how this is related to Exercise 2.14.] Then we apply Lemma 5.8 with $\beta=\tilde\beta_\lambda$ and obtain from (5.51) and (5.52) that in the event $\Omega_1(r)$, $K(\tilde\beta_\lambda)-K(\bar\beta) > 0$ and

$$\|\tilde\beta_\lambda-\bar\beta\|_1 < \frac{M_0}{A_0-1}|S_{\bar\beta}|\lambda_0 = r,$$

which contradicts $\tilde\beta_\lambda-\bar\beta = r$. In other words, the case of $\|\hat\beta_\lambda-\bar\beta\|_1 > r$ does not hold, and hence $\|\hat\beta_\lambda-\bar\beta\|_1 \leq r$ as in (5.49a), in the event $\Omega_1(r)$.

In the case of $\|\hat{\beta}_\lambda - \bar{\beta}\|_1 \le r$, we apply Lemma 5.7 and obtain that in the event $\Omega_1(r)$,

$$\begin{aligned} K(\beta) - K(\bar{\beta}) + A_0\lambda_0\|\beta - \bar{\beta}\|_1 &\le 2A_0\lambda_0 \sum_{j\in S_{\bar{\beta}}} |\beta_j - \bar{\beta}| + r\lambda_0 \\ &\le (2A_0+1)r\lambda_0, \end{aligned}$$

which by the choice of r yields

$$\begin{aligned} K(\beta) - K(\bar{\beta}) + A_0\lambda_0\|\beta - \bar{\beta}\|_1 &\le \frac{2A_0+1}{A_0-1} M_0 |S_{\bar{\beta}}|\lambda_0^2 \\ &= \frac{4A_0^2(2A_0+1)}{A_0-1}\nu_0^{*-2}\kappa_0^{*-1}|S_{\bar{\beta}}|\lambda_0^2, \end{aligned}$$

that is, $\Delta \le M_1|S_{\bar{\beta}}|\lambda_0^2$ as stated in (5.49b). In summary, inequalities (5.49a) and (5.49b) hold in the event $\Omega_1(r)$ with the *chosen* r, which has a probability at least $1-2\epsilon$ by Proposition 3.1. □

The error bound (5.49b) is in terms of the population loss difference $K(\hat{\beta}_\lambda) - K(\bar{\beta})$. Error bounds of the order $|S_{\bar{\beta}}|\lambda_0^2$ can also be obtained in terms of quadratic forms of $\hat{\beta}_\lambda - \bar{\beta}$.

Corollary 5.1 (Error bounds for quadratic forms). *In the setting of Theorem 5.5, we have with probability at least* $1-2\epsilon$,

$$Q(\hat{\beta}_\lambda, \bar{\beta}) = (\hat{\beta}_\lambda - \bar{\beta})^{\mathrm{T}} H (\hat{\beta}_\lambda - \bar{\beta}) \le \kappa_0^{*-1} M_1 |S_{\bar{\beta}}|\lambda_0^2,$$

and with probability at least $1-2\epsilon-2\epsilon^2$,

$$(\hat{\beta}_\lambda - \bar{\beta})^{\mathrm{T}} \hat{H} (\hat{\beta}_\lambda - \bar{\beta}) \le \kappa_0^{*-1} M_1 |S_{\bar{\beta}}|\lambda_0^2 + 2C_0^2\zeta_0^* \frac{M_0}{A_0-1}|S_{\bar{\beta}}|\lambda_0\sqrt{\frac{\log(p/\epsilon)}{n}},$$

where $H = E(\mathbf{X}^{\mathrm{T}}\mathbf{X}/n)$ *and* $\hat{H} = \mathbf{X}^{\mathrm{T}}\mathbf{X}/n$.

Proof. Let $r = \frac{M_0}{A_0-1}|S_{\bar{\beta}}|\lambda_0$ as in the proof of Theorem 5.5. The bound on $Q(\hat{\beta}_\lambda, \bar{\beta})$ follows from the margin condition with $\|\hat{\beta}_\lambda - \bar{\beta}\| \le r \le \zeta_0^*$,

$$Q(\hat{\beta}_\lambda, \bar{\beta}) \le \kappa_0^{*-1}\{K(\hat{\beta}_\lambda) - K(\bar{\beta})\}.$$

Moreover, the second error bound follows because in the intersection of (5.49a) and the event $\Omega_2(\lambda_2)$ with $\lambda_2 = 2C_0^2\sqrt{\frac{\log(p/\epsilon)}{n}}$ in Lemma 4.4 (Concentration of Gram matrix),

$$\begin{aligned} |(\hat{\beta}_\lambda - \bar{\beta})^{\mathrm{T}}(\hat{H} - H)(\hat{\beta}_\lambda - \bar{\beta})| &\le \left\{\max_{j,k=1,\ldots,p} |\hat{H}_{jk} - H_{jk}|\right\} \|\hat{\beta}_\lambda - \bar{\beta}\|_1^2 \\ &\le \lambda_2 r^2 \le \lambda_2\zeta_0^* r, \end{aligned}$$

where the first inequality follows similarly as in (4.27). □

To relate to the analysis of Lasso estimation in generalized linear regression (Theorems 5.3 and 5.4), it is instructive to consider the situation where the margin and theoretical compatibility conditions are modified such that the expected (scaled) Gram matrix H is replaced by the Hessian of the population loss function at $\bar{\beta}$, $H(\bar{\beta}) = \frac{\partial^2}{\partial\beta\partial\beta^{\mathrm{T}}} K(\bar{\theta})$. In other words, consider the following modification of conditions (iv) and (v):

(iv†) [Margin condition] The quadratic margin condition holds for some constants $\zeta_0^* > 0$ and $\kappa_0^* > 0$: $K(\beta) - K(\bar{\beta}) \geq \kappa_0^*(\beta - \bar{\beta})^{\mathrm{T}} H(\bar{\beta})(\beta - \bar{\beta})$ for any β satisfying $\|\beta - \bar{\beta}\|_1 \leq \zeta_0^*$.

(v†) [Theoretical compatibility condition] The compatibility condition $(S_{\bar{\beta}}, \xi_0^*, \nu_0^*)$ holds for $H(\bar{\beta})$, where $S_{\bar{\beta}} = \{1 \leq j \leq p : \bar{\beta}_j \neq 0\}$ and $\xi_0^* > 1$ and $\nu_0^* > 0$ are constants.

The margin condition (iv†) can be justified from primitive conditions as follows. The proof is left as Exercise 5.11.

Lemma 5.9 (Sufficient conditions for the modified margin condition). *Suppose that condition (iii) is satisfied.*

(i) For τ-quantile regression, if there exist constants $B_1, B_2, B_3 > 0$ such that the conditional distribution of Y_i given X_i admits density $f_{Y_i|X_i}(\cdot)$ satisfying $f_{Y_i|X_i}(u) \geq B_2$ for $|u - X_i^{\mathrm{T}}\bar{\beta}| \leq B_1$ and $f_{Y_i|X_i}(X_i^{\mathrm{T}}\bar{\beta}) \leq B_3$ for $i = 1, \ldots, n$, then the margin condition (iv†) holds with $\zeta_0^* = B_1/C_0$ and $\kappa_0^* = B_2/B_3$. [Question: Can you find a weaker condition concerning $f_{Y_i|X_i}(\cdot)$ in a similar manner as the Hessian condition (5.23)*?]*

(ii) For logistic regression with maximum likelihood estimation, the margin condition (iv†) holds with $\zeta_0^* = 1/C_0$ and $\kappa_0^* = 1/\mathrm{e}$. [Question: Can you find a larger κ_0^* given the stated ζ_0^*?]

An examination of the proof of Theorem 5.5 shows that the result still holds after the modification.

Theorem 5.6 (Fast rate in random design). *Suppose that margin and theoretical compatibility conditions (iv) and (v) are replaced by (iv†) and (v†) respectively. Then Theorem 5.5 remains valid.*

For logistic regression with maximum likelihood estimation, Theorem 5.6 is comparable to Theorem 5.4 (Fast rate in random design with model misspecification). The same theoretical compatibility condition on $H(\bar{\beta})$ is used in both results, and the margin condition (iv†) in Theorem 5.6 and the Hessian condition (iv) in Theorem 5.4 are always satisfied. Moreover, the earlier result Theorem 5.5 with the theoretical compatibility condition on H and the boundedness of $|X_i^{\mathrm{T}}\bar{\beta}|$, $i = 1, \ldots, n$, [which implies the margin condition (iv)] is comparable to Theorem 5.4 with the same compatibility condition on

H and the bounded $|X_i^{\mathrm{T}}\bar{\beta}|$ condition [which together imply the compatibility condition on $H(\bar{\beta})$ and the Hessian condition (iv)]. Nevertheless, Theorem 5.6 requires a Lipschitz loss function and hence does not apply to linear or log-linear regression, which is covered by Theorem 5.4.

Exercises

Ex 5.1. Consider generalized linear regression (5.1) with a canonical link based on exponential family (5.2). Show that the Bregman divergence at $(\tilde{\beta}, \beta)$ associated with the average negative log likelihood function (5.4) can be expressed as the average KL divergence:

$$D_{K_n}(\tilde{\beta}, \beta) = \frac{1}{n} \sum_{i=1}^{n} \mathrm{KL}(p_i(\cdot; \beta), p_i(\cdot; \tilde{\beta})),$$

where $p_i(\cdot; \beta)$ denotes the probability density (5.2) with $\theta_i^* = X_i^{\mathrm{T}}\beta$.

Ex 5.2. Lemma 5.1 (Basic inequality) is proved in Section 5.3.1 by comparing the Lasso objective function at $\hat{\beta}_\lambda$ and $(1-t)\hat{\beta}_\lambda + t\beta$ for $0 \le t \le 1$. Give an alternative proof of Lemma 5.1 using the stationary condition for $\hat{\beta}_\lambda$.

Ex 5.3. Show that equality (5.25) holds in the proof of Lemma 5.3 (Lower bound of Bregman divergence).

Ex 5.4.

(i) For $c \ge 0$ and $\eta \in (0, 1)$, show that if $c \int_0^1 \mathrm{e}^{-ct}\, \mathrm{d}t \le \eta$, then $\int_0^1 \mathrm{e}^{-ct}\, \mathrm{d}t \ge 1 - \eta$.

(ii) Complete the derivation from (5.32) to (5.35) using result (i) with the integral expression of the Bregman divergence lower bound (5.24a), but not directly the closed-form expression (5.24b).

Ex 5.5. Consider fixed-design logistic regression (5.5), correctly specified, as in Section 5.3.1. For the Lasso-penalized likelihood estimator $\hat{\beta}_\lambda$ defined as (5.9) with K_n in (5.6), show that under some conditions,

$$D_{K_n}(\hat{\beta}_\lambda, \beta^*) + D_{K_n}(\beta^*, \hat{\beta}_\lambda) = O_p(1)|S_{\beta^*}|\lambda_0^2.$$

where $\lambda_0 = \sqrt{\log(n \vee p)/n}$. Present your conditions and proof of the result. [Hint: This is covered by Theorem 5.2. You are asked to present your conditions and proof in a concrete and clean manner.]

Ex 5.6. Consider fixed-design logistic regression as in Exercise 5.5. Let $K_n(\beta)$ be the calibration loss function in (5.11). See also Exercise 2.18 where the non-penalized calibrated estimator is studied.

(i) Derive the Bregman divergence associated with $K_n(\beta)$. What relationship can you find about the Bregman divergence associated with the calibration loss function $K_n(\beta)$ and that associated with the likelihood loss function in (5.6)?

(ii) For the Lasso-penalized calibrated estimator $\hat{\beta}_\lambda$ defined as (5.9) with K_n in (5.11), show that under some conditions,

$$D_{K_n}(\hat{\beta}_\lambda, \beta^*) + D_{K_n}(\beta^*, \hat{\beta}_\lambda) = O_p(1)|S_{\beta^*}|\lambda_0^2.$$

where $\lambda_0 = \sqrt{\log(n \vee p)/n}$. Present your conditions and proof of the result.

(iii) Compare your conditions and results from part (ii) and those from Exercise 5.5.

Ex 5.7. In the setting of Theorem 5.2 (Fast rate in fixed design), replace condition (iv) by the following condition

(iv†) [Hessian and bounded mean condition] $a''(\tilde{u}) \le a''(u)\mathrm{e}^{C_2(|\tilde{u}-u|+|\tilde{u}^2-u^2|)}$ for any $u, \tilde{u} \in \mathbb{R}$ with a constant $C_2 > 0$, and $|X_i^{\mathrm{T}}\beta^*| \le B_0$ for $i = 1, \ldots, n$ with a constant $B_0 > 0$.

(i) Under conditions (iii) and (iv†), show that for any $\tilde{\beta} \in \mathbb{R}^p$,

$$D_{K_n}(\tilde{\beta}, \beta^*) + D_{K_n}(\beta^*, \tilde{\beta}) \ge \int_0^1 \mathrm{e}^{-(\tilde{C}_2\|b\|_1 t + \tilde{C}_2^2\|b\|_1^2 t^2)}\,\mathrm{d}t \cdot \{b^{\mathrm{T}}\hat{H}(\beta^*)b\},$$

where $b = \tilde{\beta} - \beta^*$ and some constant $\tilde{C}_2$. State your $\tilde{C}_2$ explicitly.

(ii) Let $c_0 = \int_0^\infty \mathrm{e}^{-(s+s^2)}\,\mathrm{d}s$. For $c \ge 0$ and $\eta \in (0, c_0)$, show that if $c\int_0^1 \mathrm{e}^{-(ct+c^2t^2)}\,\mathrm{d}t \le \eta$, then $\int_0^1 \mathrm{e}^{-(ct+c^2t^2)}\,\mathrm{d}t \ge c_0 - \eta$.

(iii) Show that a similar error bound holds as in Theorem 5.2 with condition (iv) replaced by (iv†) above. State the result with technical conditions explicitly, and provide a proof.

(iv) Can you prove a similar error bound as in Theorem 5.2 by retaining the boundedness of $X_i^{\mathrm{T}}\beta^*$ and further relaxing the condition on $a''(\cdot)$ in condition (iv†)? Provide your reasoning.

Ex 5.8.

(i) Show that conditions (iv) and (v) are implied by conditions (iv†) and (v†) jointly in Section 5.3.2 (Random-design generalized linear regression). State the associated constants explicitly.

(ii) Can you find a simpler proof of similar error bounds as in Theorem 5.3 but under conditions (i), (ii), (iii), (iv†), and (v†) than that of Theorem 5.3 outlined in Section 5.3.2?

Ex 5.9. Consider random-design logistic regression (5.5), correctly specified, as in Section 5.3.2. For the Lasso-penalized likelihood estimator $\hat{\beta}_\lambda$ defined as (5.9) with K_n in (5.6), show that under some conditions,

$$D_{K_n}(\hat{\beta}_\lambda, \beta^*) + D_{K_n}(\beta^*, \hat{\beta}_\lambda) = O_p(1)|S_{\beta^*}|\lambda_0^2.$$

where $\lambda_0 = \sqrt{\log(n \vee p)/n}$. Present your conditions and proof of the result. [Hint: This is covered by Theorem 5.3. You are asked to present your conditions and proof in a concrete and clean manner.]

Ex 5.10 (Slow rate with non-smooth M-estimation)**.** Let $\hat{\beta}_\lambda$ be the Lasso-penalized estimator in (5.47) with $\lambda = A_0\lambda_0$ in Section 5.5.2 (Penalized non-smooth M-estimation). Show that under conditions (i)–(iii), we have with probability at least $1 - 2\epsilon$,

$$K(\hat{\beta}_\lambda) - K(\bar{\beta}) + (A_0 - 1)\lambda_0\|\hat{\beta}_\lambda - \bar{\beta}\|_1 \leq M_0\lambda_0$$

for some constant M_0. State your M_0 explicitly.

Ex 5.11.

(i) Provide a proof of Lemma 5.6 (Sufficient conditions for the margin condition).

(ii) Provide a proof Lemma 5.9 (Sufficient conditions for the modified margin condition)

6

High-dimensional inference for regression coefficients

In this chapter, we study statistical inference such as confidence intervals about individual regression coefficients in high-dimensional regression models. For linear regression, we derive debiased Lasso estimation as an extension of the ideas underlying classical Wald confidence intervals, and discuss the associated asymptotic expansion which enables Wald confidence intervals in high dimensions. Moreover, we demonstrate how debiased Lasso estimation achieves doubly robust inference in partially linear regression. For generalized linear regression, we also derive debiased Lasso estimation by leveraging ideas from classical Wald confidence intervals, and establish the associated asymptotic expansion in high-dimensional settings. However, debiased Lasso estimation does not in general lead to double robustness in partially generalized linear regression.

In Chapters 4 and 5, we study (non-asymptotic) error bounds for the Lasso estimator of the coefficient vector, including the prediction error and the L_1 estimation error, in linear regression and generalized linear regression respectively. However, asymptotic distributions of the Lasso estimator may be non-normal and complex, especially in high-dimensional settings. See, for example, Knight and Fu (2000) for asymptotic results in the classical setting with p fixed and $n \to \infty$. In this chapter, we discuss debiased Lasso estimators for individual coefficients such that under suitable sparsity conditions, $n^{-1/2}$ asymptotic expansions (hence $n^{-1/2}$ asymptotic normality) are satisfied similarly as in classical asymptotic theory in Chapter 2. From these asymptotic expansions together with consistent variance estimation, valid Wald confidence intervals can be obtained for inference about individual regression coefficients.

Debiased Lasso methods are proposed and studied for linear regression in Zhang and Zhang (2014) and Javanmard and Montanari (2014) and for generalized linear regression in van de Geer et al. (2014). Our discussion about debiased Lasso estimation achieving doubly robust inference in linear regression is based on Tan (2020a) and Ghosh and Tan (2022).

DOI: 10.1201/9781003680734-6

6.1 High-dimensional inference in linear regression

Consider a linear regression model as in (4.1):

$$Y_i = X_i^{\mathrm{T}}\beta^* + \varepsilon_i, \quad i = 1, \ldots, n, \tag{6.1}$$

where $\beta^* = (\beta_1^*, \ldots, \beta_p^*)^{\mathrm{T}}$ is an unknown coefficient vector and ε_i is a noise variable with mean 0 given X_i. It is of interest to draw inference about β^* or, as to be focused on later, individual components of β^*, in high-dimensional settings where Lasso estimation is used.

6.1.1 Classical inference with least-squares estimation

To motivate ideas, we review Wald confidence intervals in the classical setting with fixed p and large n (Chapter 2). For concreteness, assume that linear regression is correctly specified in a random design, with IID $(X_1, Y_1), \ldots, (X_n, Y_n)$ as (X, Y). We use the notation as in Chapter 4.

For the Wald method, we construct or choose a point estimator, for example, the least-squares estimator

$$\hat\beta_{\mathrm{LS}} = (\sum_{i=1}^n X_i X_i^{\mathrm{T}})^{-1}(\sum_{i=1}^n X_i Y_i) = (\mathbf{X}^{\mathrm{T}}\mathbf{X})^{-1}(\mathbf{X}^{\mathrm{T}}\mathbf{Y}),$$

provided $\mathbf{X}^{\mathrm{T}}\mathbf{X}/n$ is non-singular. Then we show that $\hat\beta_{\mathrm{LS}}$ is consistent for β^* and satisfies asymptotic normality as $n \to \infty$ and p is fixed:

$$\sqrt{n}(\hat\beta_{\mathrm{LS}} - \beta^*) \xrightarrow[\mathcal{D}]{} \mathrm{N}(0, \Sigma(\beta^*)), \tag{6.2}$$

where $\Sigma(\beta) = H^{-1}G(\beta)H^{-1}$, $H = E(XX^{\mathrm{T}})$, assumed to be non-singular, and $G(\beta) = E\{(Y - X^{\mathrm{T}}\beta)^2 XX^{\mathrm{T}}\}$. Moreover, we show that the asymptotic variance $\Sigma(\beta^*)$ can be consistently estimated by $\hat\Sigma(\hat\beta_{\mathrm{LS}}) = \hat H^{-1}\hat G(\hat\beta_{\mathrm{LS}})\hat H^{-1}$, where $\hat H = \sum_{i=1}^n X_i X_i^{\mathrm{T}}/n$ and $\hat G(\beta) = \sum_{i=1}^n (Y_i - X_i^{\mathrm{T}}\beta)^2 X_i X_i^{\mathrm{T}}/n$. From these results, the Wald confidence interval for β_j^* is defined as

$$\hat\beta_{\mathrm{LS},j} \pm z_{c/2}\sqrt{\frac{\{\hat\Sigma(\hat\beta_{\mathrm{LS}})\}_{jj}}{n}}, \tag{6.3}$$

and can be shown to achieve an asymptotic size $(1-c)$ in covering β_j^*, where $\{\hat\Sigma(\beta)\}_{jj}$ is the (j,j)th element of $\hat\Sigma(\beta)$, and $z_{c/2}$ is the $(1-c/2)$-quantile of $\mathrm{N}(0,1)$.

At the center of the argument leading to (6.2) is the following $n^{-1/2}$ asymptotic expansion:

$$\hat\beta_{\mathrm{LS}} - \beta^* = \hat H^{-1}\mathbf{X}^{\mathrm{T}}(\mathbf{Y} - \mathbf{X}\beta^*)/n \tag{6.4a}$$

$$= \Gamma \mathbf{X}^{\mathrm{T}} \epsilon / n + (\hat{H}^{-1} - \Gamma) \mathbf{X}^{\mathrm{T}} \epsilon / n \tag{6.4b}$$

$$= \Gamma \mathbf{X}^{\mathrm{T}} \epsilon / n + o_p(n^{-1/2}), \tag{6.4c}$$

where $\hat{H} = \mathbf{X}^{\mathrm{T}}\mathbf{X}/n$, $\epsilon = \mathbf{Y} - \mathbf{X}\beta^*$, and $\Gamma = H^{-1}$. The last step relies on the fact that $\hat{H}^{-1}$ is a consistent estimator of Γ, i.e., $\hat{H}^{-1} = \Gamma + o_p(1)$. By the central limit theorem, $\sqrt{n}(\mathbf{X}^{\mathrm{T}}\epsilon/n) \xrightarrow[\mathcal{D}]{} \mathrm{N}(0, G(\beta^*))$ and hence $\sqrt{n}(\Gamma\mathbf{X}^{\mathrm{T}}\epsilon/n) \xrightarrow[\mathcal{D}]{} \mathrm{N}(0, \Sigma(\beta^*))$, with $\Sigma(\beta^*) = \Gamma G(\beta^*)\Gamma$. Then (6.2) follows by application of Slutsky's theorem.

The reasoning above can also be applied to inference about individual coefficients. Suppose that X, β^*, and $\hat{\beta}_{\mathrm{LS}}$ are partitioned as

$$X = \begin{pmatrix} Z \\ W \end{pmatrix}, \quad \beta^* = \begin{pmatrix} \theta^* \\ \alpha^* \end{pmatrix}, \quad \hat{\beta}_{\mathrm{LS}} = \begin{pmatrix} \hat{\theta}_{\mathrm{LS}} \\ \hat{\alpha}_{\mathrm{LS}} \end{pmatrix},$$

where Z is scalar and W is a (column) vector of dimension $p-1$. Accordingly, the covariate matrix $\mathbf{X}$ can be partitioned as

$$\mathbf{X} = (\mathbf{Z}, \mathbf{W}), \quad \mathbf{Z} = \begin{pmatrix} Z_1 \\ \vdots \\ Z_n \end{pmatrix}, \quad \mathbf{W} = \begin{pmatrix} W_1^{\mathrm{T}} \\ \vdots \\ W_n^{\mathrm{T}} \end{pmatrix}.$$

Left-multiplying the expressions by $e_1^{\mathrm{T}} = (1, 0, \ldots, 0)$ in (6.4) gives

$$\hat{\theta}_{\mathrm{LS}} - \theta^* = e_1^{\mathrm{T}} \hat{H}^{-1} \mathbf{X}^{\mathrm{T}} \epsilon / n \tag{6.5a}$$

$$= \Gamma_1^{\mathrm{T}} \mathbf{X}^{\mathrm{T}} \epsilon / n + (\hat{\Gamma}_1^{\mathrm{T}} - \Gamma_1^{\mathrm{T}}) \mathbf{X}^{\mathrm{T}} \epsilon / n \tag{6.5b}$$

$$= \Gamma_1^{\mathrm{T}} \mathbf{X}^{\mathrm{T}} \epsilon / n + o_p(n^{-1/2}), \tag{6.5c}$$

where $\Gamma_1^{\mathrm{T}} = e_1^{\mathrm{T}}\Gamma$ or $\Gamma_1 = \Gamma e_1$, the first row or column of Γ, and $\hat{\Gamma}_1^{\mathrm{T}} = e_1^{\mathrm{T}}\hat{H}^{-1}$ or $\hat{\Gamma}_1 = \hat{H}^{-1}e_1$, the first row or column of $\hat{H}^{-1}$. The last step follows because $\hat{\Gamma}_1$ can be shown to be a consistent estimator of Γ_1, i.e., $\hat{\Gamma}_1 = \Gamma_1 + o_p(1)$. An asymptotic expansion in the form (6.4c) or (6.5c) is called a first-order or $n^{-1/2}$ asymptotic expansion, where the leading term is the average of n IID variables and the remainder is of order $o_p(n^{-1/2})$.

To gain further understanding, it is helpful to notice that $\hat{\Gamma}_1$ can be determined through linear regression of $\mathbf{Z}$ on $\mathbf{W}$ as

$$\hat{\Gamma}_1 = \begin{pmatrix} 1/\hat{\tau}_{\mathrm{LS}} \\ -\hat{\gamma}_{\mathrm{LS}}/\hat{\tau}_{\mathrm{LS}} \end{pmatrix}, \tag{6.6}$$

where

$$\begin{aligned} \hat{\gamma}_{\mathrm{LS}} &= (\mathbf{W}^{\mathrm{T}}\mathbf{W})^{-1}(\mathbf{W}^{\mathrm{T}}\mathbf{Z}), \\ \hat{\tau}_{\mathrm{LS}} &= \mathbf{Z}^{\mathrm{T}}\mathbf{Z}/n - \mathbf{Z}^{\mathrm{T}}\mathbf{W}(\mathbf{W}^{\mathrm{T}}\mathbf{W})^{-1}\mathbf{W}^{\mathrm{T}}\mathbf{Z}/n \\ &= \|\mathbf{Z} - \mathbf{W}\hat{\gamma}_{\mathrm{LS}}\|_2^2/n = (\mathbf{Z} - \mathbf{W}\hat{\gamma}_{\mathrm{LS}})^{\mathrm{T}}\mathbf{Z}/n. \end{aligned}$$

In fact, (6.6) can be shown by writing

$$\hat{H} = \mathbf{X}^{\mathrm{T}}\mathbf{X}/n = \begin{pmatrix} \mathbf{Z}^{\mathrm{T}}\mathbf{Z}/n & \mathbf{Z}^{\mathrm{T}}\mathbf{W}/n \\ \mathbf{W}^{\mathrm{T}}\mathbf{Z}/n & \mathbf{W}^{\mathrm{T}}\mathbf{W}/n \end{pmatrix},$$

and applying the inverse formula of a partitioned 2×2 matrix as in Tan (2025), Section 1.6.1. Alternatively, (6.6) can be shown by letting $\hat{\Gamma}_1 = (u_1, u_2^{\mathrm{T}})^{\mathrm{T}}$ and solving the equation $\hat{H}(u_1, u_2^{\mathrm{T}})^{\mathrm{T}} = e_1$, i.e.,

$$\begin{pmatrix} (\mathbf{Z}^{\mathrm{T}}\mathbf{Z}/n)u_1 + (\mathbf{Z}^{\mathrm{T}}\mathbf{W}/n)u_2 \\ (\mathbf{W}^{\mathrm{T}}\mathbf{Z}/n)u_1 + (\mathbf{W}^{\mathrm{T}}\mathbf{W}/n)u_2 \end{pmatrix} = \begin{pmatrix} 1 \\ 0 \end{pmatrix}.$$

Details of the calculation are left as Exercise 6.1.

Similarly, the population counterpart Γ_1 of $\hat{\Gamma}_1$ can be determined as

$$\Gamma_1 = \begin{pmatrix} 1/\bar{\tau} \\ -\bar{\gamma}/\bar{\tau} \end{pmatrix}, \tag{6.7}$$

where

$$\bar{\gamma} = E(WW^{\mathrm{T}})^{-1}E(WZ),$$
$$\bar{\tau} = E((Z - W^{\mathrm{T}}\bar{\gamma})Z).$$

Substituting the expression (6.7) of Γ_1 into (6.5c) yields

$$\hat{\theta}_{\mathrm{LS}} - \theta^* = \frac{(\mathbf{Z} - \mathbf{W}\bar{\gamma})^{\mathrm{T}}(\mathbf{Y} - \mathbf{X}\beta^*)/n}{E((Z - W^{\mathrm{T}}\bar{\gamma})Z)} + o_p(n^{-1/2}) \tag{6.8a}$$

$$= \frac{(\mathbf{Z} - \mathbf{W}\bar{\gamma})^{\mathrm{T}}(\mathbf{Y} - \mathbf{X}\beta^*)/n}{(\mathbf{Z} - \mathbf{W}\bar{\gamma})^{\mathrm{T}}\mathbf{Z}/n} + o_p(n^{-1/2}). \tag{6.8b}$$

The leading term in (6.8a) is a sample mean of IID variables, whereas that in (6.8b) is a ratio of two such sample means. The two asymptotic expansions follow from each other because

$$\frac{(\mathbf{Z} - \mathbf{W}\bar{\gamma})^{\mathrm{T}}\epsilon/n}{(\mathbf{Z} - \mathbf{W}\bar{\gamma})^{\mathrm{T}}\mathbf{Z}/n} = \frac{(\mathbf{Z} - \mathbf{W}\bar{\gamma})^{\mathrm{T}}\epsilon/n}{E((Z - W^{\mathrm{T}}\bar{\gamma})Z)} + o_p(n^{-1/2}), \tag{6.9}$$

provided $E((Z - W^{\mathrm{T}}\bar{\gamma})Z) > 0$ [Exercise]. The asymptotic expansion (6.8b) corresponds to the doubly robust asymptotic expansion of $\hat{\theta}_{\mathrm{LS}}$ in Tan (2025), Section 1.7.2, in the case of correctly specified linear regression (6.1).

The asymptotic expansion (6.8b) can be related to the residual-on-residual regression indicating that $\hat{\theta}_{\mathrm{LS}}$ can be equivalently obtained as

$$\hat{\theta}_{\mathrm{LS}} = \frac{(\mathbf{Z} - \mathbf{W}\hat{\gamma}_{\mathrm{LS}})^{\mathrm{T}}(\mathbf{Y} - \mathbf{W}\hat{\eta}_{\mathrm{LS}})/n}{(\mathbf{Z} - \mathbf{W}\hat{\gamma}_{\mathrm{LS}})^{\mathrm{T}}\mathbf{Z}/n}, \tag{6.10}$$

where $\hat{\eta}_{\mathrm{LS}}$ is the least-squares estimator of the coefficient vector in linear regression of $\mathbf{Y}$ on $\mathbf{W}$, i.e., $\hat{\eta}_{\mathrm{LS}} = (\mathbf{W}^{\mathrm{T}}\mathbf{W})^{-1}(\mathbf{W}^{\mathrm{T}}\mathbf{Y})$. The expression (6.10) says that the LS coefficient $\hat{\theta}_{\text{LS}}$ associated with $\mathbf{Z}$ in linear regression of $\mathbf{Y}$ on $\mathbf{X} = (\mathbf{Z}, \mathbf{W})$ can be obtained as the LS coefficient in linear regression of the residual $\mathbf{Y} - \mathbf{W}\hat{\eta}_{\text{LS}}$ (from linear regression of $\mathbf{Y}$ on $\mathbf{W}$) on the residual $\mathbf{Z} - \mathbf{W}\hat{\gamma}_{\text{LS}}$ (from linear regression of $\mathbf{Z}$ on $\mathbf{W}$). The asymptotic expansion (6.8b) can be rewritten as (by moving θ^* to the right-hand side)

$$\hat{\theta}_{\text{LS}} = \frac{(\mathbf{Z} - \mathbf{W}\bar{\gamma})^{\text{T}}(\mathbf{Y} - \mathbf{W}\alpha^*)/n}{(\mathbf{Z} - \mathbf{W}\bar{\gamma})^{\text{T}}\mathbf{Z}/n} + o_p(n^{-1/2}).$$

After replacing α^* and $\bar{\gamma}$ by $\hat{\alpha}_{\text{LS}}$ and $\hat{\gamma}_{\text{LS}}$, the above turns out to yield an identity for $\hat{\theta}_{\text{LS}}$ (not just an asymptotic expansion):

$$\hat{\theta}_{\text{LS}} = \frac{(\mathbf{Z} - \mathbf{W}\hat{\gamma}_{\text{LS}})^{\text{T}}(\mathbf{Y} - \mathbf{W}\hat{\alpha}_{\text{LS}})/n}{(\mathbf{Z} - \mathbf{W}\hat{\gamma}_{\text{LS}})^{\text{T}}\mathbf{Z}/n}, \tag{6.11}$$

which can be verified by using the expression (6.10) for $\hat{\theta}_{\text{LS}}$ and the stationary condition $(\mathbf{Z} - \mathbf{W}\hat{\gamma}_{\text{LS}})^{\text{T}}\mathbf{W} = 0$ for $\hat{\gamma}_{\text{LS}}$. In other words, the asymptotic expansion (6.8b) amounts to saying that replacing $\hat{\alpha}_{\text{LS}}$ and $\hat{\gamma}_{\text{LS}}$ by α^* and $\bar{\gamma}$ in (6.11) only incurs an error of order $o_p(n^{-1/2})$. This property, however, does not extend to the usual residual-on-residual expression (6.10). Replacing $\hat{\eta}_{\text{LS}}$ and $\hat{\gamma}_{\text{LS}}$ by $\bar{\eta}$ and $\bar{\gamma}$ in (6.10) incurs an error of order $O_p(n^{-1/2})$, where $\bar{\eta} = E(WW^{\text{T}})^{-1}E(WY)$ (Exercise 6.2).

From the asymptotic expansion (6.8b), the asymptotic variance for $\hat{\theta}_{\text{LS}}$ after scaled by n can be estimated by

$$\hat{V}(\hat{\theta}_{\text{LS}}, \hat{\alpha}_{\text{LS}}, \hat{\gamma}_{\text{LS}}) = \frac{\|(\mathbf{Z} - \mathbf{W}\hat{\gamma}_{\text{LS}})^{\text{T}}(\mathbf{Y} - \mathbf{X}^{\text{T}}\hat{\beta}_{\text{LS}})\|_n^2}{\{(\mathbf{Z} - \mathbf{W}\hat{\gamma}_{\text{LS}})^{\text{T}}\mathbf{Z}/n\}^2}.$$

It fact, $\{\hat{\Sigma}(\hat{\beta}_{\text{LS}})\}_{11} = \hat{V}(\hat{\theta}_{\text{LS}}, \hat{\alpha}_{\text{LS}}, \hat{\gamma}_{\text{LS}})$ algebraically, by properties of residual-on-residual regression as in Tan (2025), Proposition 1.8. The Wald confidence interval (6.3) for θ^* can be equivalently written as

$$\hat{\theta}_{\text{LS}} \pm z_{c/2}\sqrt{\frac{\hat{V}(\hat{\theta}_{\text{LS}}, \hat{\alpha}_{\text{LS}}, \hat{\gamma}_{\text{LS}})}{n}}. \tag{6.12}$$

See Tan (2025), Section 1.6, for a detailed discussion on residual-on-residual regression.

6.1.2 Debiased Lasso inference

For linear regression, consider the Lasso estimator as in (4.7):

$$\hat{\beta}_\lambda = \text{argmin}_{\beta \in \mathbb{R}^p}\left\{\frac{1}{2}\|\mathbf{Y} - \mathbf{X}\beta\|_n^2 + \lambda_\beta\|\beta\|_1\right\}, \tag{6.13}$$

where $\lambda_\beta > 0$ is a tuning parameter. Suppose that X, β^*, and $\hat{\beta}_\lambda$ are partitioned as

$$X = \begin{pmatrix} Z \\ W \end{pmatrix}, \quad \beta^* = \begin{pmatrix} \theta^* \\ \alpha^* \end{pmatrix}, \quad \hat{\beta}_\lambda = \begin{pmatrix} \hat{\theta}_\lambda \\ \hat{\alpha}_\lambda \end{pmatrix}.$$

It is of interest draw inference about θ^*, the first component of β^*.

The Lasso estimator $\hat\beta_\lambda$ or the first component $\hat\theta_\lambda$ does not admit a closed form, which makes it difficult to derive any asymptotic expansion such as (6.4) or (6.5) for least-squares estimation. Nevertheless, a careful look at the stationary condition (4.8) for $\hat\beta_\lambda$ provides important information about the behavior of $\hat\theta_\lambda$ and a possible direction to obtaining asymptotic expansions similar to (6.4) and (6.5).

A direct implication of the stationary condition (4.8) is that $\|\hat\zeta_\lambda\|_\infty \le \lambda_\beta$, where

$$\hat\zeta_\lambda = \mathbf{X}^{\mathrm{T}}(\mathbf{Y} - \mathbf{X}\hat\beta_\lambda)/n.$$

To capture $\hat\beta_\lambda - \beta^*$, we rearrange the preceding equation as

$$\mathbf{X}^{\mathrm{T}}\mathbf{X}(\hat\beta_\lambda - \beta^*)/n = \mathbf{X}^{\mathrm{T}}(\mathbf{Y} - \mathbf{X}\beta^*)/n - \hat\zeta_\lambda,$$

or equivalently

$$\hat H(\hat\beta_\lambda - \beta^*) = \mathbf{X}^{\mathrm{T}}\epsilon/n - \hat\zeta_\lambda. \tag{6.14}$$

Equation (6.14) for $\hat\beta_\lambda$ is related to (6.4a) for $\hat\beta_{\mathrm{LS}}$, but involves $\hat H$ on the left-hand right. In the case of $\lambda_\beta = 0$ and hence $\hat\zeta_\lambda = 0$, left-multiplying (6.14) by $\hat H^{-1}$ recovers (6.4a). To capture $\hat\theta_\lambda - \theta^*$, let Γ_1^{T} be the first row of $\Gamma = H^{-1}$ provided $H = E(XX^{\mathrm{T}})$ is non-singular, or more generally be any row vector such that $\Gamma_1^{\mathrm{T}} H = e_1^{\mathrm{T}}$. Then we left-multiply both sides by Γ_1^{T} and apply the approximation $\Gamma_1^{\mathrm{T}}\hat H \approx e_1^{\mathrm{T}}$:

$$\hat\theta_\lambda - \theta^* \approx \Gamma_1^{\mathrm{T}}\hat H(\hat\beta_\lambda - \beta^*) = \Gamma_1^{\mathrm{T}}\mathbf{X}^{\mathrm{T}}\epsilon/n - \Gamma_1^{\mathrm{T}}\hat\zeta_\lambda.$$

For an exact relationship, we rearrange the equation above using $\hat\theta_\lambda - \theta^* = e_1^{\mathrm{T}}(\hat\beta_\lambda - \beta^*)$:

$$\hat\theta_\lambda - \theta^* = (e_1^{\mathrm{T}} - \Gamma_1^{\mathrm{T}}\hat H)(\hat\beta_\lambda - \beta^*) + \Gamma_1^{\mathrm{T}}\mathbf{X}^{\mathrm{T}}\epsilon/n - \Gamma_1^{\mathrm{T}}\hat\zeta_\lambda. \tag{6.15}$$

On the right-hand side of (6.15), we observe that

- $\Gamma_1^{\mathrm{T}}\mathbf{X}^{\mathrm{T}}\epsilon/n$ is the average of IID variables and, after scaled by $\sqrt{n}$, admits an asymptotic normal distribution by CLT;
- $(e_1^{\mathrm{T}} - \Gamma_1^{\mathrm{T}}\hat H)(\hat\beta_\lambda - \beta^*)$ is the product of two terms, $e_1^{\mathrm{T}} - \Gamma_1^{\mathrm{T}}\hat H$ and $\hat\beta_\lambda - \beta^*$, and can be expected to be $o_p(n^{-1/2})$ under suitable conditions;
- $\Gamma_1^{\mathrm{T}}\hat\zeta_\lambda$ is expected to be $\sqrt{\frac{\log(n\vee p)}{n}}O_p(1)$ for λ_β in the order $\sqrt{\frac{\log(n\vee p)}{n}}$, and can be considered a bias term, non-negligible to the order $O_p(n^{-1/2})$, due to Lasso penalization.

Equation (6.15) for $\hat{\theta}_\lambda$ is related to (6.5b) for $\hat{\theta}_{\text{LS}}$, but involves $\hat{\beta}_\lambda - \beta^*$ on the right-hand side. In the case of $\lambda_\beta = 0$, the term $\Gamma_1^{\mathrm{T}} \hat{\zeta}_\lambda$ vanishes and (6.15) can be shown to be equivalent to (6.5b), after calculating $\hat{\beta}_{\text{LS}} - \beta^*$ by (6.4a), provided $\hat{H}$ is non-singular.

To correct for the bias represented by $\Gamma_1^{\mathrm{T}} \hat{\zeta}_\lambda$, a debiased Lasso estimator for θ^* is defined as

$$\tilde{\theta}_\lambda = \hat{\theta}_\lambda + \hat{\Gamma}_{1,\lambda}^{\mathrm{T}} \hat{\zeta}_\lambda = \hat{\theta}_\lambda + \hat{\Gamma}_{1,\lambda}^{\mathrm{T}} \mathbf{X}^{\mathrm{T}} (\mathbf{Y} - \mathbf{X}\hat{\beta}_\lambda)/n, \tag{6.16}$$

where $\hat{\Gamma}_{1,\lambda}$ is an estimator of Γ_1. By the expression (6.7) of Γ_1, it is natural to take

$$\hat{\Gamma}_{1,\lambda} = \begin{pmatrix} 1/\hat{\tau}_\lambda \\ -\hat{\gamma}_\lambda/\hat{\tau}_\lambda \end{pmatrix},$$

with $\hat{\tau}_\lambda = (\mathbf{Z} - \mathbf{W}\hat{\gamma}_\lambda)^{\mathrm{T}} \mathbf{Z}/n$ and $\hat{\gamma}_\lambda$ a Lasso estimator of $\bar{\gamma}$:

$$\hat{\gamma}_\lambda = \operatorname{argmin}_{\gamma \in \mathbb{R}^p} \left\{ \frac{1}{2} \|\mathbf{Z} - \mathbf{W}\gamma\|_n^2 + \lambda_\gamma \|\gamma\|_1 \right\}, \tag{6.17}$$

where $\lambda_\gamma > 0$ is a tuning parameter. The subscript λ in $\hat{\gamma}_\lambda$ signifies the use of Lasso penalty, not the tuning parameter being λ. The associated linear regression, $E(Z|W) = W^{\mathrm{T}} \gamma^*$, is allowed to be misspecified. It can be shown that (Exercise 6.3)

$$\hat{\tau}_\lambda = (\mathbf{Z} - \mathbf{W}\hat{\gamma}_\lambda)^{\mathrm{T}} \mathbf{Z}/n = \|\mathbf{Z} - \mathbf{W}\hat{\gamma}_\lambda\|_n^2 + \lambda_\gamma \|\hat{\gamma}_\lambda\|_1. \tag{6.18}$$

Substituting the expression of $\hat{\Gamma}_{1,\lambda}$ into (6.16) gives

$$\tilde{\theta}_\lambda = \hat{\theta}_\lambda + \frac{(\mathbf{Z} - \mathbf{W}\hat{\gamma}_\lambda)^{\mathrm{T}} (\mathbf{Y} - \mathbf{X}\hat{\beta}_\lambda)/n}{(\mathbf{Z} - \mathbf{W}\hat{\gamma}_\lambda)^{\mathrm{T}} \mathbf{Z}/n} \tag{6.19a}$$

$$= \frac{(\mathbf{Z} - \mathbf{W}\hat{\gamma}_\lambda)^{\mathrm{T}} (\mathbf{Y} - \mathbf{W}\hat{\alpha}_\lambda)/n}{(\mathbf{Z} - \mathbf{W}\hat{\gamma}_\lambda)^{\mathrm{T}} \mathbf{Z}/n}, \tag{6.19b}$$

where the second line follows because $\mathbf{X}\hat{\beta}_\lambda = \mathbf{Z}\hat{\theta}_\lambda + \mathbf{W}\hat{\alpha}_\lambda$. The expression (6.19b) for the debiased Lasso estimator $\tilde{\theta}_\lambda$ is similar to the expression (6.11) for the least-squares estimator $\hat{\theta}_{\text{LS}}$ by residual-on-residual regression. It is interesting that the expression (6.19b) does not directly involve the original Lasso estimator $\hat{\theta}_\lambda$. We will return to the expression (6.19b) for further discussion of debiased Lasso estimation in Section 6.1.3.

To derive an asymptotic expansion for $\tilde{\theta}_\lambda$, we note that the calculation from (6.14) to (6.15) with Γ_1 replaced by $\hat{\Gamma}_{1,\lambda}$ gives the decomposition

$$\tilde{\theta}_\lambda - \theta^* = \hat{\Gamma}_{1,\lambda}^{\mathrm{T}} \mathbf{X}^{\mathrm{T}} \epsilon/n + (e_1^{\mathrm{T}} - \hat{\Gamma}_{1,\lambda}^{\mathrm{T}} \hat{H})(\hat{\beta}_\lambda - \beta^*).$$

Using the definition of $\hat{\Gamma}_{1,\lambda}$ and $\hat{\tau}_\lambda$, direct calculation gives

$$\hat{\Gamma}_{1,\lambda}^{\mathrm{T}} \mathbf{X}^{\mathrm{T}} \epsilon/n = \frac{(\mathbf{Z} - \mathbf{W}\hat{\gamma}_\lambda)^{\mathrm{T}} \epsilon/n}{\hat{\tau}_\lambda},$$

$$e_1^{\mathrm{T}} - \hat{\Gamma}_{1,\lambda}^{\mathrm{T}} \hat{H} = e_1^{\mathrm{T}} - \frac{(\mathbf{Z} - \mathbf{W}\hat{\gamma}_\lambda)^{\mathrm{T}}(\mathbf{Z}, \mathbf{W})/n}{\hat{\tau}_\lambda}$$
$$= (0, -\frac{(\mathbf{Z} - \mathbf{W}\hat{\gamma}_\lambda)^{\mathrm{T}}\mathbf{W}/n}{\hat{\tau}_\lambda}).$$

Then the preceding decomposition becomes

$$\begin{aligned}\tilde{\theta}_\lambda - \theta^* &= \frac{(\mathbf{Z} - \mathbf{W}\hat{\gamma}_\lambda)^{\mathrm{T}}\epsilon/n - \{(\mathbf{Z} - \mathbf{W}\hat{\gamma}_\lambda)^{\mathrm{T}}\mathbf{W}/n\}(\hat{\alpha}_\lambda - \alpha^*)}{\hat{\tau}_\lambda} \\ &= \frac{(\mathbf{Z} - \mathbf{W}\hat{\gamma}_\lambda)^{\mathrm{T}}\epsilon/n - \{(\mathbf{Z} - \mathbf{W}\hat{\gamma}_\lambda)^{\mathrm{T}}\mathbf{W}/n\}(\hat{\alpha}_\lambda - \alpha^*)}{(\mathbf{Z} - \mathbf{W}\hat{\gamma}_\lambda)^{\mathrm{T}}\mathbf{Z}/n}.\end{aligned} \tag{6.20}$$

Alternatively, (6.20) can also be deduced by direct calculation using the expression (6.19b) [Exercise]. From (6.20), it can be shown that under suitable sparsity conditions, the debiased Lasso estimator achieves an $n^{-1/2}$ asymptotic expansion similar to (6.5c) or (6.8a)–(6.8b) for the least-squares estimator $\hat{\theta}_{\mathrm{LS}}$ in the low-dimensional setting.

For theoretical analysis, consider the following conditions:

(i) [Linear regression in random design] Linear regression (6.1) is correctly specified in a random design with IID $(X_1, Y_1), \ldots, (X_n, Y_n)$;

(ii) [Lasso estimation with Y given W] As $n \to \infty$ and p may increase as n increases,

$$\|\mathbf{W}^{\mathrm{T}}(\mathbf{Y} - \mathbf{X}^{\mathrm{T}}\beta^*)/n\|_\infty = O_p(1)\frac{\log(p)}{n}, \tag{6.21a}$$

$$\|\mathbf{X}^{\mathrm{T}}(\hat{\beta}_\lambda - \beta^*)\|_n^2 = O_p(1)|S_{\beta^*}|\frac{\log(n \vee p)}{n}, \tag{6.21b}$$

$$\|\hat{\beta}_\lambda - \beta^*\|_1 = O_p(1)|S_{\beta^*}|\sqrt{\frac{\log(n \vee p)}{n}}; \tag{6.21c}$$

(iii) [Lasso estimation with Z given W] As $n \to \infty$ and p may increase as n increases,

$$\|\mathbf{W}^{\mathrm{T}}(\mathbf{Z} - \mathbf{W}\bar{\gamma})\|_\infty = O_p(1)\frac{\log(p)}{n}, \tag{6.22a}$$

$$\|\mathbf{W}^{\mathrm{T}}(\hat{\gamma}_\lambda - \bar{\gamma})\|_n^2 = O_p(1)|S_{\bar{\gamma}}|\frac{\log(n \vee p)}{n}, \tag{6.22b}$$

$$\|\hat{\gamma}_\lambda - \bar{\gamma}\|_1 = O_p(1)|S_{\bar{\gamma}}|\sqrt{\frac{\log(n \vee p)}{n}}, \tag{6.22c}$$

where $\bar{\gamma} = \mathrm{argmin}_{\gamma \in \mathbb{R}^{p-1}} E\{(Z - W^{\mathrm{T}}\gamma)^2\}$;

(iv) [Non-degenerate Z given W] $E(Z^2) < \infty$ and $\bar{\tau} = E\{(Z - W^{\mathrm{T}}\bar{\gamma})Z\} > 0$ or equivalently $\bar{\tau} = E\{(Z - W^{\mathrm{T}}\bar{\gamma})^2\} > 0$.

The choice of β^* satisfying (6.1) is unique even if $E(XX^{\mathrm{T}})$ may be singular but a compatibility condition is assumed to hold for $E(XX^{\mathrm{T}})$, and similarly the choice of $\bar{\gamma}$ is unique even if $E(WW^{\mathrm{T}})$ may be singular but a compatibility condition is assumed to hold for $E(WW^{\mathrm{T}})$. See Section 5.5.1. The error bounds for the Lasso estimators $\hat{\beta}_\lambda$ and $\hat{\gamma}_\lambda$ can be deduced from primitive conditions as shown in Chapter 4. Note that the rate conditions in Theorems 4.3 and 4.5 are implied by the sparsity condition $(|S_{\beta^*}| + |S_{\bar{\gamma}}|)\frac{\log(n\vee p)}{\sqrt{n}} = o(1)$ below.

Theorem 6.1 (Asymptotic expansion of debiased Lasso)**.** *Assume that conditions (i)–(iv) above are satisfied, and* $(|S_{\beta^*}| + |S_{\bar{\gamma}}|)\frac{\log(n\vee p)}{\sqrt{n}} = o(1)$. *Then as* $n \to \infty$ *and* p *may increase as* n *increases,*

$$\tilde{\theta}_\lambda - \theta^* = \Gamma_1^{\mathrm{T}} \mathbf{X}^{\mathrm{T}} \epsilon / n + o_p(n^{-1/2}),$$

where Γ_1 *is defined as in* (6.7), *with* $\bar{\gamma} = \operatorname{argmin}_{\gamma \in \mathbb{R}^{p-1}} E\{(Z - W^{\mathrm{T}}\gamma)^2\}$. *Equivalently, the asymptotic expansion can be written as*

$$\tilde{\theta}_\lambda - \theta^* = \frac{(\mathbf{Z} - \mathbf{W}\bar{\gamma})^{\mathrm{T}}(\mathbf{Y} - \mathbf{X}\beta^*)/n}{(\mathbf{Z} - \mathbf{W}\bar{\gamma})^{\mathrm{T}}\mathbf{Z}/n} + o_p(n^{-1/2}), \tag{6.23a}$$

$$= \frac{(\mathbf{Z} - \mathbf{W}\bar{\gamma})^{\mathrm{T}}(\mathbf{Y} - \mathbf{X}\beta^*)/n}{E((Z - W^{\mathrm{T}}\bar{\gamma})Z)} + o_p(n^{-1/2}). \tag{6.23b}$$

Note that the leading $O_p(n^{-1/2})$ *term in* (6.23a) *is the same as the correction term in* (6.19b) *with* $(\hat{\beta}_\lambda, \hat{\gamma})$ *replaced by* $(\beta^*, \bar{\gamma})$.

Proof. We complete the proof in several steps starting from the decomposition (6.20). We write (6.20) as

$$\tilde{\theta}_\lambda - \theta^* = \frac{\mathbf{I} - \mathbf{II}}{\mathbf{III}},$$

where **I**, **II**, and **III** are defined below. First, we note

$$\begin{aligned}\mathbf{I} = (\mathbf{Z} - \mathbf{W}\hat{\gamma}_\lambda)^{\mathrm{T}}\epsilon/n &= (\mathbf{Z} - \mathbf{W}\bar{\gamma})^{\mathrm{T}}\epsilon/n - (\hat{\gamma}_\lambda - \bar{\gamma})^{\mathrm{T}}\mathbf{W}^{\mathrm{T}}\epsilon/n \\ &= (\mathbf{Z} - \mathbf{W}\bar{\gamma})^{\mathrm{T}}\epsilon/n + O_p(1)|S_{\bar{\gamma}}|\frac{\log(n\vee p)}{n}.\end{aligned}$$

This follows from the Hölder inequality and convergence bounds (6.21a) and (6.22c):

$$\begin{aligned}|(\hat{\gamma}_\lambda - \bar{\gamma})^{\mathrm{T}}\mathbf{W}^{\mathrm{T}}\epsilon/n| &\le \|\hat{\gamma}_\lambda - \bar{\gamma}\|_1 \cdot \|\mathbf{W}^{\mathrm{T}}\epsilon/n\|_\infty \\ &= O_p(1)|S_{\bar{\gamma}}|\sqrt{\frac{\log(n\vee p)}{n}} \cdot O_p(1)\sqrt{\frac{\log p}{n}} \\ &= O_p(1)|S_{\bar{\gamma}}|\frac{\log(n\vee p)}{n}.\end{aligned}$$

Second, we show

$$\mathbf{II} = \{(\mathbf{Z} - \mathbf{W}\hat{\gamma}_\lambda)^{\mathrm{T}}\mathbf{W}/n\}(\hat{\alpha}_\lambda - \alpha^*)$$

$$= O_p(1)(|S_{\beta^*}| + |S_{\beta^*}|_1^{1/2}|S_{\bar{\gamma}}|_1^{1/2})\frac{\log(n \vee p)}{n}.$$

In fact, by the triangle inequality,

$$\begin{aligned}
&|\{(\mathbf{Z} - \mathbf{W}\hat{\gamma}_\lambda)^{\mathrm{T}}\mathbf{W}/n\}(\hat{\alpha}_\lambda - \alpha^*)| \\
&\le |\{(\mathbf{Z} - \mathbf{W}\bar{\gamma})^{\mathrm{T}}\mathbf{W}/n\}(\hat{\alpha}_\lambda - \alpha^*)| + |(\hat{\gamma}_\lambda - \bar{\gamma})^{\mathrm{T}}\mathbf{W}^{\mathrm{T}}\mathbf{W}(\hat{\alpha}_\lambda - \alpha^*)/n|.
\end{aligned}$$

The two terms can be upper bounded as follows: by the Hölder inequality and convergence bounds (6.22a) and (6.21c):

$$\begin{aligned}
|\{(\mathbf{Z} - \mathbf{W}\bar{\gamma})^{\mathrm{T}}\mathbf{W}/n\}(\hat{\alpha}_\lambda - \alpha^*)| &\le \|(\mathbf{Z} - \mathbf{W}\bar{\gamma})^{\mathrm{T}}\mathbf{W}/n\|_\infty \cdot \|\hat{\alpha}_\lambda - \alpha^*\|_1 \\
&= O_p(1)\sqrt{\frac{\log p}{n}} \cdot O_p(1)|S_{\beta^*}|\sqrt{\frac{\log(n \vee p)}{n}} \\
&= O_p(1)|S_{\beta^*}|\frac{\log(n \vee p)}{n}.
\end{aligned}$$

and by the Cauchy–Schwartz inequality and convergence bounds (6.21b) and (6.22b):

$$\begin{aligned}
&|(\hat{\gamma}_\lambda - \bar{\gamma})^{\mathrm{T}}\mathbf{W}^{\mathrm{T}}\mathbf{W}(\hat{\alpha}_\lambda - \alpha^*)/n| \\
&\le \|\mathbf{W}(\hat{\gamma}_\lambda - \bar{\gamma})\|_n \cdot \|\mathbf{W}(\hat{\alpha}_\lambda - \alpha^*)\|_n \\
&\le O_p(1)|S_{\bar{\gamma}}|^{1/2}\sqrt{\frac{\log(n \vee p)}{n}} \cdot O_p(1)|S_{\beta^*}|^{1/2}\sqrt{\frac{\log(n \vee p)}{n}} \\
&= O_p(1)|S_{\bar{\gamma}}|^{1/2}|S_{\beta^*}|^{1/2}\frac{\log(n \vee p)}{n}.
\end{aligned}$$

Alternatively, $\{(\mathbf{Z} - \mathbf{W}\hat{\gamma}_\lambda)^{\mathrm{T}}\mathbf{W}/n\}(\hat{\alpha}_\lambda - \alpha^*)$ can be bounded as follows:

$$\begin{aligned}
|\{(\mathbf{Z} - \mathbf{W}\hat{\gamma}_\lambda)^{\mathrm{T}}\mathbf{W}/n\}(\hat{\alpha}_\lambda - \alpha^*)| &\le \|(\mathbf{Z} - \mathbf{W}\hat{\gamma}_\lambda)^{\mathrm{T}}\mathbf{W}/n\|_\infty \cdot \|\hat{\alpha}_\lambda - \alpha^*\|_1 \\
&= O_p(1)\sqrt{\frac{\log p}{n}} \cdot O_p(1)|S_{\beta^*}|\sqrt{\frac{\log(n \vee p)}{n}} \\
&= O_p(1)|S_{\beta^*}|\frac{\log(n \vee p)}{n},
\end{aligned}$$

where by the stationary condition for $\hat{\gamma}_\lambda$, $\|(\mathbf{Z} - \mathbf{W}\hat{\gamma}_\lambda)^{\mathrm{T}}\mathbf{W}/n\|_\infty \le \lambda_\gamma$ for λ_γ in the order $\sqrt{\frac{\log(n \vee p)}{n}}$. Such a choice is implicit in condition (iii) for $\hat{\gamma}_\lambda$.

Third, we show

$$\begin{aligned}
\mathbf{III} = \hat{\tau}_\lambda = (\mathbf{Z} - \mathbf{W}\hat{\gamma}_\lambda)^{\mathrm{T}}\mathbf{Z}/n &= (\mathbf{Z} - \mathbf{W}\bar{\gamma})^{\mathrm{T}}\mathbf{Z}/n - (\hat{\gamma}_\lambda - \bar{\gamma})^{\mathrm{T}}\mathbf{W}^{\mathrm{T}}\mathbf{Z}/n \\
&= (\mathbf{Z} - \mathbf{W}\bar{\gamma})^{\mathrm{T}}\mathbf{Z}/n + O_p(1)|S_{\bar{\gamma}}|^{1/2}\sqrt{\frac{\log(n \vee p)}{n}} \\
&= \bar{\tau} + o_p(1) + O_p(1)|S_{\bar{\gamma}}|^{1/2}\sqrt{\frac{\log(n \vee p)}{n}}.
\end{aligned}$$

In fact, by the Cauchy–Schwartz inequality and convergence bound (6.22b):

$$\begin{aligned}|(\hat{\gamma}_\lambda - \bar{\gamma})^{\mathrm{T}}\mathbf{W}^{\mathrm{T}}\mathbf{Z}/n| &\le \|\mathbf{W}(\hat{\gamma}_\lambda - \bar{\gamma})\|_n \cdot \|\mathbf{Z}\|_n \\ &= O_p(1)|S_{\bar{\gamma}}|^{1/2}\sqrt{\frac{\log(n \vee p)}{n}} \cdot O_p(1),\end{aligned}$$

where $\|\mathbf{Z}\|_n^2 \xrightarrow[\mathcal{P}]{} E(Z^2) < \infty$ and hence $\|\mathbf{Z}\|_n = O_p(1)$. Moreover, $(\mathbf{Z} - \mathbf{W}\bar{\gamma})^{\mathrm{T}}\mathbf{Z}/n \xrightarrow[\mathcal{P}]{} E\{(Z - W^{\mathrm{T}}\bar{\gamma})Z\}$, i.e., $(\mathbf{Z} - \mathbf{W}\bar{\gamma})^{\mathrm{T}}\mathbf{Z}/n = E\{(Z - W^{\mathrm{T}}\bar{\gamma})Z\} + o_p(1) = \bar{\tau} + o_p(1)$.

Finally, combining the first two steps gives

$$\begin{aligned}&\underbrace{(\mathbf{Z} - \mathbf{W}\hat{\gamma}_\lambda)^{\mathrm{T}}\epsilon/n - \{(\mathbf{Z} - \mathbf{W}\hat{\gamma}_\lambda)^{\mathrm{T}}\mathbf{W}/n\}(\hat{\alpha}_\lambda - \alpha^*)}_{\mathbf{I}-\mathbf{II}} \\ &= (\mathbf{Z} - \mathbf{W}\bar{\gamma})^{\mathrm{T}}\epsilon/n + O_p(1)(|S_{\beta^*}| + |S_{\bar{\gamma}}|)\frac{\log(n \vee p)}{n}.\end{aligned}$$

If $|S_{\bar{\gamma}}|\frac{\log(n\vee p)}{n} = o(1)$, which is implied by $|S_{\bar{\gamma}}|\frac{\log(n\vee p)}{\sqrt{n}} = o(1)$, then the third step indicates

$$\begin{aligned}\underbrace{(\mathbf{Z} - \mathbf{W}\hat{\gamma}_\lambda)^{\mathrm{T}}\mathbf{Z}/n}_{\mathbf{III}} &= (\mathbf{Z} - \mathbf{W}\bar{\gamma})^{\mathrm{T}}\mathbf{Z}/n + o_p(1) \\ &= E((Z - W^{\mathrm{T}}\bar{\gamma})Z) + o_p(1).\end{aligned}$$

By applying the preceding two bounds to (6.20), it can be shown that (Exercise 6.4)

$$\begin{aligned}\tilde{\theta}_\lambda - \theta^* &= \frac{\mathbf{I} - \mathbf{II}}{\mathbf{III}} \\ &= \frac{(\mathbf{Z} - \mathbf{W}\bar{\gamma})^{\mathrm{T}}\epsilon/n}{(\mathbf{Z} - \mathbf{W}\bar{\gamma})^{\mathrm{T}}\mathbf{Z}/n} + O_p(1)(|S_{\beta^*}| + |S_{\bar{\gamma}}|)\frac{\log(n \vee p)}{n} + o_p(n^{-1/2}) \qquad (6.24) \\ &= \frac{(\mathbf{Z} - \mathbf{W}\bar{\gamma})^{\mathrm{T}}\epsilon/n}{E((Z - W^{\mathrm{T}}\bar{\gamma})Z)} + O_p(1)(|S_{\beta^*}| + |S_{\bar{\gamma}}|)\frac{\log(n \vee p)}{n} + o_p(n^{-1/2}).\end{aligned}$$

If $(|S_{\beta^*}| + |S_{\bar{\gamma}}|)\frac{\log(n\vee p)}{\sqrt{n}} = o(1)$, then the second line reduces to the asymptotic expansion (6.23a), and the third line reduces to the asymptotic expansion (6.23b). The two asymptotic expansions (6.23a) and (6.23b) can also be directly related as shown in (6.9). □

From Theorem 6.1, asymptotic normality of $\tilde{\beta}_\lambda$ can be obtained. Moreover, consistent estimation of the asymptotic variance can also be derived. Let

$$\begin{aligned}V(\theta^*, \alpha^*, \bar{\gamma}) &= \frac{E\{(Z - W^{\mathrm{T}}\bar{\gamma})^2(Y - X^{\mathrm{T}}\beta^*)^2\}}{E^2\{(Z - W^{\mathrm{T}}\bar{\gamma})Z\}}, \\ \hat{V}(\tilde{\theta}_\lambda, \hat{\alpha}_\lambda, \hat{\gamma}_\lambda) &= \frac{\|(\mathbf{Z} - \mathbf{W}\hat{\gamma}_\lambda)^{\mathrm{T}}(\mathbf{Y} - \mathbf{X}^{\mathrm{T}}\hat{\beta}_\lambda)\|_n^2}{\{(\mathbf{Z} - \mathbf{W}\hat{\gamma}_\lambda)^{\mathrm{T}}\mathbf{Z}/n\}^2}.\end{aligned}$$

The proof of Corollary 6.1 is left as Exercise 6.5.

Corollary 6.1 (Asymptotic normality and consistent variance estimation). *In the setting of Theorem 6.1, the following results hold:*

$$\sqrt{n}(\tilde{\theta}_\lambda - \theta^*) \xrightarrow[\mathcal{D}]{} \mathrm{N}(0, V(\theta^*, \alpha^*, \bar{\gamma})),$$
$$\hat{V}(\tilde{\theta}_\lambda, \hat{\alpha}_\lambda, \hat{\gamma}_\lambda) \xrightarrow[\mathcal{P}]{} V(\theta^*, \alpha^*, \bar{\gamma}).$$

as $n \to \infty$ and p may increase as n increases.

From Corollary 6.1, the Wald confidence interval, in a similar form as (6.12),

$$\tilde{\theta}_\lambda \pm z_{c/2}\sqrt{\frac{\hat{V}(\tilde{\theta}_\lambda, \hat{\alpha}_\lambda, \hat{\gamma}_\lambda)}{n}} \tag{6.25}$$

achieves asymptotic size $1 - c$ in covering θ^* in linear regression (6.1). Therefore, the Wald confidence interval (6.25) based on $\tilde{\theta}_\lambda$ represents a high-dimensional version of the classical Wald confidence interval (6.12) based on $\hat{\theta}_{\mathrm{LS}}$ in low-dimensional settings.

Finally, the development of debiased Lasso inference with correct model specification can be extended to the scenario where linear regression (6.1) may be misspecified. The proof of the following result is left as an exercise.

Theorem 6.2 (Asymptotic expansion of debiased Lasso with model misspecification). *Assume that conditions (ii)–(iv) in Theorem 6.1 are satisfied, except the true value β^* is replaced by a target value $\bar{\beta} = (\bar{\theta}, \bar{\alpha}^{\mathrm{T}})^{\mathrm{T}}$ defined as follows:*

$$\begin{aligned}\bar{\beta} &= \mathrm{argmin}_{\beta \in \mathbb{R}^p}\, E\{(Y - X^{\mathrm{T}}\beta)^2\} \\ &= \mathrm{argmin}_{\theta \in \mathbb{R},, \alpha \in \mathbb{R}^{p-1}}\, E\{(Y - Z\theta - W^{\mathrm{T}}\alpha)^2\}.\end{aligned} \tag{6.26}$$

Assume further that $(|S_{\bar{\beta}}| + |S_{\bar{\gamma}}|)\frac{\log(n \vee p)}{\sqrt{n}} = o(1)$. Then the asymptotic expansions in (6.23) hold with β^* replaced by $\bar{\beta}$, i.e.,

$$\tilde{\theta}_\lambda - \bar{\theta} = \frac{(\mathbf{Z} - \mathbf{W}\bar{\gamma})^{\mathrm{T}}(\mathbf{Y} - \mathbf{X}\bar{\beta})/n}{(\mathbf{Z} - \mathbf{W}\bar{\gamma})^{\mathrm{T}}\mathbf{Z}/n} + o_p(n^{-1/2}) \tag{6.27a}$$

$$= \frac{(\mathbf{Z} - \mathbf{W}\bar{\gamma})^{\mathrm{T}}(\mathbf{Y} - \mathbf{X}\bar{\beta})/n}{E((Z - W^{\mathrm{T}}\bar{\gamma})Z)} + o_p(n^{-1/2}). \tag{6.27b}$$

Similarly, Corrollary 6.1 can also be extended with β^* replaced by $\bar{\beta}$. Then the Wald confidence interval (6.25) achieves asymptotic size $1 - c$ in covering $\bar{\theta}$, the first component of $\bar{\beta}$, with possible misspecification of linear regression (6.1).

6.1.3 Partially linear regression and doubly robust inference

We study debiased Lasso estimation from the perspective of doubly robust estimation in partially linear regression. The main finding is that debiased Lasso inference in Section 6.1.2 achieves doubly robust inference in partially linear regression. See a related discussion for least-squares estimation in low-dimensional settings in Tan (2025), Section 1.7.

To start, we note that from the expression (6.19b), the debiased Lasso estimator $\tilde{\theta}_\lambda$ is a solution to the estimating equation

$$0 = \frac{1}{n} \sum_{i=1}^{n} \varphi(Z_i, W_i, Y_i; \theta, \hat{\alpha}_\lambda, \hat{\gamma}_\lambda), \tag{6.28}$$

where $\hat{\alpha}_\lambda$ and $\hat{\gamma}_\lambda$ are as in Section 6.1.2, and

$$\varphi(Z, W, Y; \theta, \alpha, \gamma) = (Z - W^{\mathrm{T}}\gamma)(Y - W^{\mathrm{T}}\alpha - Z\theta). \tag{6.29}$$

In other words, $\tilde{\theta}_\lambda$ can be determined in two stages. First, the Lasso estimator $\hat{\beta}_\lambda = (\hat{\theta}_\lambda, \hat{\alpha}_\lambda^{\mathrm{T}})^{\mathrm{T}}$ is computed by (6.13), and the Lasso estimator $\hat{\gamma}_\lambda$ is computed by (6.17). Second, $\tilde{\theta}_\lambda$ is computed by solving equation (6.28). The Lasso estimator $\hat{\theta}_\lambda$ is computed together with $\hat{\alpha}_\lambda$ in the first stage, but only $\hat{\alpha}_\lambda$ is passed to the second stage. As described below, $\varphi(\cdot)$ is a doubly robust estimating function in a partially linear regression model.

Consider a random design as in Section 6.1.2, where $(X_1, Y_1), \ldots, (X_n, Y_n)$ are IID as (X, Y), with $X = (Z, W^{\mathrm{T}})^{\mathrm{T}}$. A partially linear regression model postulates that

$$E(Y|Z, W) = Z\theta^* + g^*(W), \tag{6.30}$$

where θ^* is an unknown (scalar) parameter and $g^*(w)$ is an unknown function of w. To estimate θ^*, a doubly robust approach introduces two working models as follows. The first is a linear model for $g^*(W)$:

$$g^*(W) = W^{\mathrm{T}}\alpha^*,$$

or equivalently a linear regression model of Y on (Z, W):

$$E(Y|Z, W) = X^{\mathrm{T}}\beta^* = Z\theta^* + W^{\mathrm{T}}\alpha^*, \tag{6.31}$$

where $\beta^* = (\theta^*, \alpha^{*\mathrm{T}})^{\mathrm{T}}$ and α^* is an unknown parameter vector, in addition to the parameter θ^*. The second is a linear regression model for $f^*(W) = E(Z|W)$:

$$E(Z|W) = W^{\mathrm{T}}\gamma^*, \tag{6.32}$$

where γ^* is an unknown parameter vector. It can be directly shown that $\varphi(\cdot)$ satisfies the following two properties of unbiasedness, where θ^* is well-defined by (6.30) even if model (6.31) is misspecified. See the proof of Proposition 1.9 in Tan (2025).

- If model (6.31) is correctly specified, then for any γ,

$$0 = E\{\varphi(Z, W, Y; \theta^*, \alpha^*, \gamma)\}. \tag{6.33}$$

- If model (6.32) is correctly specified, then for any α,

$$0 = E\{\varphi(Z, W, Y; \theta^*, \alpha, \gamma^*)\}. \tag{6.34}$$

In this sense, $\varphi(\cdot)$ is a doubly robust estimating function for θ^* with respect to working models (6.31) and (6.32) for nuisance functions $g^*(w)$ and $f^*(w)$.

From the double robustness of estimating function $\varphi(\cdot)$, the debiased Lasso estimator $\tilde{\theta}_\lambda$ is pointwise doubly robust for θ^* in partially linear regression (6.30), i.e., being consistent for θ^* if either model (6.31) or model (6.32) is correctly specified. The following result further shows that $\tilde{\theta}_\lambda$ admits a doubly robust, $n^{-1/2}$ asymptotic expansion about θ^* if either model (6.31) or model (6.32) is correctly specified.

Theorem 6.3 (Doubly robust asymptotic expansion). *Suppose that partially linear regression (6.30) is correctly specified. In addition, assume that conditions (ii)–(iv) in Theorem 6.1 are satisfied with β^* replaced by $\bar{\beta}$ in* (6.26) *and that $(|S_{\bar{\beta}}| + |S_{\bar{\gamma}}|)\frac{\log(n \vee p)}{\sqrt{n}} = o(1)$. If either model* (6.31) *or model* (6.32) *is correctly specified, then, with θ^* defined in (6.30),*

$$\tilde{\theta}_\lambda - \theta^* = \frac{(\mathbf{Z} - \mathbf{W}\bar{\gamma})^{\mathrm{T}}(\mathbf{Y} - \mathbf{Z}\theta^* - \mathbf{W}\bar{\alpha})/n}{(\mathbf{Z} - \mathbf{W}\bar{\gamma})^{\mathrm{T}}\mathbf{Z}/n} + o_p(n^{-1/2}) \tag{6.35a}$$

$$= \frac{(\mathbf{Z} - \mathbf{W}\bar{\gamma})^{\mathrm{T}}(\mathbf{Y} - \mathbf{Z}\theta^* - \mathbf{W}\bar{\alpha})/n}{E((Z - W^{\mathrm{T}}\bar{\gamma})Z)} + o_p(n^{-1/2}). \tag{6.35b}$$

Note that the asymptotic expansions can also be written as

$$\begin{aligned}\tilde{\theta}_\lambda - \theta^* &= \frac{n^{-1}\sum_{i=1}^n \varphi(Z_i, W_i, Y_i; \theta^*, \bar{\alpha}, \bar{\gamma})}{n^{-1}\sum_{i=1}^n \frac{\partial \varphi}{\partial \theta}(Z_i, W_i, Y_i; \theta^*, \bar{\alpha}, \bar{\gamma})} + o_p(n^{-1/2}) \\ &= \frac{n^{-1}\sum_{i=1}^n \varphi(Z_i, W_i, Y_i; \theta^*, \bar{\alpha}, \bar{\gamma})}{E\{\frac{\partial \varphi}{\partial \theta}(Z, W, Y; \theta^*, \bar{\alpha}, \bar{\gamma})\}} + o_p(n^{-1/2}),\end{aligned}$$

in terms of the estimating function $\varphi(\cdot)$ in (6.29).

Proof. By Tan (2025), Proposition 1.9, we note that if either model (6.31) or model (6.32) is correctly specified, then $\bar{\theta} = \theta^*$, where $\bar{\theta}$ is from the target value $\bar{\beta} = (\bar{\theta}, \bar{\alpha}^{\mathrm{T}})^{\mathrm{T}}$ in (6.26), and θ^* is the true value in partially linear regression (6.30). Then the asymptotic expansions in (6.35) follow from (6.27) in Theorem 6.3.

For pedagogical purposes, we give a direct proof of the fact that if either model (6.31) or model (6.32) is correctly specified, then $\bar{\theta} = \theta^*$. On one hand,

$$\bar{\theta} = \frac{E\{(Z - W^{\mathrm{T}}\bar{\gamma})(Y - W^{\mathrm{T}}\bar{\alpha})\}}{E\{(Z - W^{\mathrm{T}}\bar{\gamma})Z\}},$$

from the equation $E\{(Z - W^{\mathrm{T}}\bar{\gamma})(Y - W^{\mathrm{T}}\bar{\alpha} - Z\bar{\theta})\} = 0$, which holds because $E\{Z(Y - W^{\mathrm{T}}\bar{\alpha} - Z\bar{\theta})\} = 0$ and $E\{W(Y - W^{\mathrm{T}}\bar{\alpha} - Z\bar{\theta})\} = 0$ by the definition of $\bar{\beta} = (\bar{\theta}, \bar{\alpha}^{\mathrm{T}})^{\mathrm{T}}$ in (6.26). On the other hand, if either model (6.31) or model (6.32) is correctly specified, then

$$\theta^* = \frac{E\{(Z - W^{\mathrm{T}}\bar{\gamma})(Y - W^{\mathrm{T}}\bar{\alpha})\}}{E\{(Z - W^{\mathrm{T}}\bar{\gamma})Z\}},$$

from the equation $E\{(Z - W^{\mathrm{T}}\bar{\gamma})(Y - W^{\mathrm{T}}\bar{\alpha} - Z\theta^*)\} = 0$, which holds due to the double robustness of $\varphi(\cdot)$ discussed above. □

Similarly as in Corollary 6.1, asymptotic normality of $\tilde{\beta}_\lambda$ and consistent variance estimation can be obtained for doubly robust inference.

Corollary 6.2 (Doubly robust asymptotic normality and variance estimation). *In the setting of Theorem 6.3, If either model* (6.31) *or model* (6.32) *is correctly specified, then, with θ^* defined in (6.30),*

$$\sqrt{n}(\tilde{\theta}_\lambda - \theta^*) \xrightarrow[D]{} \mathrm{N}(0, V(\theta^*, \bar{\alpha}, \bar{\gamma})),$$
$$\hat{V}(\tilde{\theta}_\lambda, \hat{\alpha}_\lambda, \hat{\gamma}_\lambda) \xrightarrow[P]{} V(\theta^*, \bar{\alpha}, \bar{\gamma}).$$

as $n \to \infty$ and p may increase as n increases.

From Corollary 6.2, the Wald confidence interval (6.25), i.e.,

$$\tilde{\theta}_\lambda \pm z_{c/2}\sqrt{\frac{\hat{V}(\tilde{\theta}_\lambda, \hat{\alpha}_\lambda, \hat{\gamma}_\lambda)}{n}},$$

achieves asymptotic size $1-c$ in covering θ^* in partially linear regression (6.30) if either model (6.31) or model (6.32) is correctly specified. Hence the Wald confidence interval (6.25) using debiased Lasso estimation is valid in three distinct manners:

- [Heteroscedasticity robust] being valid in covering the true value θ^* in correctly specified linear regression with possibly heteroscedastic noises (Theorem 6.1 and Corollary 6.1);
- [Model robust] being valid in covering the target value $\bar{\theta}$ in possibly misspecified linear regression (Theorem 6.2);
- [Doubly robust] being valid in covering the true value θ^* in the associated partially linear regression, if either model (6.31) or model (6.32) is correctly specified (Theorem 6.3 and Corollary 6.2).

These properties provide high-dimensional extensions of the corresponding properties of Wald confidence intervals using least-squares estimation as discussed in Tan (2025), Section 1.7.

The double robustness of the debiased Lasso estimator $\hat{\theta}_\lambda$ and associated confidence interval (6.25) depends on the fact that linear regression (6.32) is used as a working model for the nuisance function $f^*(W) = E(Z|W)$. Alternatively, for a binary covariate Z, a more appropriate working model for $f^*(W)$ is logistic regression:

$$P(Z=1|W) = \text{expit}(W^{\mathrm{T}}\gamma^*), \tag{6.36}$$

where $\text{expit}(u) = (1+\mathrm{e}^{-u})^{-1}$. The debiased Lasso estimator $\hat{\theta}_\lambda$ is no longer doubly robust for θ^* in partially linear regression (6.30) with respect to working models (6.31) and (6.36). In fact, instead of φ in (6.29), a doubly robust estimating function for θ^* with respect to working models (6.31) and (6.36) is

$$\varphi(Z,W,Y;\theta,\alpha,\gamma) = (Z - \text{expit}(W^{\mathrm{T}}\gamma))(Y - W^{\mathrm{T}}\alpha - Z\theta), \tag{6.37}$$

satisfying the two properties (6.33) and (6.34) with model (6.32) replaced by (6.36) [Exercise]. A doubly robust estimator for θ^* can be defined in two stages. First, some estimators $(\hat{\alpha},\hat{\gamma})$ are derived from linear regression (6.31) and logistic regression (6.36) respectively. For example, $\hat{\alpha}$ is a regularized LS estimator, i.e., $\hat{\alpha}_\lambda$ in the earlier discussion, and $\hat{\gamma}$ is a regularized likelihood estimator defined as

$$\hat{\gamma}_{\lambda,\text{logit}} = \text{argmin}_{\gamma\in\mathbb{R}^{p-1}} \left[\frac{-1}{n}\sum_{i=1}^{n}\{Z_i W_i^{\mathrm{T}}\gamma - \log(1+\mathrm{e}^{W_i^{\mathrm{T}}\gamma})\} + \lambda\|\gamma\|_1 \right],$$

with Lasso tuning parameter λ_γ. Then an estimator for θ^*, denoted as $\hat{\theta}(\hat{\alpha},\hat{\gamma})$, is defined as a solution to (6.28) with the estimating function φ in (6.37) and (α,γ) replaced by $(\hat{\alpha},\hat{\gamma})$. By simple algebra, the estimator $\hat{\theta}(\hat{\alpha},\hat{\gamma})$ is of a closed-form:

$$\hat{\theta}(\hat{\alpha},\hat{\gamma}) = \frac{(\mathbf{Z} - \text{expit}(\mathbf{W}^{\mathrm{T}}\hat{\gamma}))^{\mathrm{T}}(\mathbf{Y} - \mathbf{W}^{\mathrm{T}}\hat{\alpha})/n}{(\mathbf{Z} - \text{expit}(\mathbf{W}^{\mathrm{T}}\hat{\gamma}))^{\mathrm{T}}\mathbf{Z}/n},$$

which is similar to (6.19b) with $\mathbf{W}\hat{\gamma}_\lambda$ replaced by $\text{expit}(\mathbf{W}\hat{\gamma})$.

The estimator $\hat{\theta}(\hat{\alpha},\hat{\gamma})$ is pointwise doubly robust for θ^*: $\hat{\theta}\xrightarrow{P}\theta^*$ if either model (6.31) or model (6.36) is correctly specified, provided that $\hat{\alpha}$ is consistent for α^*, $\hat{\alpha}\xrightarrow{P}\alpha^*$, if model (6.31) is correctly specified and $\hat{\gamma}$ is consistent for γ^*, $\hat{\gamma}\xrightarrow{P}\gamma^*$, if model (6.36) is correctly specified. In particular, with $(\hat{\alpha}_\lambda,\hat{\gamma}_{\lambda,\text{logit}})$ above, the estimator $\hat{\theta}(\hat{\alpha}_\lambda,\hat{\gamma}_{\lambda,\text{logit}})$ is pointwise doubly robust for θ^*. However, for the usual choice of estimators $(\hat{\alpha}_\lambda,\hat{\gamma}_{\lambda,\text{logit}})$, the estimator $\hat{\theta}(\hat{\alpha}_\lambda,\hat{\gamma}_{\lambda,\text{logit}})$ does not admit a doubly robust, asymptotic expansion similar to (6.35a) with $\mathbf{W}\bar{\gamma}$ replaced by $\text{expit}(\mathbf{W}\bar{\gamma})$. Nevertheless, when $(\hat{\alpha},\hat{\gamma})$ are constructed using a method called *calibrated estimation*, such an asymptotic expansion can be achieved for $\hat{\theta}(\hat{\alpha},\hat{\gamma})$, and hence the associated Wald confidence intervals are doubly robust for θ^* with respect to working models (6.31) and (6.36). See Tan (2025), Section 1.10.4, for a discussion in low-dimensional settings and Ghosh and Tan (2022) for a general discussion including high-dimensional settings.

For completeness, we point out (Exercise 6.9) that the estimator

$$\hat{\theta}(\hat{\alpha}_\lambda, \hat{\gamma}_{\lambda,\text{logit}}) = \frac{(\mathbf{Z} - \text{expit}(\mathbf{W}\hat{\gamma}_{\lambda,\text{logit}}))^{\mathrm{T}}(\mathbf{Y} - \mathbf{W}\hat{\alpha}_\lambda)/n}{(\mathbf{Z} - \text{expit}(\mathbf{W}\hat{\gamma}_{\lambda,\text{logit}}))^{\mathrm{T}}\mathbf{Z}/n}, \tag{6.38}$$

although not achieving a doubly robust asymptotic expansion, can be shown (under suitable sparsity conditions) to achieve a (singly robust) asymptotic expansion similar to (6.23a) with $\mathbf{W}\bar{\gamma}$ replaced by $\text{expit}(\mathbf{W}\bar{\gamma}_{\text{logit}})$ or equivalently (6.35a) with further $\bar{\alpha}$ replaced by α^* if linear regression (6.31) is correctly specified, where $\bar{\gamma}_{\text{logit}} = \text{argmin}_{\gamma \in \mathbb{R}^{p-1}} E\{-ZW^{\mathrm{T}}\gamma + \log(1 + \mathrm{e}^{W^{\mathrm{T}}\gamma})\}$. Therefore, $\hat{\theta}(\hat{\alpha}_\lambda, \hat{\gamma}_{\lambda,\text{logit}})$ provides an alternative to the debiased Lasso estimator $\tilde{\theta}$ in (6.19a)–(6.19b) in achieving $n^{-1/2}$ asymptotic expansion and enabling valid Wald confidence intervals about θ^* in correctly specified linear regression (6.31).

6.2 High-dimensional inference in generalized linear regression

Consider generalized linear regression with mean restriction in the form

$$E(Y_i|X_i) = a'(X_i^{\mathrm{T}}\beta^*), \quad i = 1, \ldots, n, \tag{6.39}$$

where $\beta^* = (\beta_1^*, \ldots, \beta_p^*)^{\mathrm{T}}$ is an unknown coefficient vector, and $a(\cdot)$ is a *known*, strictly convex, twice differentiable function. It is of interest to draw inference about individual components of β^* in high-dimensional settings where Lasso-penalized M-estimation is used.

6.2.1 Classical inference with M-estimation

We review Wald confidence intervals based on non-penalized M-estimation in the classical setting with fixed p and large n as studied in Chapter 2. For concreteness, assume that generalized linear regression is correctly specified in a random design, with IID $(X_1, Y_1), \ldots, (X_n, Y_n)$ as (X, Y).

For generalized linear regression (6.39), consider an M-estimator defined as

$$\hat{\beta}_{\mathrm{M}} = \text{argmin}_{\beta \in \mathbb{R}^p} K_n(\beta),$$

where $K_n(\beta)$ is an objective function in the form (5.14), i.e.

$$K_n(\beta) = \frac{-1}{n} \sum_{i=1}^{n} \{Y_i X_i^{\mathrm{T}}\beta - a(X_i^{\mathrm{T}}\beta)\},$$

Then under suitable regularity conditions, as $n \to \infty$ and p is fixed, $\hat{\beta}_{\mathrm{M}}$ is consistent for β^* and satisfies the asymptotic expansion

$$\begin{aligned}\hat{\beta}_{\mathrm{M}} - \beta^* &= H(\beta^*)^{-1}\frac{1}{n}\sum_{i=1}^{n}\{Y_i - a'(X_i^{\mathrm{T}}\beta^*)\}X_i + o_p(n^{-1/2}) \\ &= H(\beta^*)^{-1}\mathbf{X}^{\mathrm{T}}(\mathbf{Y} - a'(\mathbf{X}\beta^*))/n + o_p(n^{-1/2}),\end{aligned} \tag{6.40}$$

and the asymptotic normality

$$\sqrt{n}(\hat{\beta}_{\mathrm{M}} - \beta^*) \xrightarrow[\mathcal{D}]{} \mathrm{N}(0, \Sigma(\beta^*)),$$

where $\Sigma(\beta) = H(\beta)^{-1}G(\beta)H(\beta)^{-1}$ with $H(\beta) = E\{a''(X^{\mathrm{T}}\beta)XX^{\mathrm{T}}\}$, assumed to be non-singular, and $G(\beta) = E\{(Y - a'(X^{\mathrm{T}}\beta))^2XX^{\mathrm{T}}\}$. Moreover, the asymptotic variance $\Sigma(\beta^*)$ can be consistently estimated by $\hat{\Sigma}(\hat{\beta}_{\mathrm{M}}) = \hat{H}(\hat{\beta}_{\mathrm{M}})^{-1}\hat{G}(\hat{\beta}_{\mathrm{M}})\hat{H}(\hat{\beta}_{\mathrm{M}})^{-1}$, where

$$\hat{H}(\beta) = \frac{1}{n}\sum_{i=1}^{n} a''(X_i^{\mathrm{T}}\beta)X_iX_i^{\mathrm{T}} = \mathbf{X}^{\mathrm{T}}\mathrm{diag}(a''(\mathbf{X}\beta))\mathbf{X},$$

$$\hat{G}(\beta) = \frac{1}{n}\sum_{i=1}^{n}(Y_i - a'(X_i^{\mathrm{T}}\beta))^2X_iX_i^{\mathrm{T}} = \mathbf{X}^{\mathrm{T}}\mathrm{diag}^2(\mathbf{Y} - a'(\mathbf{X}^{\mathrm{T}}\beta))\mathbf{X}.$$

From these results, the Wald confidence interval for β_j^* is defined as

$$\hat{\beta}_{\mathrm{M},j} \pm z_{c/2}\sqrt{\frac{\{\hat{\Sigma}(\hat{\beta}_{\mathrm{M}})\}_{jj}}{n}}, \tag{6.41}$$

and can be shown to achieve an asymptotic size $(1-c)$ in covering β_j^*.

Asymptotic expansions can be calculated for inference about individual coefficients. Similarly as in Section 6.1.1, suppose that X, β^*, and $\hat{\beta}_{\mathrm{M}}$ are partitioned as

$$X = \begin{pmatrix} Z \\ W \end{pmatrix}, \quad \beta^* = \begin{pmatrix} \theta^* \\ \alpha^* \end{pmatrix}, \quad \hat{\beta}_{\mathrm{M}} = \begin{pmatrix} \hat{\theta}_{\mathrm{M}} \\ \hat{\alpha}_{\mathrm{M}} \end{pmatrix}.$$

Left-multiplying the expressions by $e_1^{\mathrm{T}} = (1, 0, ..., 0)$ in (6.40) gives

$$\hat{\theta}_{\mathrm{M}} - \theta^* = \Gamma_1(\beta^*)^{\mathrm{T}}\mathbf{X}^{\mathrm{T}}(\mathbf{Y} - a'(\mathbf{X}\beta^*))/n + o_p(n^{-1/2}),$$

where $\Gamma_1(\beta)$ is the first column of $H(\beta)^{-1}$ and can be determined as follows:

$$\Gamma_1(\beta) = \begin{pmatrix} 1/\bar{\tau}(\beta) \\ -\bar{\gamma}(\beta)/\bar{\tau}(\beta) \end{pmatrix}, \tag{6.42}$$

with

$$\bar{\gamma}(\beta) = E\{a''(X^{\mathrm{T}}\beta)WW^{\mathrm{T}}\}^{-1}E\{a''(X^{\mathrm{T}}\beta)WZ\},$$

$$\bar{\tau}(\beta) = E\{a''(X^{\mathrm{T}}\beta)(Z - W^{\mathrm{T}}\bar{\gamma}(\beta))Z\}.$$

Substituting the expression of $\Gamma_1(\beta^*)$ from (6.42) into (6.40) yields

$$\hat{\theta}_{\mathrm{M}} - \theta^* = \frac{(\mathbf{Z} - \mathbf{W}\bar{\gamma}(\beta^*))^{\mathrm{T}}(\mathbf{Y} - a'(\mathbf{X}\beta^*))/n}{E\{a''(X^{\mathrm{T}}\beta^*)(Z - W^{\mathrm{T}}\bar{\gamma}(\beta^*))Z\}} + o_p(n^{-1/2}) \tag{6.43a}$$

$$= \frac{(\mathbf{Z} - \mathbf{W}\bar{\gamma}(\beta^*))^{\mathrm{T}}(\mathbf{Y} - a'(\mathbf{X}\beta^*))/n}{(\mathbf{Z} - \mathbf{W}\bar{\gamma}(\beta^*))^{\mathrm{T}}A(\beta^*)\mathbf{Z}/n} + o_p(n^{-1/2}), \tag{6.43b}$$

where $A(\beta) = \mathrm{diag}(a''(\mathbf{X}\beta))$. In the case of $a(u) = \frac{u^2}{2}$, the asymptotic expansions (6.43a)–(6.43b) reduce to respectively (6.8a)–(6.8b) for least-squares estimation in linear regression.

From the asymptotic expansion (6.43b), the asymptotic variance for $\hat{\theta}_{\mathrm{M}}$ after scaled by n can be estimated by

$$\hat{V}(\hat{\theta}_{\mathrm{M}}, \hat{\alpha}_{\mathrm{M}}, \hat{\gamma}_{\mathrm{M}}) = \frac{\|(\mathbf{Z} - \mathbf{W}\hat{\gamma}_{\mathrm{M}})^{\mathrm{T}}(\mathbf{Y} - a'(\mathbf{X}^{\mathrm{T}}\hat{\beta}_{\mathrm{M}}))\|_n^2}{\{(\mathbf{Z} - \mathbf{W}\hat{\gamma}_{\mathrm{M}})^{\mathrm{T}}A(\hat{\beta}_{\mathrm{M}})\mathbf{Z}/n\}^2},$$

where $\hat{\gamma}_{\mathrm{M}}$ is a weighted least-squares estimator,

$$\hat{\gamma}_{\mathrm{M}} = (\mathbf{W}^{\mathrm{T}}A(\hat{\beta}_{\mathrm{M}})\mathbf{W})^{-1}(\mathbf{W}^{\mathrm{T}}A(\hat{\beta}_{\mathrm{M}})\mathbf{Z}).$$

It fact, it can be shown that $\{\hat{\Sigma}(\hat{\beta}_{\mathrm{M}})\}_{11} = \hat{V}(\hat{\theta}_{\mathrm{M}}, \hat{\alpha}_{\mathrm{M}}, \hat{\gamma}_{\mathrm{M}})$ algebraically [Exercise]. The Wald confidence interval (6.41) for θ^* can be equivalently written as

$$\hat{\theta}_{\mathrm{M}} \pm z_{c/2}\sqrt{\frac{\hat{V}(\hat{\theta}_{\mathrm{M}}, \hat{\alpha}_{\mathrm{M}}, \hat{\gamma}_{\mathrm{M}})}{n}}. \tag{6.44}$$

In the case of $a(u) = \frac{u^2}{2}$, the confidence interval (6.44) reduces to (6.12) in linear regression. In general, a notable difference from linear regression is that $\hat{\gamma}_{\mathrm{M}}$ depends on $\hat{\beta}_{\mathrm{M}}$.

6.2.2 Debiased Lasso inference

For generalized linear regression, consider the Lasso estimator as in (5.9):

$$\hat{\beta}_{\lambda} = \mathrm{argmin}_{\beta \in \mathbb{R}^p} \{K_n(\beta) + \lambda_\beta \|\beta\|_1\}, \tag{6.45}$$

where $\lambda_\beta > 0$ is a tuning parameter and $K_n(\beta)$ is an objective function in the form (5.14). Suppose that X, β^*, and $\hat{\beta}_\lambda$ are partitioned as

$$X = \begin{pmatrix} Z \\ W \end{pmatrix}, \quad \beta^* = \begin{pmatrix} \theta^* \\ \alpha^* \end{pmatrix}, \quad \hat{\beta}_\lambda = \begin{pmatrix} \hat{\theta}_\lambda \\ \hat{\alpha}_\lambda \end{pmatrix}.$$

It is of interest draw inference about θ^*, the first component of β^*.

We follow similar steps as in Section 6.1.2 to construct a debiased Lasso estimator for θ^*. First, the stationary condition (5.15) for $\hat\beta_\lambda$ implies that $\|\hat\zeta_\lambda\|_\infty \le \lambda_\beta$, where

$$\hat\zeta_\lambda = \mathbf{X}^{\mathrm{T}}(\mathbf{Y} - a'(\mathbf{X}\hat\beta_\lambda))/n.$$

To capture $\hat\beta_\lambda - \beta^*$, we rearrange the preceding equation as

$$\mathbf{X}^{\mathrm{T}}(a'(\mathbf{X}\hat\beta_\lambda) - a'(\mathbf{X}\beta^*))/n = \mathbf{X}^{\mathrm{T}}(\mathbf{Y} - a'(\mathbf{X}\beta^*))/n - \hat\zeta_\lambda,$$

and apply the approximation, $a'(\mathbf{X}\hat\beta_\lambda) - a'(\mathbf{X}\beta^*) \approx A(\beta^*)\mathbf{X}(\hat\beta_\lambda - \beta^*)$:

$$\hat H(\beta^*)(\hat\beta_\lambda - \beta^*) \approx \mathbf{X}^{\mathrm{T}}\epsilon/n - \hat\zeta_\lambda, \tag{6.46}$$

where $\epsilon = \mathbf{Y} - a'(\mathbf{X}\beta^*)$, $\hat H(\beta) = \mathbf{X}^{\mathrm{T}}A(\beta)\mathbf{X}/n$, and $A(\beta) = \mathrm{diag}(a''(\mathbf{X}\beta))$. To capture $\hat\theta_\lambda - \theta^*$, let $\Gamma_1^{\mathrm{T}}(\beta^*)$ be the first row of $H^{-1}(\beta^*)$ provided $H(\beta^*)$ is nonsingular, or more generally be any row vector such that $\Gamma_1^{\mathrm{T}}(\beta^*)H(\beta^*) = e_1^{\mathrm{T}}$. Then we left-multiply both sides by $\Gamma_1^{\mathrm{T}}(\beta^*)$ and apply the approximation $\Gamma_1^{\mathrm{T}}(\beta^*)\hat H(\beta^*) \approx e_1^{\mathrm{T}}$:

$$\hat\theta_\lambda - \theta^* \approx \Gamma_1^{\mathrm{T}}(\beta^*)\hat H(\beta^*)(\hat\beta_\lambda - \beta^*) \approx \Gamma_1^{\mathrm{T}}(\beta^*)\mathbf{X}^{\mathrm{T}}\epsilon/n - \Gamma_1^{\mathrm{T}}(\beta^*)\hat\zeta_\lambda.$$

Hence $\Gamma_1^{\mathrm{T}}(\beta^*)\hat\zeta_\lambda$ is expected to be the leading bias term, of order $\sqrt{\frac{\log(n\vee p)}{n}}O_p(1)$, for Lasso tuning parameter λ_β in the order $\sqrt{\frac{\log(n\vee p)}{n}}$.

To correct for the penalization bias, a debiased Lasso estimator for θ^* is defined as

$$\tilde\theta_\lambda = \hat\theta_\lambda + \hat\Gamma_{1,\lambda}^{\mathrm{T}}\hat\zeta_\lambda = \hat\theta_\lambda + \hat\Gamma_{1,\lambda}^{\mathrm{T}}\mathbf{X}^{\mathrm{T}}(\mathbf{Y} - \mathbf{X}\hat\beta_\lambda)/n, \tag{6.47}$$

where $\hat\Gamma_{1,\lambda}$ is an estimator of $\Gamma_1(\beta^*)$. By the expression $\Gamma_1(\beta^*)$ from (6.42), we take

$$\hat\Gamma_{1,\lambda} = \begin{pmatrix} 1/\hat\tau_\lambda \\ -\hat\gamma_\lambda/\hat\tau_\lambda \end{pmatrix},$$

with $\hat\tau_\lambda = (\mathbf{Z} - \mathbf{W}\hat\gamma_\lambda)^{\mathrm{T}}A(\hat\beta_\lambda)\mathbf{Z}/n$ and $\hat\gamma_\lambda$ a Lasso estimator of $\bar\gamma$:

$$\begin{aligned}\hat\gamma_\lambda &= \mathrm{argmin}_{\gamma\in\mathbb{R}^p}\left\{\frac{1}{2n}(\mathbf{Z} - \mathbf{W}\gamma)^{\mathrm{T}}A(\hat\beta_\lambda)(\mathbf{Z} - \mathbf{W}\gamma) + \lambda_\gamma\|\gamma\|_1\right\} \\ &= \mathrm{argmin}_{\gamma\in\mathbb{R}^p}\left\{\frac{1}{2n}\sum_{i=1}^n a''(X_i^{\mathrm{T}}\hat\beta_\lambda)(Z_i - W_i^{\mathrm{T}}\gamma)^2 + \lambda_\gamma\|\gamma\|_1\right\},\end{aligned} \tag{6.48}$$

where $\lambda_\gamma > 0$ is a tuning parameter. The associated linear regression, $E(Z|W) = W^{\mathrm{T}}\gamma^*$, is allowed to be misspecified. It can be shown that [Exercise]

$$\hat\tau_\lambda = (\mathbf{Z} - \mathbf{W}\hat\gamma_\lambda)^{\mathrm{T}}A(\hat\beta_\lambda)\mathbf{Z}/n$$

$$= (\mathbf{Z} - \mathbf{W}\hat{\gamma}_\lambda)^{\mathrm{T}} A(\hat{\beta}_\lambda)(\mathbf{Z} - \mathbf{W}\hat{\gamma}_\lambda)/n + \lambda_\gamma \|\hat{\gamma}_\lambda\|_1.$$

Substituting the expression of $\hat{\Gamma}_{1,\lambda}$ into (6.47) gives

$$\begin{aligned} \tilde{\theta}_\lambda &= \hat{\theta}_\lambda + \frac{(\mathbf{Z} - \mathbf{W}\hat{\gamma}_\lambda)^{\mathrm{T}}(\mathbf{Y} - a'(\mathbf{X}\hat{\beta}_\lambda))/n}{(\mathbf{Z} - \mathbf{W}\hat{\gamma}_\lambda)^{\mathrm{T}} A(\hat{\beta}_\lambda)\mathbf{Z}/n} \\ &= \hat{\theta}_\lambda + \frac{\sum_{i=1}^n (Z_i - W_i^{\mathrm{T}}\hat{\gamma}_\lambda)(Y_i - a'(X_i^{\mathrm{T}}\hat{\beta}_\lambda))}{\sum_{i=1}^n a''(X_i^{\mathrm{T}}\hat{\beta}_\lambda)(Z_i - W_i^{\mathrm{T}}\hat{\gamma}_\lambda)Z_i} \end{aligned} \tag{6.49}$$

In the case of $a(u) = \frac{u^2}{2}$, the definition (6.49) recovers the expression (6.19a) for the debiased Lasso estimator in linear regression. In general, the expression (6.49) cannot be simplified into a similar form as (6.19b), depending on $\hat{\alpha}_\lambda$, but not $\hat{\theta}_\lambda$, in the Lasso estimator $\hat{\beta}_\lambda$.

Example 6.1. For logistic regression (5.5), $a(u) = \log(1 + \mathrm{e}^u)$. The estimator $\hat{\gamma}_\lambda$ is defined as

$$\hat{\gamma}_\lambda = \operatorname{argmin}_{\gamma \in \mathbb{R}^p} \left\{ \frac{1}{2n} \sum_{i=1}^n \frac{\mathrm{e}^{X_i^{\mathrm{T}}\hat{\beta}_\lambda}}{(1 + \mathrm{e}^{X_i^{\mathrm{T}}\hat{\beta}_\lambda})^2} (Z_i - W_i^{\mathrm{T}}\gamma)^2 + \lambda_\gamma \|\gamma\|_1 \right\},$$

depending on $\hat{\beta}_\lambda$. The debiased estimator for θ^* is

$$\tilde{\theta}_\lambda = \hat{\theta}_\lambda + \frac{\sum_{i=1}^n (Z_i - W_i^{\mathrm{T}}\hat{\gamma}_\lambda)(Y_i - \mathrm{e}^{X_i^{\mathrm{T}}\hat{\beta}_\lambda})}{\sum_{i=1}^n \frac{\mathrm{e}^{X_i^{\mathrm{T}}\hat{\beta}_\lambda}}{(1 + \mathrm{e}^{X_i^{\mathrm{T}}\hat{\beta}_\lambda})^2} (Z_i - W_i^{\mathrm{T}}\hat{\gamma}_\lambda) Z_i},$$

depending on $\hat{\beta}_\lambda$ including $\hat{\theta}_\lambda$.

Example 6.2. For log-linear regression (5.7), $a(u) = \mathrm{e}^u$. The estimator $\hat{\gamma}_\lambda$ is defined as

$$\hat{\gamma}_\lambda = \operatorname{argmin}_{\gamma \in \mathbb{R}^p} \left\{ \frac{1}{2n} \sum_{i=1}^n \mathrm{e}^{X_i^{\mathrm{T}}\hat{\beta}_\lambda} (Z_i - W_i^{\mathrm{T}}\gamma)^2 + \lambda_\gamma \|\gamma\|_1 \right\},$$

depending on $\hat{\beta}_\lambda$. The debiased estimator for θ^* is

$$\tilde{\theta}_\lambda = \hat{\theta}_\lambda + \frac{\sum_{i=1}^n (Z_i - W_i^{\mathrm{T}}\hat{\gamma}_\lambda)(Y_i - \mathrm{e}^{X_i^{\mathrm{T}}\hat{\beta}_\lambda})}{\sum_{i=1}^n \mathrm{e}^{X_i^{\mathrm{T}}\hat{\beta}_\lambda} (Z_i - W_i^{\mathrm{T}}\hat{\gamma}_\lambda) Z_i},$$

depending on $\hat{\beta}_\lambda$ including $\hat{\theta}_\lambda$.

To derive an asymptotic expansion for $\tilde{\theta}_\lambda$, direct calculation from (6.49) gives

$$\tilde{\theta}_\lambda - \theta^* \tag{6.50}$$

$$= \frac{\sum_{i=1}^n (Z_i - W_i^{\mathrm{T}}\hat{\gamma}_\lambda)\{Y_i - \mu_i(\hat{\beta}_\lambda) + v_i(\hat{\beta}_\lambda) Z_i(\hat{\theta}_\lambda - \theta^*)\}}{\sum_{i=1}^n (Z_i - W_i^{\mathrm{T}}\hat{\gamma}_\lambda) v_i(\hat{\beta}_\lambda) Z_i}$$

$$= \frac{\sum_{i=1}^n (Z_i - W_i^{\mathrm{T}}\hat{\gamma}_\lambda)\{Y_i - \mu_i(\beta^*) - (\mu_i(\hat{\beta}_\lambda) - \mu_i(\beta^*)) + v_i(\hat{\beta}_\lambda) Z_i(\hat{\theta}_\lambda - \theta^*)\}}{\sum_{i=1}^n (Z_i - W_i^{\mathrm{T}}\hat{\gamma}_\lambda) v_i(\hat{\beta}_\lambda) Z_i}, \tag{6.51}$$

where $\mu_i(\beta) = a'(X_i^{\mathrm{T}}\beta)$ and $v_i(\beta) = a''(X_i^{\mathrm{T}}\beta)$. From (6.51), it can be shown that under suitable sparsity conditions, the debiased Lasso estimator achieves an $n^{-1/2}$ asymptotic expansion similar to (6.43a)–(6.43b) for the M-estimator $\hat{\theta}_{\mathrm{M}}$ in the low-dimensional setting.

For theoretical analysis, consider the following conditions:

(i) [Generalized linear regression in random design] Generalized linear regression (6.39) is correctly specified in a random design with IID $(X_1, Y_1), \ldots, (X_n, Y_n)$;

(ii) [Lasso estimation with Y given W] As $n \to \infty$ and p may increase as n increases,

$$\|\mathbf{W}^{\mathrm{T}}(\mathbf{Y} - a'(\mathbf{X}^{\mathrm{T}}\beta^*))/n\|_\infty = O_p(1)\frac{\log(p)}{n}, \tag{6.52a}$$

$$\|A(\beta^*)^{1/2}\mathbf{X}(\hat{\beta}_\lambda - \beta^*)\|_n^2 = O_p(1)|S_{\beta^*}|\frac{\log(n \vee p)}{n}, \tag{6.52b}$$

$$\|\hat{\beta}_\lambda - \beta^*\|_1 = O_p(1)|S_{\beta^*}|\sqrt{\frac{\log(n \vee p)}{n}}, \tag{6.52c}$$

where $A(\beta) = \mathrm{diag}(a''(\mathbf{X}\beta))$.

(iii) [Lasso estimation with Z given W] As $n \to \infty$ and p may increase as n increases,

$$\|\mathbf{W}^{\mathrm{T}}A(\beta^*)(\mathbf{Z} - \mathbf{W}\bar{\gamma})\|_\infty = O_p(1)\frac{\log(p)}{n}, \tag{6.53a}$$

$$\|A(\beta^*)^{1/2}\mathbf{W}(\hat{\gamma}_\lambda - \bar{\gamma})\|_n^2 = O_p(1)(|S_{\beta^*}| + |S_{\bar{\gamma}}|)\frac{\log(n \vee p)}{n}, \tag{6.53b}$$

$$\|\hat{\gamma}_\lambda - \bar{\gamma}\|_1 = O_p(1)(|S_{\beta^*}| + |S_{\bar{\gamma}}|)\sqrt{\frac{\log(n \vee p)}{n}}, \tag{6.53c}$$

where $\bar{\gamma} = \mathrm{argmin}_{\gamma \in \mathbb{R}^{p-1}} E\{a''(X^{\mathrm{T}}\beta^*)(Z - W^{\mathrm{T}}\gamma)^2\}$;

(iv) [Bounded covariates] $\|X\|_\infty = \max(|Z|, \|W\|_\infty) \le C_0$ almost surely with a constant $C_0 > 0$;

(v) [Hessian condition] (v.1) $a''(\tilde{u}) \le a''(u)\mathrm{e}^{C_1|\tilde{u}-u|}$ for any $u, \tilde{u} \in \mathbb{R}$ with a constant $C_1 > 0$, and (v.2) $a''(X^{\mathrm{T}}\beta^*) \le C_2$ almost surely with a constant $C_2 > 0$;

(vi) [Bounded pseudo-noise from Z given W] $|Z - W^{\mathrm{T}}\bar{\gamma}| \le C_3$ almost surely with a constant $C_3 > 0$.

(vii) [Non-degenerate Z given W] $\bar{\tau} = E\{a''(X^{\mathrm{T}}\beta^*)(Z - W^{\mathrm{T}}\bar{\gamma})Z\} > 0$ or equivalently $\bar{\tau} = E\{a''(X^{\mathrm{T}}\beta^*)(Z - W^{\mathrm{T}}\bar{\gamma})^2\} > 0$.

The error bounds (6.52b) and (6.52c) for the Lasso estimator $\hat{\beta}_\lambda$ can be deduced from primitive conditions as shown in Chapter 5. However, the error bounds (6.53b) and (6.53c) for $\hat{\gamma}_\lambda$ are more complex than in Chapter 4, depending on both $|S_{\beta^*}|$ and $|S_{\bar{\gamma}}|$ due to the data-dependent weights $a''(X_i^{\mathrm{T}}\hat{\beta}_\lambda)$. See Section 6.3.1 for formal analysis.

Theorem 6.4 (Asymptotic expansion of debiased Lasso). *Assume that conditions (i)–(vii) above are satisfied, and* $(|S_{\beta^*}| + |S_{\bar{\gamma}}|)\frac{\log(n\vee p)}{\sqrt{n}} = o(1)$. *Then as* $n \to \infty$ *and* p *may increase as* n *increases,*

$$\tilde{\theta}_\lambda - \theta^* = \frac{(\mathbf{Z} - \mathbf{W}\bar{\gamma})^{\mathrm{T}}(\mathbf{Y} - a'(\mathbf{X}\beta^*))/n}{(\mathbf{Z} - \mathbf{W}\bar{\gamma})^{\mathrm{T}}A(\beta^*)\mathbf{Z}/n} + o_p(n^{-1/2}) \tag{6.54a}$$

$$= \frac{(\mathbf{Z} - \mathbf{W}\bar{\gamma})^{\mathrm{T}}(\mathbf{Y} - a'(\mathbf{X}\beta^*))/n}{E((Z - W^{\mathrm{T}}\bar{\gamma})A(\beta^*)Z)} + o_p(n^{-1/2}). \tag{6.54b}$$

Note that the leading $O_p(n^{-1/2})$ *term in* (6.54) *is the same as the correction term in* (6.49) *with* $(\hat{\beta}_\lambda, \hat{\gamma})$ *replaced by* $(\beta^*, \bar{\gamma})$.

Proof. We proceed from (6.51) in similar steps as in the proof of Theorem 6.1, but need to deal with complications related to $\mu_i(\hat{\beta}_\lambda)$ and $v_i(\hat{\beta}_\lambda)$. We write (6.51) as

$$\tilde{\theta}_\lambda - \theta^* = \frac{\mathbf{I} - (\mathbf{II} + \mathbf{III})}{\mathbf{IV}},$$

where **I**, **II**, **III**, and **IV** are defined below.

First, we note that

$$\begin{aligned}
&\underbrace{\frac{1}{n}\sum_{i=1}^n (Z_i - W_i^{\mathrm{T}}\hat{\gamma}_\lambda)(Y_i - \mu_i(\beta^*))}_{\mathbf{I}} \\
&= \frac{1}{n}\sum_{i=1}^n (Z_i - W_i^{\mathrm{T}}\bar{\gamma})(Y_i - \mu_i(\beta^*)) - \frac{1}{n}\sum_{i=1}^n (\hat{\gamma}_\lambda - \bar{\gamma})^{\mathrm{T}}W_i(Y_i - \mu_i(\beta^*)) \\
&= \frac{1}{n}\sum_{i=1}^n (Z_i - W_i^{\mathrm{T}}\bar{\gamma})(Y_i - \mu_i(\beta^*)) + O_p(1)(|S_{\beta^*}| + |S_{\bar{\gamma}}|)\frac{\log(n\vee p)}{n},
\end{aligned}$$

using the Hölder inequality and convergence bounds (6.52a) and (6.53c).

Second, we show that

$$\underbrace{\frac{1}{n}\sum_{i=1}^n (Z_i - W_i^{\mathrm{T}}\hat{\gamma}_\lambda)\{\mu_i(\hat{\beta}_\lambda) - \mu_i(\beta^*) - v_i(\hat{\beta}_\lambda)Z_i(\hat{\theta}_\lambda - \theta^*)\}}_{\mathbf{II}+\mathbf{III}}$$

$$= \underbrace{\frac{1}{n}\sum_{i=1}^{n}(Z_i - W_i^{\mathrm{T}}\hat{\gamma}_\lambda)\{\mu_i(\hat{\beta}_\lambda) - \mu_i(\beta^*) - v_i(\hat{\beta}_\lambda)X_i^{\mathrm{T}}(\hat{\beta}_\lambda - \beta^*)\}}_{\mathbf{II}}$$

$$+ \underbrace{\frac{1}{n}\sum_{i=1}^{n}(Z_i - W_i^{\mathrm{T}}\hat{\gamma}_\lambda)v_i(\hat{\beta}_\lambda)W_i^{\mathrm{T}}(\hat{\alpha}_\lambda - \alpha^*)}_{\mathbf{III}}$$

$$= O_p(1)(|S_{\beta^*}| + |S_{\bar{\gamma}}|)\frac{\log(n \vee p)}{n}.$$

In fact, by the fundamental theorem of calculus, $\mu_i(\beta^*) - \mu_i(\hat{\beta}_\lambda) = \int_0^1 a''(X_i^{\mathrm{T}}(\hat{\beta}_\lambda + t(\beta^* - \hat{\beta}_\lambda)))\,\mathrm{d}t \cdot (X_i^{\mathrm{T}}(\beta^* - \hat{\beta}_\lambda))$, we have

$$\begin{aligned}
&|\mu_i(\beta^*) - \mu_i(\hat{\beta}_\lambda) - v_i(\hat{\beta}_\lambda)X_i^{\mathrm{T}}(\beta^* - \hat{\beta}_\lambda)| \\
&\leq \int_0^1 |a''(X_i^{\mathrm{T}}(\hat{\beta}_\lambda + t(\beta^* - \hat{\beta}_\lambda))) - a''(X_i^{\mathrm{T}}\hat{\beta}_\lambda))|\,\mathrm{d}t \cdot |X_i^{\mathrm{T}}(\beta^* - \hat{\beta}_\lambda)| \\
&\leq \int_0^1 a''(X_i^{\mathrm{T}}\beta^*)C_1\mathrm{e}^{tC_0C_1\|\hat{\beta}_\lambda - \beta^*\|_1}\,\mathrm{d}t \cdot (X_i^{\mathrm{T}}(\hat{\beta}_\lambda - \beta^*))^2 \\
&\leq C_1\mathrm{e}^{C_0C_1\|\hat{\beta}_\lambda - \beta^*\|_1}a''(X_i^{\mathrm{T}}\beta^*)(X_i^{\mathrm{T}}(\hat{\beta}_\lambda - \beta^*))^2,
\end{aligned}$$

where the third line uses $|v_i(\hat{\beta}_\lambda + t(\beta^* - \hat{\beta}_\lambda)) - v_i(\hat{\beta}_\lambda)| \leq C_1\mathrm{e}^{C_0C_1t\|\hat{\beta}_\lambda - \beta^*\|_1}v_i(\beta^*)|X_i^{\mathrm{T}}(\hat{\beta}_\lambda - \beta^*)|$ by Lemma 6.4 in Section 6.3.1, with condition (iv) and (v.1). [Exercise: Show that the conclusion above (the initial quantity being upper bounded by the final quantity) can also be obtained by applying the mean value theorem to $\mu_i(\beta^*) - \mu_i(\hat{\beta}_\lambda)$ and then Lemma 6.4.] Note that $|Z_i - W_i^{\mathrm{T}}\hat{\gamma}_\lambda| \leq |z_i - W_i^{\mathrm{T}}\bar{\gamma}| + |W_i^{\mathrm{T}}(\hat{\gamma}_\lambda - \bar{\gamma})| \leq C_3 + C_0\|\hat{\gamma}_\lambda - \bar{\gamma}\|_1$, by $\|W_i\|_\infty \leq C_0$ (condition (iv)) and $|Z_i - W_i^{\mathrm{T}}\bar{\gamma}| \leq C_3$ (condition (vi)). If $(|S_{\beta^*}| + |S_{\bar{\gamma}}|)\sqrt{\frac{\log(n \vee p)}{n}} = o(1)$, then $\|\hat{\beta}_\lambda - \beta^*\|_1 = o_p(1)$, $\|\hat{\gamma}_\lambda - \bar{\gamma}\|_1 = o_p(1)$, and

$$\begin{aligned}
&|\mathbf{II}| \\
&\leq \frac{1}{n}\sum_{i=1}^{n}\{|Z_i - W_i^{\mathrm{T}}\bar{\gamma}| + W_i^{\mathrm{T}}(\hat{\gamma}_\lambda - \bar{\gamma})|\}\cdot|\mu_i(\beta^*) - \mu_i(\hat{\beta}_\lambda) - v_i(\hat{\beta}_\lambda)X_i^{\mathrm{T}}(\beta^* - \hat{\beta}_\lambda)| \\
&\leq (C_3 + C_0\|\hat{\gamma}_\lambda - \bar{\gamma}\|_1)\cdot C_1\mathrm{e}^{C_0C_1\|\hat{\beta}_\lambda - \beta^*\|_1}\cdot\frac{1}{n}\sum_{i=1}^{n}v_i(\beta^*)(X_i^{\mathrm{T}}(\beta^* - \hat{\beta}_\lambda))^2 \\
&= O_p(1)|S_{\beta^*}|\frac{\log(n \vee p)}{n},
\end{aligned}$$

using convergence bounds (6.52b), (6.52c), and (6.53b). To control **III**, consider the decomposition

$$\frac{1}{n}\sum_{i=1}^{n}(Z_i - W_i^{\mathrm{T}}\hat{\gamma}_\lambda)v_i(\hat{\beta}_\lambda)W_i^{\mathrm{T}}(\hat{\alpha}_\lambda - \alpha^*)$$

$$= \underbrace{\frac{1}{n}\sum_{i=1}^{n}(Z_i - W_i^{\mathrm{T}}\bar{\gamma})v_i(\beta^*)W_i^{\mathrm{T}}(\hat{\alpha}_\lambda - \alpha^*)}_{\mathbf{III}_1}$$
$$+ \underbrace{\frac{1}{n}\sum_{i=1}^{n}(Z_i - W_i^{\mathrm{T}}\bar{\gamma})(v_i(\hat{\beta}_\lambda) - v_i(\beta^*))W_i^{\mathrm{T}}(\hat{\alpha}_\lambda - \alpha^*)}_{\mathbf{III}_2}$$
$$+ \underbrace{\frac{-1}{n}\sum_{i=1}^{n}W_i^{\mathrm{T}}(\hat{\gamma}_\lambda - \bar{\gamma})v_i(\hat{\beta}_\lambda)W_i^{\mathrm{T}}(\hat{\alpha}_\lambda - \alpha^*)}_{\mathbf{III}_3}.$$

The three terms, $\mathbf{III}_1$, $\mathbf{III}_2$, and $\mathbf{III}_3$, can be controlled as follows. By the Hölder inequality and convergence bounds (6.53a) and (6.52c),

$$|\mathbf{III}_1| = O_p(1)|S_{\beta^*}|\frac{\log(n \vee p)}{n}.$$

Note that $|v_i(\hat{\beta}_\lambda) - v_i(\beta^*)| \leq C_1 \mathrm{e}^{C_0 C_1 \|\hat{\beta}_\lambda - \beta^*\|_1} v_i(\beta^*)|X_i^{\mathrm{T}}(\hat{\beta}_\lambda - \beta^*)|$ by Lemma 6.4 in Section 6.3.1. By using this together with $|Z_i - W_i^{\mathrm{T}}\bar{\gamma}| \leq C_3$ (condition (vi)), the Cauchy–Schwartz inequality, and convergence bounds (6.52b) and (6.53b), we have

$$\begin{aligned}
&|\mathbf{III}_2| \\
&\leq \frac{1}{n}\sum_{i=1}^{n}|Z_i - W_i^{\mathrm{T}}\bar{\gamma}| \cdot C_1 \mathrm{e}^{C_0 C_1 \|\hat{\beta}_\lambda - \beta^*\|_1} v_i(\beta^*)|X_i^{\mathrm{T}}(\hat{\beta}_\lambda - \beta^*)| \cdot |W_i^{\mathrm{T}}(\hat{\alpha}_\lambda - \alpha^*)| \\
&\leq C_3 C_1 \mathrm{e}^{C_0 C_1 \|\hat{\beta}_\lambda - \beta^*\|_1} \\
&\quad \times \left\{\frac{1}{n}\sum_{i=1}^{n} v_i(\beta^*)(X_i^{\mathrm{T}}(\hat{\beta}_\lambda - \beta^*))^2\right\}^{1/2} \left\{\frac{1}{n}\sum_{i=1}^{n} v_i(\beta^*)(W_i^{\mathrm{T}}(\hat{\alpha}_\lambda - \alpha^*))^2\right\}^{1/2} \\
&= O_p(1)|S_{\beta^*}|\frac{\log(n \vee p)}{n}.
\end{aligned}$$

Note that $v_i(\hat{\beta}_\lambda) \leq v_i(\beta^*)\mathrm{e}^{C_1|X_i^{\mathrm{T}}(\hat{\beta}_\lambda - \beta^*)|} \leq v_i(\beta^*)\mathrm{e}^{C_0 C_1 \|\hat{\beta}_\lambda - \beta^*\|_1}$ by conditions (v.1) and $\|X_i\|_\infty \leq C_0$ (condition (iv)). By using this together with the Cauchy–Schwartz inequality, and convergence bounds (6.52b), (6.52c), and (6.53b), if $|S_{\beta^*}|\sqrt{\frac{\log(n \vee p)}{n}} = o(1)$, then

$$\begin{aligned}
&|\mathbf{III}_3| \\
&\leq \frac{1}{n}\sum_{i=1}^{n} v_i(\beta^*)\mathrm{e}^{C_0 C_1 \|\hat{\beta}_\lambda - \beta^*\|_1}|W_i^{\mathrm{T}}(\hat{\gamma}_\lambda - \bar{\gamma})| \cdot |W_i^{\mathrm{T}}(\hat{\alpha}_\lambda - \alpha^*)| \\
&\leq \mathrm{e}^{C_0 C_1 \|\hat{\beta}_\lambda - \beta^*\|_1}
\end{aligned}$$

$$\times \{\frac{1}{n}\sum_{i=1}^{n} v_i(\beta^*)(W_i^{\mathrm{T}}(\hat{\gamma}_\lambda - \bar{\gamma}))^2\}^{1/2}\{\frac{1}{n}\sum_{i=1}^{n} v_i(\beta^*)(W_i^{\mathrm{T}}(\hat{\alpha}_\lambda - \alpha^*))^2\}^{1/2}$$
$$= O_p(1)|S_{\beta^*}|^{1/2}(|S_{\beta^*}| + |S_{\bar{\gamma}}|)^{1/2}\frac{\log(n \vee p)}{n}.$$

Combining the preceding bounds gives

$$|\mathbf{III}| = O_p(1)(|S_{\beta^*}| + |S_{\bar{\gamma}}|)\frac{\log(n \vee p)}{n}.$$

Third, we show that

$$\underbrace{\frac{1}{n}\sum_{i=1}^{n}(Z_i - W_i^{\mathrm{T}}\hat{\gamma}_\lambda)v_i(\hat{\beta}_\lambda)Z_i}_{\mathbf{IV}}$$
$$= \frac{1}{n}\sum_{i=1}^{n}(Z_i - W_i^{\mathrm{T}}\bar{\gamma})v_i(\beta^*)Z_i + O_p(1)(|S_{\beta^*}|^{1/2} + |S_{\bar{\gamma}}|^{1/2})\sqrt{\frac{\log(n \vee p)}{n}}.$$

In fact, consider the decomposition

$$\frac{1}{n}\sum_{i=1}^{n}(Z_i - W_i^{\mathrm{T}}\hat{\gamma}_\lambda)v_i(\hat{\beta}_\lambda)Z_i$$
$$= \underbrace{\frac{1}{n}\sum_{i=1}^{n}(Z_i - W_i^{\mathrm{T}}\bar{\gamma})v_i(\beta^*)Z_i}_{\mathbf{IV}_1}$$
$$+ \underbrace{\frac{1}{n}\sum_{i=1}^{n}(Z_i - W_i^{\mathrm{T}}\bar{\gamma})(v_i(\hat{\beta}_\lambda) - v_i(\beta^*))Z_i}_{\mathbf{IV}_2} + \underbrace{\frac{-1}{n}\sum_{i=1}^{n} W_i^{\mathrm{T}}(\hat{\gamma}_\lambda - \bar{\gamma})v_i(\hat{\beta}_\lambda)Z_i}_{\mathbf{IV}_3}.$$

The two terms, $\mathbf{IV}_2$ and $\mathbf{IV}_3$, can be controlled similarly as $\mathbf{III}_2$ and $\mathbf{III}_3$ respectively:

$$|\mathbf{IV}_2| \leq \frac{1}{n}\sum_{i=1}^{n}|Z_i - W_i^{\mathrm{T}}\bar{\gamma}| \cdot \mathrm{e}^{C_0C_1\|\hat{\beta}_\lambda - \beta^*\|_1}v_i(\beta^*)|X_i^{\mathrm{T}}(\hat{\beta}_\lambda - \beta^*)| \cdot |Z_i|$$
$$\leq C_3\{\frac{1}{n}\sum_{i=1}^{n} v_i(\beta^*)(X_i^{\mathrm{T}}(\hat{\beta}_\lambda - \beta^*))^2\}^{1/2}\{\frac{1}{n}\sum_{i=1}^{n} v_i(\beta^*)Z_i^2\}^{1/2}$$
$$= O_p(1)|S_{\beta^*}|^{1/2}\sqrt{\frac{\log(n \vee p)}{n}},$$

and

$$|\mathbf{IV}_3| \leq \frac{1}{n}\sum_{i=1}^{n} v_i(\beta^*)\mathrm{e}^{C_0C_1\|\hat{\beta}_\lambda - \beta^*\|_1}|W_i^{\mathrm{T}}(\hat{\gamma}_\lambda - \bar{\gamma})| \cdot |Z_i|$$

$$\leq e^{C_0 C_1 \|\hat{\beta}_\lambda - \beta^*\|_1} \left\{ \frac{1}{n} \sum_{i=1}^n v_i(\beta^*)(W_i^{\mathrm{T}}(\hat{\gamma}_\lambda - \bar{\gamma}))^2 \right\}^{1/2} \left\{ \frac{1}{n} \sum_{i=1}^n v_i(\beta^*) Z_i^2 \right\}^{1/2}$$

$$= O_p(1)(|S_{\beta^*}| + |S_{\bar{\gamma}}|)^{1/2} \sqrt{\frac{\log(n \vee p)}{n}},$$

where $\frac{1}{n} \sum_{i=1}^n v_i(\beta^*) Z_i^2 \leq C_2 C_0^2$ by condition (iv) and (v.2). Combining the preceding bounds gives

$$|\mathbf{IV}_2| + |\mathbf{IV}_3| = O_p(1)(|S_{\beta^*}|^{1/2} + |S_{\bar{\gamma}}|^{1/2}) \sqrt{\frac{\log(n \vee p)}{n}}.$$

Fourth, by applying the results from the first three steps to (6.51), it can be shown that [Exercise] if $(|S_{\beta^*}| + |S_{\bar{\gamma}}|) \frac{\log(n \vee p)}{n} = o(1)$, then

$$\begin{aligned} &\tilde{\theta}_\lambda - \theta^* \\ &= \frac{\sum_{i=1}^n (Z_i - W_i^{\mathrm{T}} \bar{\gamma})(Y_i - \mu_i(\beta^*))}{\sum_{i=1}^n (Z_i - W_i^{\mathrm{T}} \bar{\gamma}) v_i(\beta^*) Z_i} \\ &\quad + O_p(1)(|S_{\beta^*}| + |S_{\bar{\gamma}}|) \frac{\log(n \vee p)}{n} + o_p(n^{-1/2}). \end{aligned}$$

If $(|S_{\beta^*}| + |S_{\bar{\gamma}}|) \frac{\log(n \vee p)}{\sqrt{n}} = o(1)$, this reduces to the asymptotic expansion (6.54a).

Finally, the asymptotic expansion (6.54b) can be deduced from (6.54a) [Exercise]. □

From Theorem 6.4, asymptotic normality of $\tilde{\beta}_\lambda$ can be obtained. Moreover, consistent estimation of the asymptotic variance can also be derived. Let

$$V(\theta^*, \alpha^*, \bar{\gamma}) = \frac{E\{(Z - W^{\mathrm{T}} \bar{\gamma})^2 (Y - a'(X^{\mathrm{T}} \beta^*))^2\}}{E^2\{a''(X^{\mathrm{T}} \beta^*)(Z - W^{\mathrm{T}} \bar{\gamma}) Z\}},$$

$$\hat{V}(\tilde{\theta}_\lambda, \hat{\alpha}_\lambda, \hat{\gamma}_\lambda) = \frac{\|(\mathbf{Z} - \mathbf{W}\hat{\gamma}_\lambda)^{\mathrm{T}} (\mathbf{Y} - a'(\mathbf{X}^{\mathrm{T}} \hat{\beta}_\lambda))\|_n^2}{\{(\mathbf{Z} - \mathbf{W}\hat{\gamma}_\lambda)^{\mathrm{T}} A(\hat{\beta}_\lambda) \mathbf{Z}/n\}^2}.$$

The proof of Corollary 6.3 is left as Exercise 6.7

Corollary 6.3 (Asymptotic normality and consistent variance estimation). *In the setting of Theorem 6.4, the following results hold:*

$$\sqrt{n}(\tilde{\theta}_\lambda - \theta^*) \xrightarrow[\mathcal{D}]{} \mathrm{N}(0, V(\theta^*, \alpha^*, \bar{\gamma})),$$

$$\hat{V}(\tilde{\theta}_\lambda, \hat{\alpha}_\lambda, \hat{\gamma}_\lambda) \xrightarrow[\mathcal{P}]{} V(\theta^*, \alpha^*, \bar{\gamma}).$$

as $n \to \infty$ and p may increase as n increases.

From Corollary 6.3, the Wald confidence interval, in a similar form as (6.44),

$$\tilde{\theta}_\lambda \pm z_{c/2} \sqrt{\frac{\hat{V}(\tilde{\theta}_\lambda, \hat{\alpha}_\lambda, \hat{\gamma}_\lambda)}{n}} \tag{6.55}$$

achieves asymptotic size $1 - c$ in covering θ^* in generalized linear regression (6.39). Therefore, the Wald confidence interval (6.55) based on $\tilde{\theta}_\lambda$ represents a high-dimensional version of the classical Wald confidence interval (6.44) based on $\hat{\theta}_{\mathrm{M}}$.

Finally, the development of debiased Lasso inference with correct model specification can be extended to the scenario where generalized linear regression (6.39) may be misspecified. The proof of the following result is left as an exercise.

Theorem 6.5 (Asymptotic expansion of debiased Lasso with model misspecification). *Assume that conditions (ii)–(iv) in Theorem 6.4 are satisfied, except the true value β^* is replaced by a target value $\bar{\beta} = (\bar{\theta}, \bar{\alpha}^{\mathrm{T}})^{\mathrm{T}}$ defined as follows:*

$$\begin{aligned}\bar{\beta} &= \mathrm{argmin}_{\beta \in \mathbb{R}^p} E\{-YX^{\mathrm{T}}\beta + a(X^{\mathrm{T}}\beta)\} \\ &= \mathrm{argmin}_{\theta \in \mathbb{R},, \alpha \in \mathbb{R}^{p-1}} E\{-Y(Z\theta + W^{\mathrm{T}}\alpha) + a(Z\theta + W^{\mathrm{T}}\alpha)\}. \qquad (6.56)\end{aligned}$$

Assume further that $(|S_{\bar{\beta}}| + |S_{\bar{\gamma}}|)\frac{\log(n \vee p)}{\sqrt{n}} = o(1)$. *Then the asymptotic expansions in* (6.54) *hold with β^* replaced by $\bar{\beta}$, i.e.,*

$$\tilde{\theta}_\lambda - \bar{\theta} = \frac{(\mathbf{Z} - \mathbf{W}\bar{\gamma})^{\mathrm{T}}(\mathbf{Y} - a'(\mathbf{X}\bar{\beta}))/n}{(\mathbf{Z} - \mathbf{W}\bar{\gamma})^{\mathrm{T}}A(\bar{\beta})\mathbf{Z}/n} + o_p(n^{-1/2}) \qquad (6.57a)$$

$$= \frac{(\mathbf{Z} - \mathbf{W}\bar{\gamma})^{\mathrm{T}}(\mathbf{Y} - a'(\mathbf{X}\bar{\beta}))/n}{E((Z - W^{\mathrm{T}}\bar{\gamma})A(\bar{\beta})Z)} + o_p(n^{-1/2}). \qquad (6.57b)$$

Similarly, Corrollary 6.3 can also be extended with β^* replaced by $\bar{\beta}$. Then the Wald confidence interval (6.25) achieves asymptotic size $1 - c$ in covering $\bar{\theta}$, the first component of $\bar{\beta}$, with possible misspecification of generalized linear regression (6.39).

6.2.3 Partially generalized linear regression

We discuss how debiased Lasso estimation can be related to doubly robust estimation in partially generalized linear regression, in a similar manner as in linear regression. By simple algebra, the debiased Lasso estimator $\tilde{\theta}_\lambda$ in (6.49) is a solution to the estimating equation

$$0 = \frac{1}{n}\sum_{i=1}^{n} \varphi(Z_i, W_i, Y_i; \theta, \hat{\beta}, \hat{\gamma}),$$

where $\hat{\beta}_\lambda = (\hat{\theta}_\lambda, \hat{\alpha}_\lambda^{\mathrm{T}})^{\mathrm{T}}$ and $\hat{\gamma}_\lambda$ are as in Section 6.2.2 and

$$\varphi(Z, W, Y; \theta, \hat{\beta}, \hat{\gamma}) = (Z - W^{\mathrm{T}}\gamma)\{Y - a'(X^{\mathrm{T}}\hat{\beta}) + a''(X^{\mathrm{T}}\hat{\beta})Z(\hat{\theta} - \theta)\}, \quad (6.58)$$

with $\hat{\beta} = (\hat{\theta}, \hat{\alpha}^{\mathrm{T}})^{\mathrm{T}}$. Compared with $\varphi(\cdot)$ in (6.29), $\varphi(\cdot)$ in (6.58) depends on $\hat{\beta}$ including both $\hat{\theta}$ and $\hat{\alpha}$, hence the need to distinguish $\hat{\theta}$ and θ in the notation.

Consider a random design as in Section 6.1.2, where $(X_1, Y_1), \ldots, (X_n, Y_n)$ are IID as (X, Y), with $X = (Z, W^{\mathrm{T}})^{\mathrm{T}}$. A partially generalized linear regression model postulates that

$$E(Y|Z, W) = a\{Z\theta^* + g^*(W)\}, \tag{6.59}$$

where $a(\cdot)$ is a *known* convex differentiable function, θ^* is an unknown (scalar) parameter and $g^*(w)$ is an unknown function of w. For doubly robust estimation of θ^*, a direct extension of the ideas from Section 6.1.3 involves two working models as follows. The first is a linear model for $g^*(W)$:

$$g^*(W) = W^{\mathrm{T}}\alpha^*,$$

or equivalently a generalized linear regression model of Y on (Z, W):

$$E(Y|Z, W) = a(X^{\mathrm{T}}\beta^*) = a(Z\theta^* + W^{\mathrm{T}}\alpha^*) \tag{6.60}$$

where $\beta^* = (\theta^*, \alpha^{*\mathrm{T}})^{\mathrm{T}}$ and α^* is an unknown parameter vector, in addition to the parameter θ^*. The second is a linear regression model for $f^*(W) = E(Z|W)$:

$$E(Z|W) = W^{\mathrm{T}}\gamma^*, \tag{6.61}$$

where γ^* is an unknown parameter vector. However, it can be shown that $\varphi(\cdot)$ in (6.58) satisfies the first unbiasedness property, but in general not the second one, below.

- If model (6.60) is correctly specified, then for any γ,

$$0 = E\{\varphi(Z, W, Y; \theta^*, \beta^*, \gamma)\}.$$

- If model (6.61) is correctly specified, then for any $\beta = (\theta, \alpha^{\mathrm{T}})^{\mathrm{T}}$,

$$0 = E\{\varphi(Z, W, Y; \theta^*, \beta, \gamma^*)\}.$$

In this sense $\varphi(\cdot)$ in (6.58) is *not* a doubly robust estimating function for θ^* with respect to working models (6.60) and (6.61) for nuisance functions $g^*(w)$ and $f^*(w)$. The debiased Lasso estimator $\tilde{\theta}_\lambda$ is in general *not* doubly robust for θ^* in partially generalized linear regression (6.59) with respect to models (6.60) and (6.61). Instead, $\tilde{\theta}_\lambda$ is singly robust (or consistent) for θ^* if model (6.60) is correctly specified but model (6.61) may be misspecified.

It remains in general an open question whether and how a doubly robust estimating function for θ^* can be constructed in partially generalized linear regression (6.59). Positive answers are known in several specific cases, depending on both the link-related function $a(\cdot)$ and the working models involved. For partially log-linear regression with $a'(u) = \mathrm{e}^u$, a doubly robust estimating function for θ^* is available with respect to working models for $g^*(W)$ and $f^*(W) = E(Z|W)$. For partially logistic regression with $a'(u) = \frac{\mathrm{e}^u}{1+\mathrm{e}^u}$, a doubly robust estimating function for θ^* is available with respect to working models for $g^*(W)$ and $f^*(W) = E(Z|Y=0,W)$. However, for partially probit regression with $a'(u) = \Phi(u)$, no doubly robust estimating function for θ^* seems to be available. See Robins and Rotnitzky (2001) and Ghosh and Tan (2022) and references therein for more information.

6.3 Appendices

6.3.1 Analysis of weighted Lasso estimation

We provide formal analysis to show that the convergence bounds (6.53b) and (6.53c) are satisfied under primitive conditions by the weighted Lasso estimator $\hat{\gamma}_\lambda$, where the linear regression model of Z on W may be misspecified. Compared with related analysis in Chapter 4, a complication is that the weights $a''(X_i^{\mathrm{T}}\hat{\beta}_\lambda)$, $i = 1, \ldots, n$, depend on the Lasso estimator $\hat{\beta}_\lambda$ in generalized linear regression of Y given (Z, W), and this data-dependency needs to be carefully tackled. For concreteness, generalized linear regression (6.39) is assumed to be correctly specified. But the analysis can be readily extended to allow misspecification of (6.39) as in Theorem 6.4.

For theoretical analysis, consider the following conditions:

(i) [Generalized linear regression in random design] Generalized linear regression (6.39) is correctly specified in a random design with IID $(X_1, Y_1), \ldots, (X_n, Y_n)$;

(ii) [Lasso estimation with Y given W] For $0 < \epsilon < 1$, the Lasso estimator $\hat{\beta}_\lambda$ with suitable choice of λ_β in (6.45) satisfies that with probability at least $1 - 4\epsilon$,

$$\|A(\beta^*)^{1/2}\mathbf{X}(\hat{\beta}_\lambda - \beta^*)\|_n^2 \le M_{01}|S_{\beta^*}|\lambda_{\beta 0}^2, \tag{6.62a}$$

$$\|\hat{\beta}_\lambda - \beta^*\|_1 \le M_{02}|S_{\beta^*}|\lambda_{\beta 0}, \tag{6.62b}$$

where $\lambda_{\beta 0} = O(1)\sqrt{\frac{\log(p/\epsilon)}{n}}$, and $M_{01} > 0$ and $M_{02} > 0$ are constants;

(iii) [Theoretical compatibility condition] The theoretical compatibility condition $(S_{\bar{\gamma}}, \xi_0^*, \nu_0^*)$ holds for the matrix $D(\beta^*) = E\{a''(X^{\mathrm{T}}\beta^*)WW^{\mathrm{T}}\}$, where $S_{\bar{\gamma}} = \{1 \le j \le p-1 : \bar{\gamma}_j \ne 0\}$ and $\xi_0^* > 1$ and $\nu_0^* > 0$ are constants;

(iv) [Bounded covariates] $\|X\|_\infty = \max(|Z|, \|W\|_\infty) \le C_0$ almost surely with a constant $C_0 > 0$;

(v) [Hessian condition] (v.1) $a''(\tilde{u}) \le a''(u)\mathrm{e}^{C_1|\tilde{u}-u|}$ for any $u, \tilde{u} \in \mathbb{R}$ with a constant $C_1 > 0$, and (v.2) $a''(X^{\mathrm{T}}\beta^*) \le C_2$ almost surely with a constant $C_2 > 0$;

(vi) [Bounded pseudo-noise from Z given W] $|Z - W^{\mathrm{T}}\bar{\gamma}| \le C_3$ almost surely with a constant $C_3 > 0$.

Conditions (i), (iv), (v), and (vi) are the same as in Section 6.2.2. Condition (ii) state high-probability bounds leading to condition (ii) in Section 6.2.2.

For any $0 < \epsilon < 1$, suppose that the tuning parameter λ_γ in (6.48) is specified as $\lambda_\gamma = A_1\lambda_{\gamma 0}$ with a constant $A_1 > 1$ and

$$\lambda_{\gamma 0} = \max\left\{ C_0C_2C_3\sqrt{\frac{2\log(p/\epsilon)}{n}},\ 2C_0^2C_2\sqrt{\frac{\log(p/\epsilon)}{n}} \right\},$$

depending on constants (C_0, C_2, C_3) from conditions (iv), (v.2), and (vi). The following result extends the fast-rate error bound in Theorem 4.5 to allow data-dependent weights.

Theorem 6.6 (Fast rate in random design with data-dependent weights). *Assume that conditions (i)–(vi) above are satisfied. For any* $0 < \epsilon < 1$, *suppose further that*

(vii) [Rate condition] $(1+\xi_0^*)^2 \nu_0^{*-2} |S_{\bar{\gamma}}| \lambda_{\gamma 0} \le \eta_0$ *for a constant* $\eta_0 \in (0,1)$, *and* $C_0C_1M_{02}|S_{\beta^*}|\lambda_{\beta 0} \le \eta_1$ *for a constant* $\eta_1 \in (0,1)$.

Then it holds with probability at least $1 - 6\epsilon - 2\epsilon^2$ *uniformly over* $A_1 > \frac{\xi_0^*+1}{\xi_0^*-1}$ *that the Lasso estimator* $\hat{\gamma}_\lambda$ *with* $\lambda_\gamma = A_1\lambda_{\gamma 0}$ *in* (6.48) *satisfies*

$$\begin{aligned} &\|A(\beta^*)^{1/2}\mathbf{W}(\hat{\gamma}_\lambda - \bar{\gamma})\|_n^2 + (A_1 - 1)\lambda_{\gamma 0}\|\hat{\gamma}_\lambda - \bar{\gamma}\|_1 \\ &\le M_{11}|S_{\beta^*}|\lambda_{\beta 0}^2 + M_{12}|S_{\bar{\gamma}}|\lambda_{\gamma 0}^2, \end{aligned} \tag{6.63}$$

where $M_{11} = \xi_1^{-2}C_3^2C_1^2 e^{3\eta_1} M_{01}$ *and* $M_{12} = \xi_2^2 \nu_0^{-2} e^{\eta_1}$, *with* $\xi_1 = 1 - 2A_1/\{(\xi_0^* + 1)(A_1 - 1)\} \in (0,1)$, $\xi_2 = (\xi_0^* + 1)(A_1 - 1)$, *and* $\nu_0 = \nu_0^*(1-\eta_0)^{1/2}$.

Theorem 6.6 can be shown by carefully extending the analyses in Chapter 4 to deal with the data-dependency of the weights $a''(X_i^{\mathrm{T}}\hat{\beta}_\lambda)$. The following three lemmas are similar to their counterparts in Chapter 4, and the proofs are left as exercises.

Lemma 6.1 (Basic inequality). *For any* $\gamma \in \mathbb{R}^{p-1}$,

$$\|A(\hat{\beta}_\lambda)^{1/2}\mathbf{W}(\hat{\gamma}_\lambda - \gamma)\|_n^2 - \langle \mathbf{Z} - \mathbf{W}\gamma, A(\hat{\beta}_\lambda)\mathbf{W}(\hat{\gamma}_\lambda - \gamma)\rangle_n + \lambda_\gamma\|\hat{\gamma}_\lambda\|_1 \le \lambda_\gamma\|\gamma\|_1,$$

where $A(\beta) = \mathrm{diag}(a''(\mathbf{X}\beta))$.

Lemma 6.2 (Concentration of weighted inner products in random design). *For* $0 < \epsilon < 1$, *let* $\Omega_1(\lambda_1)$ *be the event that*

$$\max_{j=1,\ldots,p} |\langle \mathbf{Z} - \mathbf{W}\bar{\gamma}, A(\beta^*)\mathbf{W}_j\rangle_n| \le \lambda_1.$$

Under condition (iv), (v.2), and (vi), if $\lambda_1 \ge C_0C_2C_3\sqrt{\frac{2\log(p/\epsilon)}{n}}$, *then* $P(\Omega_1(\lambda_1)) \ge 1 - 2\epsilon$.

Lemma 6.3 (Concentration of Hessian matrix). *For* $0 < \epsilon < 1$, *let* $\Omega_2(\lambda_2)$ *be the event that*

$$\max_{j,k=1,\ldots,p-1} |\hat{D}_{jk}(\beta^*) - D_{jk}(\beta^*)| \leq \lambda_2,$$

where $\hat{D}_{jk}(\beta^*)$ *is the* (j,k)*th element of* $\hat{D}(\beta^*) = \mathbf{W}^{\mathrm{T}} A(\beta^*)\mathbf{W}/n$, *and* $D_{jk}(\beta^*)$ *is the* (j,k)*th element of* $D(\beta^*) = E\{a''(X^{\mathrm{T}}\beta^*)WW^{\mathrm{T}}\}$. *Under condition (iv) and (v.2), if* $\lambda_2 \geq 2C_0^2 C_2 \sqrt{\frac{\log(p/\epsilon)}{n}}$, *then* $P(\Omega_2(\lambda_2)) \geq 1 - 2\epsilon^2$.

Proof of Theorem 6.6. Let Ω_0 be the event that (6.62a) and (6.62b) hold. Let $\Omega_1 = \Omega_1(\lambda_{\gamma 0})$ with $\lambda_1 = \lambda_{\gamma 0}$ in Lemma 6.2, and $\Omega_2 = \Omega_2(\lambda_{\gamma 0})$ with $\lambda_2 = \lambda_{\gamma 0}$ in Lemma 6.3. Then $P(\Omega_0) \geq 1-4\epsilon$, $P(\Omega_1) \geq 1-2\epsilon$, and $P(\Omega_2) \geq 1-2\epsilon^2$.

Taking $\gamma = \bar{\gamma}$ in the basic inequality with $\lambda_\gamma = A_1\lambda_{\gamma 0}$ gives

$$\begin{aligned}&\|A(\hat{\beta}_\lambda)^{1/2}\mathbf{W}(\hat{\gamma}_\lambda - \bar{\gamma})\|_n^2 - \langle \mathbf{Z} - \mathbf{W}\bar{\gamma}, A(\hat{\beta}_\lambda)\mathbf{W}(\hat{\gamma}_\lambda - \bar{\gamma})\rangle_n + A_1\lambda_{\gamma 0}\|\hat{\gamma}_\lambda\|_1 \\ &\leq A_1\lambda_{\gamma 0}\|\bar{\gamma}\|_1. \end{aligned} \tag{6.64}$$

In the event Ω_0, the following two inequalities can be obtained. First, by condition (v.1) and $|X_i^{\mathrm{T}}(\hat{\beta}_\lambda - \beta^*)| \leq C_0\|\hat{\beta}_\lambda - \beta^*\|_1$,

$$\begin{aligned}\|A(\hat{\beta}_\lambda)^{1/2}\mathbf{W}(\hat{\gamma}_\lambda - \bar{\gamma})\|_n^2 &= \frac{1}{n}\sum_{i=1}^n a''(X_i^{\mathrm{T}}\hat{\beta}_\lambda)(W_i^{\mathrm{T}}(\hat{\gamma}_\lambda - \bar{\gamma}))^2 \\ &\geq \frac{1}{n}\sum_{i=1}^n \mathrm{e}^{-C_0C_1\|\hat{\beta}_\lambda - \beta^*\|_1} a''(X_i^{\mathrm{T}}\beta^*)(W_i^{\mathrm{T}}(\hat{\gamma}_\lambda - \bar{\gamma}))^2 \\ &\geq \frac{1}{n}\sum_{i=1}^n \mathrm{e}^{-\eta_1} a''(X_i^{\mathrm{T}}\beta^*)(W_i^{\mathrm{T}}(\hat{\gamma}_\lambda - \bar{\gamma}))^2 = \mathrm{e}^{-\eta_1}\|A(\beta^*)^{1/2}\mathbf{W}(\hat{\gamma}_\lambda - \bar{\gamma})\|_n^2,\end{aligned}$$

where the third line uses $C_0C_1\|\hat{\beta}_\lambda - \beta^*\|_1 \leq C_0C_1M_{02}|S_{\beta^*}|\lambda_{\beta 0} \leq \eta_1$ in the event Ω_0. Second, by the triangle inequality,

$$\begin{aligned}&|\langle \mathbf{Z} - \mathbf{W}\bar{\gamma}, A(\hat{\beta}_\lambda)\mathbf{W}(\hat{\gamma}_\lambda - \bar{\gamma})\rangle_n| - |\langle \mathbf{Z} - \mathbf{W}\bar{\gamma}, A(\beta^*)\mathbf{W}(\hat{\gamma}_\lambda - \bar{\gamma})\rangle_n| \\ &\leq \left|\frac{1}{n}\sum_{i=1}^n (Z_i - W_i^{\mathrm{T}}\bar{\gamma})(a''(X_i^{\mathrm{T}}\hat{\beta}_\lambda) - a''(X_i^{\mathrm{T}}\beta^*))W_i^{\mathrm{T}}(\hat{\gamma}_\lambda - \bar{\gamma})\right| \\ &\leq C_3C_1\mathrm{e}^{\eta_1}\frac{1}{n}\sum_{i=1}^n a''(X_i^{\mathrm{T}}\beta^*)|X_i^{\mathrm{T}}(\hat{\beta}_\lambda - \beta^*)|\cdot|W_i^{\mathrm{T}}(\hat{\gamma}_\lambda - \bar{\gamma})| \\ &\leq C_3C_1\mathrm{e}^{\eta_1}\|A(\beta^*)^{1/2}\mathbf{X}(\hat{\beta}_\lambda - \beta^*)\|_n\|A(\beta^*)^{1/2}\mathbf{W}(\hat{\gamma}_\lambda - \gamma^*)\|_n \\ &\leq C_3C_1\mathrm{e}^{\eta_1}(M_{01}|S_{\beta^*}|\lambda_{\beta 0}^2)^{1/2}\|A(\beta^*)^{1/2}\mathbf{W}(\hat{\gamma}_\lambda - \gamma^*)\|_n,\end{aligned}$$

where the third line uses $|Z_i - W_i^{\mathrm{T}}\bar{\gamma}| \leq C_3$ (condition (vi)) and $|v_i(\hat{\beta}_\lambda) - v_i(\beta^*)| \leq C_1\mathrm{e}^{C_0C_1\|\hat{\beta}_\lambda - \beta^*\|_1}v_i(\beta^*)|X_i^{\mathrm{T}}(\hat{\beta}_\lambda - \beta^*)| \leq C_1\mathrm{e}^{\eta_1}v_i(\beta^*)|X_i^{\mathrm{T}}(\hat{\beta}_\lambda - \beta^*)|$ by Lemma 6.4 in Section 6.3.1, the fourth lines uses the Cauchy–Schwartz inequality, and the fifth line uses $\|A(\beta^*)^{1/2}\mathbf{X}(\hat{\beta}_\lambda - \beta^*)\|_n^2 \le M_{01}|S_{\beta^*}|\lambda_{\beta 0}^2$ in the event Ω_0. In the event $\Omega_0 \cap \Omega_1$, applying the preceding two inequalities and the Hölder inequality to (6.64) yields

$$\begin{aligned}
&\mathrm{e}^{-\eta_1}\|A(\beta^*)^{1/2}\mathbf{W}(\hat{\gamma}_\lambda - \bar{\gamma})\|_n^2 + A_1\lambda_{\gamma 0}\|\hat{\gamma}_\lambda\|_1 \\
&\le |\langle \mathbf{Z} - \mathbf{W}\bar{\gamma}, A(\beta^*)\mathbf{W}(\hat{\gamma}_\lambda - \bar{\gamma})\rangle_n| + A_1\lambda_{\gamma 0}\|\bar{\gamma}\|_1 \\
&\quad + C_3C_1\mathrm{e}^{\eta_1}(M_1|S_{\beta^*}|\lambda_{\gamma 0}^2)^{1/2}\|A(\beta^*)^{1/2}\mathbf{W}(\hat{\gamma}_\lambda - \gamma^*)\|_n \\
&\le \lambda_{\gamma 0}\|\hat{\gamma}_\lambda - \bar{\gamma}\|_1 + A_1\lambda_{\gamma 0}\|\bar{\gamma}\|_1 \\
&\quad + C_3C_1\mathrm{e}^{\eta_1}(M_{01}|S_{\beta^*}|\lambda_{\gamma 0}^2)^{1/2}\|A(\beta^*)^{1/2}\mathbf{W}(\hat{\gamma}_\lambda - \gamma^*)\|_n
\end{aligned}$$

and further applying the triangle inequality gives

$$\begin{aligned}
&\mathrm{e}^{-\eta_1}\|A(\beta^*)^{1/2}\mathbf{W}(\hat{\gamma}_\lambda - \bar{\gamma})\|_n^2 + (A_1 - 1)\lambda_{\gamma 0}\|\hat{\gamma}_\lambda - \bar{\gamma}\|_1 \\
&\le 2A_1\lambda_{\gamma 0}\sum_{j\in S_{\bar{\gamma}}}|\hat{\gamma}_{\lambda,j} - \bar{\gamma}_j| \\
&\quad + C_3C_1\mathrm{e}^{\eta_1}(M_1|S_{\beta^*}|\lambda_{\beta 0}^2)^{1/2}\|A(\beta^*)^{1/2}\mathbf{W}(\hat{\gamma}_\lambda - \gamma^*)\|_n.
\end{aligned} \tag{6.65}$$

Denote by Δ_n the left-hand side of (6.65).

It suffices to deduce (6.63) from (6.65). Consider two complementary cases: either

$$\xi_1\Delta_n \le C_3C_1\mathrm{e}^{\eta_1}(M_1|S_{\beta^*}|\lambda_{\beta 0}^2)^{1/2}\|A(\beta^*)^{1/2}\mathbf{W}(\hat{\gamma}_\lambda - \gamma^*)\|_n, \tag{6.66}$$

or $(1-\xi_1)\Delta_n \le 2A_1\lambda_{\gamma 0}\sum_{j\in S_{\bar{\gamma}}}|\hat{\gamma}_{\lambda,j} - \bar{\gamma}_j|$, that is,

$$\Delta_n \le \xi_2\lambda_{\gamma 0}\sum_{j\in S_{\bar{\gamma}}}|\hat{\gamma}_{\lambda,j} - \bar{\gamma}_j|, \tag{6.67}$$

where $\xi_1 = 1 - 2A_1/\{(\xi_0^* + 1)(A_1 - 1)\} \in (0,1)$ with $\xi_0^* > \frac{A_1+1}{A_1-1}$ or equivalently $A_1 > \frac{\xi_0^*+1}{\xi_0^*-1}$, and $\xi_2 = (\xi_0^* + 1)(A_1 - 1)$. In the case of (6.66), $\Delta_n \le \xi_1^{-1}C_3C_1\mathrm{e}^{\eta_1}(M_1|S_{\beta^*}|\lambda_{\beta 0}^2)^{1/2}(\mathrm{e}^{\eta_1}\Delta_n)^{1/2}$ and hence

$$\Delta_n \le \xi_1^{-2}C_3^2C_1^2\mathrm{e}^{3\eta_1}(M_1|S_{\beta^*}|\lambda_{\beta 0}^2). \tag{6.68}$$

In the case of (6.67), $\sum_{j\notin S_{\bar{\gamma}}}|\hat{\gamma}_{\lambda,j} - \bar{\gamma}_j| \le \xi_0^*\sum_{j\in S_{\bar{\gamma}}}|\hat{\gamma}_{\lambda,j} - \bar{\gamma}_j|$, that is, $\hat{\gamma}_\lambda - \bar{\gamma}$ satisfies the cone condition. By similar reasoning as in Lemma 4.5, in the event Ω_2, the empirical compatibility condition $(S_{\bar{\gamma}}, \xi_0, \nu_0)$ holds for $\hat{D}(\beta^*)$ with $\xi_0 = \xi_0^*$ and $\nu_0 = \nu_0^*(1-\eta_0)^{1/2}$. Then

$$\begin{aligned}
&\nu_0^2\Big(\sum_{j\in S_{\bar{\gamma}}}|\hat{\gamma}_{\lambda,j} - \bar{\gamma}_j|\Big)^2 \\
&\le |S_{\bar{\gamma}}|(\hat{\gamma}_\lambda - \bar{\gamma})^{\mathrm{T}}\hat{D}(\beta^*)(\hat{\gamma}_\lambda - \bar{\gamma}) = |S_{\bar{\gamma}}|\cdot\|A(\beta^*)^{1/2}\mathbf{W}(\hat{\gamma}_\lambda - \bar{\gamma})\|_n^2.
\end{aligned}$$

Returning to (6.67) gives

$$\begin{aligned}\Delta_n &\le \xi_2 \lambda_{\gamma 0} \sum_{j \in S_{\bar{\gamma}}} |\hat{\gamma}_{\lambda,j} - \bar{\gamma}_j| \\ &\le \xi_2 \lambda_{\gamma 0} \cdot \nu_0^{-1} |S_{\bar{\gamma}}|^{1/2} \| A(\beta^*)^{1/2} \mathbf{W}(\hat{\gamma}_\lambda - \bar{\gamma}) \|_n \\ &\le \xi_2 \lambda_{\gamma 0} \nu_0^{-1} |S_{\bar{\gamma}}|^{1/2} (\mathrm{e}^{\eta_1} \Delta_n)^{1/2}\end{aligned}$$

and hence

$$\Delta_n \le \xi_2^2 \nu_0^{-2} \mathrm{e}^{\eta_1} |S_{\bar{\gamma}}| \lambda_{\gamma 0}^2. \tag{6.69}$$

Combining (6.68) and (6.69) leads to the desired inequality (6.63) in the event $\Omega_0 \cap \Omega_1 \cap \Omega_2$ with probability at least $1 - 6\epsilon - 2\epsilon^2$. □

Lemma 6.4 (Bound on Hessian difference). *Under condition (iv) and (v.1), for any* $\beta, \tilde{\beta} \in \mathbb{R}^p$,

$$|a''(X^{\mathrm{T}} \tilde{\beta}) - a''(X^{\mathrm{T}} \beta)| \le C_1 \mathrm{e}^{C_0 C_1 \|\tilde{\beta} - \beta\|_1} |X^{\mathrm{T}} (\tilde{\beta} - \beta)| a''(X^{\mathrm{T}} \beta).$$

Proof. By condition (v.1),

$$\begin{aligned}|a''(X^{\mathrm{T}} \tilde{\beta}) - a''(X^{\mathrm{T}} \beta)| &\le \{\mathrm{e}^{C_1 |X^{\mathrm{T}}(\tilde{\beta} - \beta)|} - 1\} a''(X^{\mathrm{T}} \beta) \\ &\le \mathrm{e}^{C_1 |X^{\mathrm{T}}(\tilde{\beta} - \beta)|} C_1 |X^{\mathrm{T}} (\tilde{\beta} - \beta)| a''(X^{\mathrm{T}} \beta),\end{aligned}$$

where the second line uses $\mathrm{e}^{C_1 u} - 1 = \int_0^{C_1 u} \mathrm{e}^t \, \mathrm{d}t \le \mathrm{e}^{C_1 u} C_1 u$ for $u \ge 0$. [This can also be obtained from the equality $1 - \mathrm{e}^{-C_1 u} \le C_1 u$, i.e., $1 - \mathrm{e}^{-x} \le x$ with $x = C_1 u$.] The desired result follows because $|X^{\mathrm{T}}(\tilde{\beta} - \beta)| \le C_0 \|\tilde{\beta} - \beta\|_1$ by condition (iv). □

Exercises

Ex 6.1. Provide a proof of expression (6.6) for $\hat{\Gamma}_1$.

Ex 6.2. Assume that linear regression is correctly specified as in Section 6.1.1. From expression (6.10) for $\hat{\theta}_{\mathrm{LS}}$, show that as p is fixed and $n \to \infty$,

$$\begin{aligned}\hat{\theta}_{\mathrm{LS}} &= \frac{(\mathbf{Z} - \mathbf{W}\bar{\gamma}^{\mathrm{T}})(\mathbf{Y} - \mathbf{W}\bar{\eta})/n}{(\mathbf{Z} - \mathbf{W}\bar{\gamma}^{\mathrm{T}})\mathbf{Z}/n} + \theta^* \frac{(\mathbf{Z} - \mathbf{W}\bar{\gamma}^{\mathrm{T}})\mathbf{W}\bar{\gamma}/n}{(\mathbf{Z} - \mathbf{W}\bar{\gamma}^{\mathrm{T}})\mathbf{Z}/n} + o_p(n^{-1/2}) \\ &= \frac{(\mathbf{Z} - \mathbf{W}\bar{\gamma}^{\mathrm{T}})(\mathbf{Y} - \mathbf{W}\bar{\eta})/n}{(\mathbf{Z} - \mathbf{W}\bar{\gamma}^{\mathrm{T}})\mathbf{Z}/n} + O_p(n^{-1/2}).\end{aligned}$$

How can this expansion be extended when linear regression is misspecified?

Ex 6.3. Provide a proof of expression (6.18) for $\hat{\tau}_\lambda$.

Ex 6.4. In the proof of Theorem 6.1 (Asymptotic expansion of debiased Lasso), show that (6.24) holds. [Note that $o_p(n^{-1/2})$ is needed on the right-hand side before introducing $(|S_{\beta^*}| + |S_{\bar{\gamma}}|)\frac{\log(n \vee p)}{\sqrt{n}} = o(1)$.]

Ex 6.5. Provide a proof of the consistency of $\hat{V}$ in Corollary 6.1. Present your proof in full details but as simple as possible.

Ex 6.6. Comment on how the results in Theorem 6.1 and Corollary 6.1 are compared with the results in van de Geer et al. (2014), Section 2.

Ex 6.7. Provide a proof of the consistency of $\hat{V}$ in Corollary 6.3. Present your proof in full details but as simple as possible.

Ex 6.8. Comment on how the results in Theorem 6.4 and Corollary 6.3 are compared with the results in van de Geer et al. (2014), Section 3.

Ex 6.9 (Alternative to debiased Lasso).

(i) Show that if linear regression (6.31) for Y given (Z, W) is correctly specified but logistic regression (6.36) for binary Z given W may be misspecified, then the estimator $\hat{\theta}(\hat{\alpha}_\lambda, \hat{\gamma}_{\lambda,\text{logit}})$ in (6.38) achieves an $n^{-1/2}$ asymptotic expansion similar as the debiased Lasso estimator in Theorem 6.1,

$$\hat{\theta}(\hat{\alpha}_\lambda, \hat{\gamma}_{\lambda,\text{logit}}) - \theta^* = \frac{(\mathbf{Z} - \text{expit}(\mathbf{W}\bar{\gamma}_{\text{logit}}))^{\mathrm{T}}(\mathbf{Y} - \mathbf{X}\beta^*)/n}{(\mathbf{Z} - \text{expit}(\mathbf{W}\bar{\gamma}_{\text{logit}}))^{\mathrm{T}}\mathbf{Z}/n} + o_p(n^{-1/2}),$$

where $\bar{\gamma}_{\text{logit}} = \text{argmin}_{\gamma \in \mathbb{R}^{p-1}} E\{-ZW^{\mathrm{T}}\gamma + \log(1 + \mathrm{e}^{W^{\mathrm{T}}\gamma})\}$. Present your conditions and proof in full details but as simple as possible.

[Hint: From the definition of $\hat{\theta}(\hat{\alpha}_\lambda, \hat{\gamma}_{\lambda,\text{logit}})$, direct calculation gives

$$\hat{\theta}(\hat{\alpha}_\lambda, \hat{\gamma}_{\lambda,\text{logit}}) - \theta^* = \frac{(\mathbf{Z} - \text{expit}(\mathbf{W}\hat{\gamma}_{\lambda,\text{logit}}))^{\mathrm{T}}(\mathbf{Y} - \mathbf{Z}\theta^* - \mathbf{W}\hat{\alpha}_\lambda)/n}{(\mathbf{Z} - \text{expit}(\mathbf{W}\hat{\gamma}_{\lambda,\text{logit}}))^{\mathrm{T}}\mathbf{Z}/n}.$$

The desired result can be obtained by establishing

$$\begin{aligned}
&(\mathbf{Z} - \text{expit}(\mathbf{W}\hat{\gamma}_{\lambda,\text{logit}}))^{\mathrm{T}}(\mathbf{Y} - \mathbf{Z}\theta^* - \mathbf{W}\hat{\alpha}_\lambda)/n \\
&= (\mathbf{Z} - \text{expit}(\mathbf{W}\bar{\gamma}_{\text{logit}}))^{\mathrm{T}}(\mathbf{Y} - \mathbf{Z}\theta^* - \mathbf{W}\alpha^*)/n + o_p(n^{-1/2}), \\
&(\mathbf{Z} - \text{expit}(\mathbf{W}\hat{\gamma}_{\lambda,\text{logit}}))^{\mathrm{T}}\mathbf{Z}/n \\
&= (\mathbf{Z} - \text{expit}(\mathbf{W}\bar{\gamma}_{\text{logit}}))^{\mathrm{T}}\mathbf{Z}/n + o_p(1),
\end{aligned}$$

under suitable conditions.]

(ii) With possible misspecification of linear regression (6.31), can a similar result to Theorem 6.2 for the debiased Lasso estimator be established for the estimator $\hat{\theta}(\hat{\alpha}_\lambda, \hat{\gamma}_{\lambda,\text{logit}})$? Provide your reasoning.

Ex 6.10. Consider generalize linear regression (6.60) for Y given (Z, W) and, with binary Z, logistic regression (6.36) for Z given W as in Exercise 6.9. Formulate an alternative to the debiased Lasso estimator (6.49) for coefficient θ^* in (6.60), similarly to the alternative estimator $\hat{\theta}(\hat{\alpha}_\lambda, \hat{\gamma}_{\lambda,\text{logit}})$ to the debiased Lasso estimator in linear regression. For your new estimator, can you establish a similar result as Theorem 6.4 (Correct model) or as Theorem 6.5 (Misspecified model)? Provide you reasoning.

Ex 6.11. Suppose that (X_i, Y_i), $i = 1, \ldots, n$, are IID as (X, Y), where Y is a response variable and $X = (X^{(1)}, \ldots, X^{(p)})^{\mathrm{T}}$ is a $p \times 1$ covariate vector, such that each covariate has mean 0, i.e., $E(X^{(j)}) = 0$ for $j = 1, \ldots, p$. The problem of interest is to estimate the mean $\alpha^* = E(Y)$, while taking advantage of the fact that $E(X^{(j)}) = 0$ for $j = 1, \ldots, p$. Each covariate $X^{(j)}$ is called an auxiliary variable or a control variate.

(i) Let $(\hat{\alpha}, \hat{\beta})$ be the least-squares estimator defined as

$$(\hat{\alpha}, \hat{\beta}) = \text{argmin}_{\alpha \in \mathbb{R}, \beta \in \mathbb{R}^p} \ell_n(\alpha, \beta)$$

where $\ell_n(\alpha, \beta) = \frac{1}{2n} \sum_{i=1}^n (Y_i - \alpha - X_i^{\mathrm{T}} \beta)^2$. Show that $\hat{\alpha} = \bar{Y} - \bar{X}^{\mathrm{T}} \hat{\beta}$, where $\bar{Y} = n^{-1} \sum_{i=1}^n Y_i$ and $\bar{X} = n^{-1} \sum_{i=1}^n X_i$.

(ii) Assume that $\Sigma = E(XX^{\mathrm{T}})$ is non-singular. As $n \to \infty$ and p fixed, show that $\sqrt{n}(\hat{\alpha} - \alpha^*) \underset{\mathcal{D}}{\to} \mathrm{N}(0, V)$, with variance matrix V. Find $\varphi(x, y)$ such that $V = E\{\varphi^2(X, Y)\}$.

For the remaining questions, p may be greater than n as $n \to \infty$.

(iii) Let $(\tilde{\alpha}, \tilde{\beta})$ be a Lasso estimator defined as

$$(\tilde{\alpha}, \tilde{\beta}) = \text{argmin}_{\alpha \in \mathbb{R}, \beta \in \mathbb{R}^p} \left\{ \ell_n(\alpha, \beta) + \lambda \|\beta\|_1 \right\},$$

where $\lambda > 0$ is a tuning parameter. Show that $\tilde{\alpha} = \bar{Y} - \bar{X}^{\mathrm{T}} \tilde{\beta}$.

(iv) Assume that each covariate is bounded in absolute values by a constant $C_0 > 0$, i.e., $|X^{(j)}| \le C_0$ almost surely for $j = 1, \ldots, p$. Under some further conditions, $(\tilde{\alpha}, \tilde{\beta})$ is expected to converge to (α^*, β^*) in the following sense, with $(\alpha^*, \beta^*) = \text{argmin}_{\alpha \in \mathbb{R}, \beta \in \mathbb{R}^p} E\{(Y - \alpha - X^{\mathrm{T}} \beta)^2\}$:

$$\frac{1}{n} \sum_{i=1}^n (\tilde{\alpha} - \alpha^* + X_i^{\mathrm{T}} (\tilde{\beta} - \beta^*))^2 = O_p(1)(1 + |S_{\beta^*}|) \lambda_0^2, \tag{6.70a}$$

$$|\tilde{\alpha} - \alpha^*| + \|\tilde{\beta} - \beta^*\|_1 = O_p(1)(1 + |S_{\beta^*}|) \lambda_0, \tag{6.70b}$$

where $\lambda_0 = \sqrt{\log(n \vee p)/n}$, and $|S_{\beta^*}|$ is the size of the set $S_{\beta^*} = \{1 \le j \le p : \beta_j^* \ne 0\}$ with $\beta^* = (\beta_1^*, \ldots, \beta_p^*)^{\mathrm{T}}$. Please formulate a set of technical conditions (as weak as you can) under which you expect that (6.70a) and (6.70b) can be established.

(v) As one of the steps to prove (6.70a) and (6.70b), you are asked to prove the following result. For $0 < \epsilon < 1$, formulate an event Ω_0 such that $P(\Omega_0) \geq 1 - 2\epsilon$ and the following inequality holds under event Ω_0 with $\lambda = A_0 \lambda_{0,\epsilon}$ for a constant $A_0 > 1$:

$$\frac{1}{n}\sum_{i=1}^{n}(\tilde{\alpha} - \alpha^* + X_i^{\mathrm{T}}(\tilde{\beta} - \beta^*))^2 + (A_0 - 1)\lambda_{0,\epsilon}(|\tilde{\alpha} - \alpha^*| + \|\tilde{\beta} - \beta^*\|_1)$$
$$\leq 2A_0\lambda_{0,\epsilon}\big(|\tilde{\alpha} - \alpha^*| + \sum_{j \in S_{\beta^*}} |\hat{\beta}_j - \beta_j^*|\big),$$

where $\lambda_{0,\epsilon} = B_0\sqrt{\frac{\log((1+p)/\epsilon)}{n}}$ for some constant B_0. State your B_0 explicitly.

(vi) Suppose that (6.70a) and (6.70b) hold. Show that $\sqrt{n}(\tilde{\alpha} - \alpha^*) \xrightarrow[\mathcal{D}]{} \mathrm{N}(0, V)$, with variance matrix V as stated in (ii), provided $(1 + |S_{\beta^*}|)\log(n \vee p) = o(n^{1/2})$.

Bibliography

Bach, F. (2010). Self-concordant analysis for logistic regression. *Electronic Journal of Statistics*, 4, 384–414. (cit. on p. 165)

Bertsekas, D. P. (1973). Stochastic optimization problems with nondifferentiable cost functionals. *Journal of Optimization Theory and Applications*, 12, 218–231. (cit. on p. 72)

Bickel, P. J., Ritov, Y., & Tsybakov, A. B. (2009). Simultaneous analysis of Lasso and Dantzig selector. *Annals of Statistics*, 37, 1705–1732. (cit. on p. 122)

Boucheron, S., Lugosi, G., & Massart, P. (2013). *Concentration Inequalities: A Nonasymptotic Theory of Independence.* Oxford University Press. (cit. on pp. 82, 87, 103, 107, 109, 117)

Bühlmann, P., & van de Geer, S. (2011). *Statistics for High-Dimensional Data: Methods, Theory and Applications.* Springer. (cit. on pp. 82, 122, 153, 178)

Candes, E., & Tao, T. (2007). The Dantzig selector: Statistical estimation when p is much larger than n (with discussion). *Annals of Statistics*, 35, 2313–2351. (cit. on p. 122)

Casella, G., & Berger, R. L. (2001). *Statistical Inference (2nd edition).* Cengage Learning. (cit. on pp. 60, 72, 114)

Ferguson, T. S. (1996). *A Course in Large Sample Theory.* Chapman; Hall. (cit. on pp. 1, 7, 19, 28, 32, 59)

Ghosh, S., & Tan, Z. (2022). Doubly robust semiparametric inference using regularized calibrated estimation with high-dimensional data. *Bernoulli*, 28, 1675–1703. (cit. on pp. 189, 204, 218)

Hastie, T., Tibshirani, R., & Wainwright, M. J. (2015). *Statistical Learning with Sparsity: The Lasso and Generalizations.* Chapman & Hall. (cit. on pp. 122, 153)

Hiriart-Urruty, J.-B., & Lemaréchal, C. (2001). *Fundamentals of Convex Analysis.* Springer. (cit. on pp. 55, 64)

Hjort, N. L., & Pollard, D. (1993). *Asymptotics for minimisers of convex processes* (tech. rep.). University of Oslo and Yale University. (cit. on p. 19)

Huang, J., & Zhang, C.-H. (2012). Estimation and selection via absolute penalized convex minimization and its multistage adaptive applications. *Journal of Machine Learning Research*, 13, 1839–1864. (cit. on p. 136)

Javanmard, A., & Montanari, A. (2014). Confidence intervals and hypothesis testing for high-dimensional regression. *Journal of Machine Learning Research*, 15, 2869–2909. (cit. on p. 189)

Knight, K., & Fu, W. (2000). Asymptotics for Lasso-type estimators. *Annals of Statistics*, 28, 1356–1378. (cit. on p. 189)

Mallows, C. L. (1973). Some comments on Cp. *Technometrics*, 15, 661–675. (cit. on p. 125)

Massart, P. (2000). About the constants in Talagrand's concentration inequalities for empirical processes. *Annals of Probability*, 28, 863–884. (cit. on p. 103)

McCullagh, P., & Nelder, J. A. (1989). *Generalized Linear Models (2nd edition)*. Chapman & Hall. (cit. on pp. 153, 154)

Meir, R., & Zhang, T. (2003). Generalization error bounds for Bayesian mixture algorithms. *Journal of Machine Learning Research*, 4, 839–860. (cit. on p. 109)

Negahban, S. N., Ravikumar, P., Wainwright, M. J., & Yu, B. (2012). A unified framework for high-dimensional analysis of M-estimators with decomposable regularizers. *Statistical Science*, 27, 538–557. (cit. on p. 136)

Newey, W. K. (1991). Uniform convergence in probability and stochastic equicontinuity. *Econometrica*, 59, 1161–1167. (cit. on p. 62)

Newey, W. K., & McFadden, D. (1994). Large sample estimation and hypothesis testing. In R. F. Engle & D. McFadden (Eds.), *Handbook of Econometrics, Volume IV*. Elsevier. (cit. on pp. 19, 64)

Niemiro, W. (1992). Asymptotics for M-Estimators defined by convex minimization. *Annals of Statistics*, 20, 1514–1533. (cit. on pp. 19, 49, 64, 67)

Pollard, D. (1984). *Convergence of Stochastic Processes*. Springer. (cit. on p. 3)

Pollard, D. (1991). Asymptotics for least absolute deviation regression estimators. *Econometric Theory*, 7, 186–199. (cit. on pp. 19, 49, 64)

Robins, J. M., & Rotnitzky, A. (2001). Comment on "Inference for semiparametric models: Some questions and an answer" by Bickel and Kwon. *Statistica Sinica*, 11, 920–936. (cit. on p. 218)

Rockafellar, R. T. (1970). *Convex Analysis*. Princeton University Press. (cit. on pp. 64, 67)

Talagrand, M. (2005). *The Generic Chaining*. Springer. (cit. on p. 102)

Tan, Z. (2020a). Model-assisted inference for treatment effects using regularized calibrated estimation with high-dimensional data. *Annals of Statistics*, 48, 811–837. (cit. on pp. 157, 189)

Tan, Z. (2020b). Regularized calibrated estimation of propensity scores with model misspecification and high-dimensional data. *Biometrika*, 107, 137–158. (cit. on pp. 153, 157, 158, 178)

Tan, Z. (2024). Model-assisted sensitivity analysis for treatment effects under unmeasured confounding via regularized calibrated estimation. *Journal of the Royal Statistical Society, Ser. B*, 86, 1339–1363. (cit. on p. 178)

Tan, Z. (2025). *A Course in Regression and Smoothing Methods.* Book draft. (cit. on pp. viii, 55, 154, 192, 193, 201, 202, 203, 204)

Tibshirani, R. (1996). Regression shrinkage and selection via the Lasso. *Journal of the Royal Statistical Society, Ser. B*, 58, 267–288. (cit. on pp. 122, 153)

van de Geer, S. (2000). *Empirical Processes in M-Estimation.* Cambridge University Press. (cit. on p. 28)

van de Geer, S., Bühlmann, P., Ritov, Y., & Dezeure, R. (2014). On asymptotically optimal confidence regions and tests for high-dimensional models. *Annals of Statistics*, 42, 1166–1202. (cit. on pp. 189, 223)

van der Vaart, A. W. (1998). *Asymptotic Statistics.* Cambridge University Press. (cit. on pp. 1, 19)

Vershynin, R. (2018). *High-Dimensional Probability: An Introduction with Applications in Data Science.* Cambridge University Press. (cit. on p. 82)

Wainwright, M. J. (2019). *High-Dimensional Statistics: A Non-Asymptotic Viewpoint.* Cambridge University Press. (cit. on pp. 82, 122, 153)

Zhang, C.-H., & Zhang, S. S. (2014). Confidence intervals for low dimensional parameters in high dimensional linear models. *Journal of the Royal Statistical Society, Ser. B*, 76, 217–242. (cit. on p. 189)

Index

For Product Safety Concerns and Information please contact our
EU representative GPSR@taylorandfrancis.com Taylor & Francis
Verlag GmbH, Kaufingerstraße 24, 80331 München, Germany